Genetically Modified Crops and Future Agriculture

Genetically Modified Crops and Future Agriculture

Dr. Siya Ram

RANDOM PUBLICATIONS
NEW DELHI (INDIA)

Genetically Modified Crops and Future Agriculture

ISBN 978-93-5111-952-4

Published in 2016 in India by

RANDOM PUBLICATIONS

4376-A/4B, Gali Murari Lal, Ansari Road
New Delhi-110 002
Phone : +9111-43580356, 011-23289044, 011-43142548
e-mail: sales@randompublications.com,
info@randompublications.com, randomexports@gmail.com

Reprinted 2021

Type Setting by : Friends Media, Delhi-110089
Digitally Printed at : Replika Press Pvt. Ltd.

Preface

The development of GM crops may continue and they may have good impact on the future agriculture, we need to consider the apprehensions of the critics of GM technology about the health hazards and biosafety concerns arising out of the use of these gene combinations prior to their widespread use.

GM Food is a food product that has been derived from a genetically modified organism (GMO). This organism or life form can be anything starting from crop plants to animals and even microscopic organisms. The 'genetic modification' or 'genetic reconstitution' of these life forms is based on the advanced form of Genetic Engineering, Recombinant DNA technology, which attempts to improve the food-yield (from the organism) – both quality and quantity wise – by combining genes from different organisms.

GM food crops thus are high-grade crops with superior quality and taste. The genetically modified food crops are planned to be nutritionally rich. GM not only improves the quality of the yield, you get increased produce and that in less time. Genetic modification also endows the crops with greater resistivity to common diseases and harsh weather conditions. Genetic modification of animals, likewise, improves animal health, minimizing their chances of being affected with common infections. Naturally, the improved breeds give better yields of eggs, meat and milk.

Modern plant biotechnology is considered the key to what is known as a "bio-based economy". In such a scenario, agriculture is used not only for producing food and feed, but also as a major source of energy and renewable raw materials. This system takes advantage of microorganisms and enzymes for producing raw materials such as carbohydrates, fatty acids, and fibres in energy- and resource-conserving industrial systems.

The book will be of help to researcher in the field of agriculture, crop improvement, biotechnology etc. It will also be helpful to teachers and students for better understanding of the subject.

– Author

Contents

1

Significant of Agricultural Biotechnology

Every year, millions of people worldwide die from starvation and nutritional deficiencies. Although these challenges remain significant, we are making some progress. Agricultural technologies have increased caloric intake on a per-capita basis in nearly every nation of the world. technological development— particularly in the area of agricultural biotechnology — is crucial if we want to further reduce malnutrition and starvation and meet the needs of growing populations. Unfortunately, misperceptions about biotechnology have led to activist movements to slow down, if not eliminate, this critical technology via government regulation. In the United States, the advancement of biotech crops has slowed down because of attempts to regulate them differently than conventionally grown foods. In Europe, antibiotech movements are much stronger, and international bureaucrats are pushing various international agreements and trade restrictions. While Western nations may be able to absorb the costs of such policies, people in the developing world will pay a higher price as many continue to suffer from starvation.

WHAT IS AGRICULTURAL BIOTECHNOLOGY?

Broadly speaking, biotechnology is any technique that uses living organisms or substances from these organisms to make or modify a product for a practical purpose. Biotechnology can be applied to all classes of organism-from viruses and bacteria to plants and animals-and it is becoming a major feature of modern medicine, agriculture and industry. Modern agricultural biotechnology includes a range of tools that scientists employ to understand and manipulate the genetic make-up of organisms for use in the production or processing of agricultural products.

Some applications of biotechnology, such as fermentation and brewing, have been used for millennia. Other applications are newer but also well established. For example, micro-organisms have been used for decades as living factories for the production of life-saving antibiotics including penicillin, from the fungus Penicillium, and streptomycin from the bacterium Streptomyces. Modern detergents rely on enzymes produced via biotechnology, hard cheese production largely relies on rennet produced by biotech yeast and human insulin for diabetics is now produced using biotechnology.

Biotechnology is being used to address problems in all areas of agricultural production and processing. This includes plant breeding to raise and stabilize yields; to improve resistance to pests, diseases and abiotic stresses such as drought and cold; and to enhance the nutritional content of foods. Biotechnology is being used to develop low-cost disease-free planting materials for crops such as cassava, banana and potato and is creating new tools for the diagnosis and treatment of plant and animal diseases and for the measurement and conservation of genetic resources. Biotechnology is being used to speed up breeding programmes for plants, livestock and fish and to extend the range of traits that can be addressed. Animal feeds and feeding practices are being changed by biotechnology to improve animal nutrition and to reduce environmental waste. Biotechnology is used in disease diagnostics and for the production of vaccines against animal diseases.

Clearly, biotechnology is more than genetic engineering. Indeed, some of the least controversial aspects of agricultural biotechnology are potentially the most powerful and the most beneficial for the poor. Genomics, for example, is revolutionizing our understanding of the ways genes, cells, organisms and ecosystems function and is opening new horizons for marker-assisted breeding and genetic resource management. At the same time, genetic engineering is a very powerful tool whose role should be carefully evaluated. It is important to understand how biotechnology-particularly genetic engineering-complements and extends other approaches if sensible decisions are to be made about its use.

UNDERSTANDING, CHARACTERIZING AND MANAGING GENETIC RESOURCES

Farmers and pastoralists have manipulated the genetic make-up of plants and animals since agriculture began more than 10 000 years ago. Farmers managed the process of domestication over millennia, through many cycles of selection of the best adapted individuals. This exploitation of the natural variation in biological organisms has given us the crops, plantation trees, farm animals and farmed fish of today, which often differ radically from their early ancestors. The aim of modern breeders is the same as that of early farmers-to produce superior crops or animals. Conventional breeding, relying on the application of classic genetic principles based on the phenotype or physical characteristics of the organism concerned, has been very successful in introducing desirable traits into crop cultivars or livestock breeds from domesticated or wild relatives or mutants. In a conventional cross, whereby each parent donates half the genetic make-up of the progeny, undesirable traits may be passed on along with the desirable ones, and these undesirable traits may then have to be eliminated through successive generations of breeding.

With each generation, the progeny must be tested for its growth characteristics as well as its nutritional and processing traits. Many generations may be required before the desired combination of traits is found, and time lags may be very long, especially for perennial crops such as trees and some species of livestock. Such phenotype-based selection is thus a slow, demanding process and is expensive in terms of both time and money. Biotechnology can make the application of conventional breeding methods more efficient.

GENOMICS

The most significant breakthroughs in agricultural biotechnology are coming from research into the structure of genomes and the genetic mechanisms behind economically important traits. The rapidly progressing discipline of genomics is providing information on the identity, location, impact and function of genes affecting such traits-knowledge that will increasingly drive the application of biotechnology in all agricultural sectors. Genomics sets the foundation for post-genomics activities, including new disciplines such as proteomics and metabolomics to generate knowledge on gene and protein structure, as well as their functions and interactions. These disciplines seek to understand systematically the molecular biology of organisms for their practical use. A vast range of new and rapidly advancing technologies and equipment has also been developed to generate and process information about the structure and function of biological systems. The use and organization of this information is called bioinformatics. Advances in bioinformatics may allow the prediction of gene function from gene sequence data: from a listing of an organism's genes, it will become possible to build a theoretical framework of its biology. The comparison across organisms of physical and genetic maps and DNA sequences will significantly reduce the time needed to identify and select potentially useful genes.

Through the production of genetic maps that provide the precise location and sequences of genes, it is apparent that even distantly related genomes share common features. Comparative genomics assists in the understanding of many genomes based on the intensive study of just a few.

For instance, the rice genome sequence is useful for studying the genomes of other cereals with which it shares features according to its degree of relatedness, and the mouse and malaria genomes provide models for livestock and some of the diseases that affect them. There are now model species for most types of crops, livestock and diseases and knowledge of their genomes is accumulating rapidly.

BIOTECHNOLOGICAL PRINCIPAL OF AGRICULTURE

Agricultural technology can be defined as biotechnology to include 'any technique that uses living organisms or processes to make or modify products, to improve plants or animals or to develop micro-organisms for specific uses- it focuses upon recombinant DNA and cell fusion technologies'. Longworth concurs with this definition, with the significant addition of tissue-culture techniques, an aspect of biotechnology which has great economic, and therefore social and political, potential impact.

Despite debates about whether these or any other definitions are adequate, they do touch on the main aspects of biotechnology which deserve to be considered. The aspect of this new biotechnology which most captures the imagination and stirs the greatest controversy is gene splicing or recombinant DNA techniques, which inspire researchers to consider the possibilities of producing reproducible animals and plants markedly different from current existing species, and referred to as transgenic species. Already in higher animals and plants recombinant DNA has produced transgenic forms which are being commercially exploited. Their agricultural significance is, however, so far limited, and commercial application is concentrated in highly profitable pharmaceutical and horticultural markets. Genes have been introduced into several animal species which alter their protein synthesis to enable transgenic sheep to produce insulin in their milk, and rabbits to produce interferon. A completely different application has originated in Denmark for salmon, where it has proved possible to introduce germplasm which enables the salmon's physiology to handle heavy metals, which are normally toxic, so opening up new locations for farm fisheries. Much current research is directed to conferring disease immunity on animals, and holds out the prospect of widespread commercial application. In plants one achievement has been the transfer of genetic resistance to antibiotics in the petunia, and another has been the introduction of storage-protein genes from French bean plants into tobacco plants. As yet commercial progress with recombinant-DNA technology in plants appears limited, but extensive opportunities beckon, particularly because plant research is less restricted by the ethical and animal-welfare concerns which apply to research on transgenic animals. The most important commercial developments based on gene splicing have so far occurred with genetically much simpler microorganisms, and it is with these that the greatest short-to mediumrun commercial potential lies.

Already genetically engineered micro-organisms are producing a variety of hormones, vaccines, enzymes and other proteins. Important examples are the production of insulin, vaccines for neo-natal diarrhoea in calves and piglets, and the bovine growth hormone BST identical to that produced in cows which can stimulate a 20 per cent increase in milk yield. BST is already a source of problems for legislators in the European Community, and has provoked strong reactions from the media and milk consumers. As far as larger animals, and cattle in particular, are concerned, it is developments in embryo transfer and in many processes for manipulating reproduction which hold out the prospect of continuing increases in yields, feed conversion efficiency and general economic efficiency.

Already, apparently over 1 per cent of dairy calves in the USA are from embryo transplants, despite the high costs still associated with this procedure. Widespread adoption of these sophisticated technologies would further distance livestock farming from its traditional rural simplicity, and from the natural mating of animals to produce offspring. It would place technical demands upon the operators which favour large-scale company farms capable of supporting a range of highly trained specialists. Longworth identifies new techniques of tissue and cell culture as having 'the potential for enormous advances in crop improvement in the next couple of decades'. Those techniques 'can both increase the genetic diversity and greatly increase selection efficiency', and they permit innumerable plants to be reproduced asexually from single cells or small pieces of tissue. The particular technique known as 'callus culture' has been used for many years to clone highly valued horticultural plants such as orchids, and cloning of cuttings is widely practised for tree crops such as tea and palm oil as well as by millions of gardeners for garden plants. According to Longworth 'cell culture' of single cells has unexpectedly, and so far inexplicably, resulted in plants with different properties being regenerated from the same clump of parent tissue. Sugar cane, maize and potato plants regenerated in this way have been found which are resistant to important pathogens.

This application of cell culture with the capacity to generate vast numbers of seedlings rapidly has the potential to simplify greatly the hitherto labourious procedures of plant breeding and selection. However, in terms of current and immediate commercial importance, it is through micro-organisms that biotechnology has its greatest impact. Microbial fermentation processes have been of great commercial significance for centuries, for example in bread, wine and cheese production, and there is the prospect of considerable development. Two recent examples of new processes indicate the sorts of impact that such developments can have.

In the late 1960s genetically engineered bacteria were developed which were able to digest corn starch to produce high-fructose corn syrup (HFCS) and which left as a residue corn gluten, which is now an important protein feed for livestock. HFCS has made considerable inroads in the United States and in Japan. It has largely replaced sugar as a sweetener in Coca Cola as well as many other food and drink products. This has helped depress sugar prices to Third World growers. In the European Community steps have been taken to prevent HFCS and other new powerful sweeteners from being produced and from undermining the market for domestically grown beet sugar, which is supported by the Common Agricultural Policy.

A second important application of microbial fermentation has been the production of ethanol from sugar cane in Brazil and from maize in the USA.

The commercial viability of these processes is critically dependent upon the price of oil as the main non-renewable source of fuel, and, to date, massive subsidies have been required to maintain the Brazilian and USA ethanol programmes. Longworth states, however, that a new biotechnology, Sucrotech, is being patented, which will not only reduce the cost of producing ethanol from sugar cane, but will simultaneously produce fructose at a cost which will be competitive with HFCS and may thereby reclaim part of the sweetener market for sugar cane.

That such possibilities are in prospect indicates how volatile the future might be; biotechnology has tilted the competitive balance from sugar cane to maize, causing economic pressure and even disruption to sugar-cane-dependent economies and may in future switch it back again. In the long run the capacity to ferment fuels microbially from renewable agricultural feedstocks points to an important long-term reorientation of agriculture if non-renewable oil becomes uncompetitively expensive as a fuel for cars and feedstock for certain chemicals. It suggests that eventually there will be an increased emphasis on agricultural production of industrial feedstocks at the same time as continually increasing food output will be required to feed the expanding world population. All this will require considerable increases in agricultural productivity, to which biotechnology will increasingly contribute, but it will at the same time impose great strains on the natural environment and the structure of agriculture.

USES OF BIOTECHNOLOGY

Biotechnology-the sum of the technologies deriving from molecular and cell biology-will have utility in plant breeding beyond the prebreeding stages. Genetic transformation—non-sexual insertion of alien genes and/or gene regulating systems—is envisaged as eventually being applied, quickly and easily, to finished cultivars, to endow them with a series of useful new traits.

However, experience is beginning to show that genetic background affects expression of genes and gene systems that are transferred *in vitro,* as is also the case when they are transferred via regular sexual mechanisms.

Long periods of trial and error are needed to bring transformed lines into desired levels of expression. Thus, for some years to come, genetic transformation perhaps will be used only as a prebreeding technique, one in which alien genes are put into breeding pools of elite germplasm, following which standard plant breeding methods are used to extract good new varieties containing the alien genes. Cell culture techniques—particularly protoplast fusion techniques—will have utility for making new combinations of nuclear and cytoplasmic genes.

Immediate utility is seen for manufacturing new male sterility systems to make F1 hybrid cultivars, or to develop new kinds of herbicide resistance. With time and experience it may be possible to increase primary productive potential as well, by selecting and fixing optimum mitochondrial and chloroplast genotypes, balanced with proper nuclear genotypes. Such success seems far in the future, however, since at present little is known about optimization of nuclear, mitochondrial, and chloroplastic combinations. DNA technology will help plant breeding in a more traditional way, when information is further developed on maps of restriction fragment length polymorphisms (RFLPs) and their linkages with useful traits in crop plants.

In vitro tests, using RFLP markers, will allow near perfect accuracy in selecting for such useful linked traits. RFLP technology also will be used for precisely characterizing cultivar lineages and relationships, for positive identification of miss-identified or mislabeled cultivars, for characterizing gene frequencies of populations, and even to test hypotheses of landrace evolution, and global movement of crop genotypes. Perhaps the greatest contribution of molecular biology to plant breeding will be a deeper and more precise understanding of gene action and inheritance, of cellular physiology and biochemistry, and perhaps of whole plant physiology. With such deepened understanding, breeding and selection via standard sexual methods will be planned and executed with greater vision and economy. Such knowledge and execution will be, in part, a consequence of efforts to make genetic transform-nation a practical reality. The fall-out of knowledge from genetic engineering efforts, wisely utilized by observant biologists, will be of great benefit to plant breeding, and to plant culture in general.

DEVELOPMENT OF NEW CROPS

In addition to broadening the genetic base of established crops, genetic enhancement-and plant breeding in the sense of final cultivar development-can be used in two other ways: to develop new crops from heretofore uncultivated species, and to change old crops into new crops. Breeding of the first type (creating new crops where none existed before) usually is interpreted also to include development of advanced cultivars from landraces.

Usually the advanced cultivars are made for industrial countries and the landraces come from developing countries. Soybeans and amaranth thus are spoken of as "new crops" in the United States, even though they have been cultivated for centuries as landraces in China, and in Central and South America, respectively. Challenges are great and are about the same, for changing either wild species or landraces into advanced cultivars, able to withstand monocropping and intensive culture at high production levels.

Selection for productivity inevitably requires narrowing the germplasm base, in the course of selecting plant architecture and physiology able to withstand crowding and the stress of high yields, and amenable to machine planting, weeding and harvesting. Planting the narrowly selected genotypes in large areas gives excellent access to specially adapted disease and insect pests, as already noted, and so the need for tolerance or resistance always is next to be added to the list of breeding priorities for new crops. Thus, genetic enhancement first narrows and then later must broaden the genetic base, in the course of new crop development. And as soon as users of the new product get used to larger quantities of it, they will want to standardize and then to change the quality of the product-as, in oil percentage, grain storability and hardness, or flavour and texture. The rule is simple: the more successful the crop, economically, the greater the demands on (and complaints about) the plant breeders. Breeding of the second type-changing old crops into new ones-has been considered as a broad-scale possibility only since concepts of genetic engineering were developed. Ability to produce human insulin in bacteria gave rise to thoughts of producing insulin, or other useful primary gene products, in crop plants. Plants were envisaged as now able to turn sunshine, air and water (and soil nutrients) into high value specialty chemicals, instead of commodity chemicals like corn starch or soybean oil. Most of these ideas still are only ideas, but workers are proceeding to put some of the least complicated ones into action. Modification, via genetic engineering, of the ratios of fatty acid components of rapeseed oil is contemplated. Standard breeding to modify chemical components has been used in Canada already, to make rapeseed into a new crop: "canola."

Canola oil, unlike standard rapeseed oil, is essentially free of erucic acid and glucosinolates and therefore is safe for human consumption when used as a cooking oil. Another possibility derives from recent success in cloning the R gene (first identified by Mendel) in peas which provides the opportunity to identify and transfer, *in vitro*, genes giving new starch branching properties and thus, potentially, useful new kinds of specialty starches in old starch-producing crops such as maize or wheat. A word of caution is needed for those looking to replace major portions of commodity crop production areas with specialty-product new versions of the old crops. High value specialty products, by definition, require only small amounts of product to saturate the market. They won't replace very much of the production area now used for production of commodity crops. On the other hand, they will make intensive use of crop breeding and crop management skills.

They are "brain-power intensive" crops, and as such, must command higher preunit prices—premiums—or they won't be grown.

PURE LINES SELECTION

When dealing with autogamous species, plant breeders have used two broad approaches towards developing new strains. For most species (cereals, soybeans, and many vegetable crops), breeders have been concerned with the creation ofpure line cultivars that are adapted to a specific set of environments and optimized for yield and other desirable characteristics. An important exception is maize, where (for both biological and historical reasons) the goal of breeders has generally been the construction of collections of pure lines whose hybrid seed is the commercial product.

There are two broad issues when considering selection involving pure lines: choosing the best lines froma large collection and the optimal construction of such lines following the initial hybridization. Suppose we have a collection of pure-breeding lines, either because each is completely homozygous or each is asexually (clonally) propagated, and we wish to select the best fraction p. Under this setting, offspring are genetically identical to their parents (ignoring mutation), and between-line selection has the advantage that elite genotypes are propagated intact across generations. Thus, it would appear that between-line selection is very straightforward. If we choose inferior lines, we cannot recover more elite genotypes unless we cross these back to other lines. Hence, attempting to chose the best pure line (or lines) carries with it the risk that we will not improve the population or (more likely) not pick the best genotypes present in the collection of lines. The second broad issue is the actual formation of the lines, which entails two components: when to apply selection

MULTISTAGE SELECTION: FINNEY'S RULE

Lines are often measured over two (or more) growing seasons before selection, with the mean value of the line over seasons used for final selection. This approach partly accounts for genotype £ year interactions, which can have a considerable effect of $\sigma_e^{2\cdot}$ Equation gives the expected response for such a single cycle of selection. Alternatively, a breeder may choose to use several cycles, rather than a single cycle, of selection to choose the elite lines. The logic behind such multistage selection is that the initial cycle can cull out many of the inferior lines, allowing for greater replications (and hence higher precision) to discriminate among the remaining lines.

CONTROLLING THE RESIDUAL VARIANCE

A major advantage of either pure or clonal lines is that one can replicate identical genotypes, replacing individual measurements with group measurements, such as the yield in a plot. Further, individuals can be replicated over environments, such as different locations and/or different seasons/years. The use of replication at several levels (multiple individuals within a plot, plots within a major environment, plots or blocks of plots across major environments) reduces the error

CHOOSING THE BEST GENOTYPES

Selecting for the single best line from a collection of pure lines brings with it the risk that the best genotype is actually not the one choosen. We examine several issues related to this risk. First, what statistical method should be used to pick the "best" lines from a collection of lines. Second, we review the types of statistical errors that can be made. For choosing the best line, our major concern is controlling type III errors (picking inferior lines). We conclude by examining the tradeoff between adding new lines versus increasing the number of replicates per line.

Least Significant Difference (LSD) Tests

A problem that arises in selection among pure lines is determining which lines, from a potentially large number of comparisons, are significantly different from each other. A number of multiple comparison tests, such as Fisher's least significant difference (LSD), Tukey's significant difference, and Duncan's multiple range (to name a few), have been proposed to deal with this issue. Simulation studies by Carmer and Swanson find that a simple modification of Fisher's LSD test is at least as good as any other tested procedure and is the simplest to implement. Fisher's original LSD test compares the absolute value of the difference for the pair of means under consideration with the test statistic.

Advancement of Generations

A key issue in the development of pure lines is how to advance the derived lines from a cross to complete homozygosity and in particular at what stage should selection be practiced. Historically, plant breeders have employed at least some selection during the segregating generations following a hybridization, but for low heritability traits (such as yield), such selection was usually not very effective. Pedigree and bulk selection are examples of such schemes that use either artificial or natural selection to cull undesirable genotypes as the lines are moving towards fixation.

The effectiveness of such early generations selection is often questionable, leaving to advancement of generations approaches that attempt to minimize selection (such as single seed descent and doubled haploids), practicing selection only once the resulting lines are very close to completely inbred.

Outcrossing Hybrids Before Selfing

While most selection schemes start by immediately allowing the F1s to self, the alternative is to allow one or a few generations of intermating before selfing. Such interbreeding schemes are generally not done, as intermating takes time and (above all) effort, as pollinations must be carefully controlled and is often done by hand. The potential advantage of intermating is that is allows recombination to break up favourable alleles tightly linked in repulsion. Such was the situation for two studies with cotton. Miller and Rawlings found a negative genetic correlation between yield and fibre strength, and this correlation was reduced by several generations of intermating. Similar results were reported by Meredith and Bridge. Conversely, if there are favourable alleles tightly linked in coupling, intermating would reduce the potential response relative to immediate selfing.

Pedigree Selection

The idea of pedigree selection to handle the segregating generations of an autogamous species as the lines inbred to complete homozygosity has its roots at the very beginning of modern plant breeding. It represents a compromise at trying to move self-pollinating lines along towards complete homozygosity while still allowing some selection to occur. Under pedigree selection, as individuals become more inbred, selection decisions move from individual plants (during the F2 and perhaps F3 generations) to family-based decisions and finally to extensive yield traits (with extensive replication) in later generations (the F6 or higher) when resulting lines are sufficiently inbred.

There are any number of variants of pedigree selection. Pedigree selection has been historically preferred over mass selection by breeders of autogamous crops as selection can be shifted towards family, rather than individual, performance, increasing the efficiency for low heritability traits. This scheme attempts to balance between being able to locate superior genotypes during early generations of selection while still maintaining sufficient genetic variation for selection in later generations.

Individual selection in early generations requires that plants are grownunder space-planted conditions so that individuals can be scored (as opposed to the spacing used in many crops where individual scoring is essentially impossible). Family selection, on the other hand, often involves planting family members in a bulk, with the entire plot (as opposed to single individuals) being scored. One pedigree scheme turns this testing on its head — under the pure-line family method (or PLF) of Ivers and Fehr, pure lines are used to identify the superior families.

Here, individual F2 plants are harvested and bulk progeny from each plant are selfed for several generations (typically the F4 or F5). A pure line is then extracted from each of these bulks and yield tested. For superior lines, additional members from the original F2 family are chosen for yield testing. of environments, pedigree selection can involve sequential selection, where the different generations are cycled through a series of different environments to minimize the effects of G X E.

While historically pedigree selection schemes have been very widely used for a number of autogamous crops, there are two major concerns with this approach, the effectiveness of individual selection and the effectiveness of early-generation selection. Because individual selection for low heritability traits (such as yield) is very inefficient, one modification of pedigree selection is to focus on only highly heritability traits (such as height or disease resistance) in the early generations, postponing selection on other traits until later generations. For lower heritability traits, several approaches have been suggested to improve the effectiveness of individual selection. One is improved planting designs to reduce the error variance associated with soil heterogeneity. Another approach is the honeycomb design of Fasoulas where individuals are planted in a honeycomb fashion and compared with their six equidistant neighbours.

Wide spacing (*e.g.*, 60cm in wheat) is used to avoid the effects of competition and the neighbourhood comparison will reduce the effect of soil heterogeneity. Mitchell used this design for early generation selection in Durum wheat yield. While they found a slight improvement under this design, it was not deemed sufficiently large to be worth the additional resources. Even if we can alleviate the concerns of individual selection, a second and perhaps more serious problem with pedigree selection is that we are interested in the performance of completely inbred (F1 lines), and F2 or F3 lines are often very poor predictors of the final line performance. Thus, early-generations selection, whether individual- or family-based, may simply not be efficient.

Because of both potential differences in planting designs (space-planting versus more seeding densities) and in the composition of the population, high yielding genotypes in early generations may not be high yielding under normal cropping conditions. In early generations, individual performance is scored in heterogeneous populations (mixtures of different genotypes), but for cropping conditions we seek the best performance in a monoculture. Thus early generations select for performance under allogenotypic competition, while crop conditions require performance under autogenotypic competition. Pedigree selection tends to select for lines that can do well in both mixed and pure cultures, because successive lines must survive both early generations selection and yield testing in later generations. It is thus unclear if lines with the highest performance in pure cultures are those that are also most favoured by pedigree selection. While the pedigree approach has been very widely applied, one sobreing appraisal of its success is offered by Kulshrestha, who examined 630 wheat strains developed using pedigree selection by the Indian government's National Wheat Programme. These lines were tested at a total of 271 locations from 1983 to 1987, and involved tests in nine agro-climatic zones.

Single Seed Descent (SSD) Selection

The motivation behind the single seed descent (SSD) is that since in many cases early-generations selection can be misleading, the simplest approach is to inbred the lines in (as much as possible) the absence of selection. One way to do this is to propagate a single seed from each selfed plant. This is done under space planted conditions (to reduce competition) until the F5 or F6 generation, at which point the resulting lines are expanded and yield-tested. This approach was initially suggested by Goulden, and later by Grafius, Brim, who called this approach the modified pedigree method, and Kaufman, who referred to this as the random method.

The term single seed descent is due to Empig and Fehr. One subtle, but important, advantage of SSD is that it allows populations to be grown in environments quite different from those in which the final cultivar will be grown, without being unduly influenced by selection. Hence, lines can be advanced towards complete inbreeding in a greenhouse and/or in winter nurseries experiencing a very different environment than the final cultivar. SSD has been used on a number of crops – most cereals, soybeans, tomato, lettuce, and safflower to name a few. SSD requires more work than bulk selection, as single seeds must be harvested from each plant, but this is far less work than pedigree selection. When starting a set of lines for SSD, the F1 population should be as large as possible. The number of lines limits the possible range of genotypes that will be recovered. Further, the population size is expected decline over time, as some plants will die out or not seed. Roy found in wheat that SSD lines are impacted by competition but not to the extent of bulk selection. He suggests growing plants under conditions of relatively uniform plant survival to mitigate the effects of competition on the development of SSD lines. In the absence of selection, SSD and pedigree inbreeding are expected to generate the same F1 distribution of lines. This expectation was indeed observed by Jinks and Pooni, who found no differences in phenotypic and genotypic distributions of seven quantitative traits in *Nicotiana rustica* lines developed from SSD and pedigree inbreeding.

Doubled Haploid (DH) Selection

In a number of plants, cytological techniques can allow one to directly create completely homozygous line by producing doubled haploids. As with SSD, the motivation for DH is to sample a collection of completely inbred lines that have undergone as little selection as possible. The advantage of DH over SSD is that a single generation takes a line to complete homozygosity, as opposed to selfing for 5-6 generations (which only takes a line to, respectively, 96.9 per cent and 98.4 per cent homozygosity). This advantage is partially offset by the additional effort required to haploidize a line. In the absence of selection, SSD, and DH produce an identical distribution of pure lines, except when linkage is present. Since DH lines typically undergo only a single round of meiosis (with gametes from the F1's being haploidized), the resulting DH lines can be strongly influenced by parental linkage phase, even for moderately linked loci.

Favourable genes in repulsion in parents tend to remain so, reducing the joint probability of fixation relative to SSD. Conversely, favourable genes in coupling have a higher change of being jointly fixed than under SSD. Powell *et al*. found no evidence of such linkage effects for yield in spring barley, where SSD and DH lines showed the same performance. If linkage is a concern, its effect can be reduced by allowing several generations of meiosis before haploization.

Sampling of Genotypes and Selection Limits

One concern with SSD and DH lines, which holds for pedigree selection on low heritability traits as well, is that only a very small subset of the possible genotypes are represented in the final collection of lines to be tested. If the number of final lines is too small, the chances of recovering favourable genotypes are greatly diminished. Likewise, the selection limit imposed by choosing the best genotype in our sample is also greatly affected by the sample size. Suppose we wish to recover at least *r* lines with a particular genotype.

How many lines must to sampled so have probability *p* that at least *r* such lines are included? This is a binomial sampling problem, with success parameter *q*, sampling from *N* lines. The solution is given by the smallest N satisfying:

N Xj=r

(N°j)! j!

N! qj (1°q)N°j¸ p

This is usually more easily computed by considering the complementary event,

Choice of Parents

The final issue for breeding an optimal pure line is the choice of parents. In particular, what is the best approach to predict parental prepotency — finding the parents whose inbred hybrid offspring will yield the best lines. The principal difficulty is that our interest in the completely-inbred (F1) lines, and these take a number of generations to realise. Further, since our goal is extracting the *best* lines, the mean of the F1 lines from a particular cross may be below average, but this cross may also segregate out the most exceptional lines. A more difficult issue is how to chose parents such that the resulting lines show transgressive segregation for the trait of interest—with some of the derived hybrid lines exceeding the values of either parent. Such is usually the case for yield, where we wish to extract lines with higher yield that either parent. The idea is that each line contains favourable alleles that the other lacks. Recombination generates lines containing more favourable alleles that either parent, and hence a more extreme character value. Ideally, one could use just phenotypic information from parental lines (avoiding the cost and time for line crosses), and we examine such strategies first. The other approach is to actually cross at least certain lines and use this information to increase our predictable ability. One obvious approach is to pick those lines with the best mean values over the first few generations of inbreeding. While a number of workers suggested that using a bulk (measuring all descendent lines in a single batch) was sufficient for predicting the crosses that will generate a high proportion of high-yield

Using only Parental Phenotypic Information

As mentioned, when we wish to combine desirable features, such as performance in particular traits or over particular environments, the choice of parental lines can be rather obvious. At a minimum, it is often clear that certain lines should be avoided as parents, and this limits the combinations we may wish to further test. Conversely, there is no general consensus on the best approach for choosing parents to achieve transgenic segregation. Langham proposed the highlow method, suggesting that transgressive segregation would generate lines with larger yields (on average) from such crosses as opposed to high-high or low-low cross.

The support for this view is mixed. In maize, Johnson and Hayes found that high £ low and high £ high crosses gave the same proportion of high-yielding hybrids (with the most extreme lines coming for the H X L crosses), while Green found that high £ high crosses were best, as did Lonnquist. In wheat, Busch *et al.* found that while the highest performing line came from a high £ low cross, the high £ high cross gave the highest frequency of superior lines.

On theoretical grounds, Bailey and Comstock and Bailey suggest that when one of the parental lines is significantly superior to the other, then any given derived line is less likely to contain more favourable alleles than the original superior parent.

Using Information from Crosses: Diallels and DHs

While one expects that information from crossing parental lines is far more informative than simply using the phenotypic values of the parents, this is not always the case. For example, while the hybrid potential of a particular cross can be immediately assessed from the F1, our concern is the inbreeding potential of the resulting lines (the F1), and these may be poorly correlated with the F1. The most straightforward solution is to use doubled-haploid lines when these can be generated.DHlines immediately represent the completely inbred offspring from a hybrid. There is the minor complication that linkage can result in a different distribution of lines than F1 lines produced by inbreeding. However, in most cases, a set of DH lines from a hybrid provides a quick assessment of parental potential. Reinbergs *et al.* used 100 DH lines per cross, but by random elimination of the data, they found that 20 lines gave essentially the same amount of signal in predicting the most superior crosses.

Example Thurling and Ratinam compared three methods of parent selection for yield improvement in the cowpea in western Australia. These were: selection based simply on parental yield, selection based on the estimated GCA of the parents (requiring parental and F1 data), and selection based on the F2 lines. A total of 10 potential parents were tested, resulting in a total of 45 crosses. Seven of the 45 F2 lines had a higher yield that the best of the 10 original parents, while 12 had yields below the worst parent.

The F2 yield from the cross between the lines (G and C) with the two largest GCAs was exceeded by 10 of the other F2 lines. Indeed, C and G had a rather significant negative interaction, with 6 of the other 8 F2's involving C having higher values than the C £ G cross (only one of the other F2's involving

G had a higher value). Hence, GCA was a poor predictor of F2 yield. When F3 bulks (selfed collections of the F2s planted in bulk) were examined, yield was higher for bulks whose parents were chosen by their F2 values than by their GCA values (10.75 gr/plot vs. 8.78 gr/plot), but the difference was not significant. Both the GCA and F2 selection schemes gave larger F3 bulk yields than the cross chosen simply from parental yield (6.44 gr/plot).

Number of Crosses vs. Number of Lines per Cross

There are two areas of risk when trying to chose elite genotypes: not choosing the best crosses and not choosing the best lines within a cross. What is the optimal strategy for balancing these? If a total of N lines are to be used, the breeder must weight the number of crosses m against the number of lines n from each cross (with $N = mn$). In the absence of any trait information, how should the breeder chosen m and n so as to minimize the risk that no elite genotypes exist in the lines being examined? This question has been examined by Yonezama and Yamagata, Weber and (most generally) Wricke and Weber.

Wricke and Weber show that, in the absence of any phenotypic information on the crosses, the risk is minimized by maximizing the number of crosses (taking $m = N$ and hence $n = 1$). There is also a tradeoff between the within and between cross variance. Wide crosses are expected to have somewhat intermediate performance but a high within-cross segregation variance, while crosses between closely related lines may have higher mean cross values but also smaller within cross variance, reflecting fewer genetic differences than between the parents of wide-species crosses.

RECURRENT SELECTION

The goal of recurrent selection is to *improve the mean performance* of a population of plants; a secondary goal, but nevertheless very important, is to *maintain the genetic variability* present in the population to the extent possible. Progress in selection is based on the heritability of the trait and the types of genetic variation controlling the trait in the particular population under selection and on the selection differential. We will initially discuss various recurrent selection methods, and then discuss the genetic parameters that are associated with them, as a means to differentiate between them.

The generalized method consists of three parts:

1. Development of a base population with which to begin selection,
2. Evaluation of individuals from the population
3. Selection of superior individuals and intercrossing to form a new population.

Evaluation of Individuals in the Population

Individuals can be evaluated in one of two ways: *Phenotypic evaluation* considers the individual plant *per se*, so that no evaluation of the actual *breeding value* of the individual is possible. In contrast, *genotypic evaluation* is based on the performance of the progeny of the plant under evaluation, and therefore, provides information on the individual's breeding value.

Three types of progeny may be evaluated:

1. Half-sib progeny,
2. Full-sib progeny, and
3. Selfed progeny.

Half-sibs are formed by crossing the individuals to be evaluated to a common parent (which can be a population, an inbred line, or a specific genotype), called a *tester*; thus, all half-sib families have the tester in common. To produce full-sib families, the individuals to be evaluated are crossed in pair wise combinations–thus, full sib families do not have any parents in common, although all the parents come from the same population.

Finally self-fertilizing the individuals to be evaluated for one or two generations produces selfed progeny. These partially inbred progeny are evaluated.

Methods of Intrapopulation Improvement

Recurrent Phenotypic Selection

Recurrent phenotypic selection (RPS), also termed mass selection is the simplest form of recurrent selection.

Typically, RPS has three stages:

1. A population of plants is planted in a nursery in a manner that allows notes to be taken on individual plants,
2. The population is evaluated for the trait/traits of interest and the best individuals are identified,
3. Seed is harvested from the selected individuals to constitute a new population from which the next cycle will begin.

Evaluation can only occur after pollination, obviously, so when the best ears are selected from individual plants, clearly the pollen has derived from the entire population, and only the female parent is controlled.

Perennial plants, in contrast, can be dug up from their location and all selections can be transplanted to a isolated crossing nursery (called a *polycross* nursery) where the plants are typically replicated to ensure random intercrossing among all the selections, or to the greenhouse to enable hand crossing among them.

Recurrent half-sib Selection

The general scheme for recurrent half-sib selection (RHSS) is to cross individuals being evaluated to a common tester, evaluate the half-sib progeny of each plant, select the best individuals, and intercross them. RHSS can be conducted for either *general* or *specific combining ability*. General combining ability (GCA) refers to the average performance of an individual's progeny, which are developed by crossing that individual to a variety of other individuals.

If the tester being used is a population of plants, then the progeny performance represents the parental plant's GCA. If the tester were an inbred line, then the progeny performance would represent specific combining ability (SCA), because it would represent the ability of a particular plant to combine with a specific inbred rather than with a broad selection of genotypes. The ears on the tester, then, represent the half-sib family to be tested.

Recombining selfed progeny will require three seasons:

1. Selfing and crossing to the tester,
2. Evaluation,
3. Intercrossing selfed progeny.

This will lengthen a cycle of selection. The benefit gained from using selfed rather than half-sib seed may be offset by the longer selection cycle

Recurrent Full-sib Selection

Recurrent full-sib selection (RFSS) tests paired plant crosses rather than crosses of individual plants with a population (or inbred) tester. For this method, paired crosses are made between individuals in the population in season one. Evaluations are conducted in the field the next season and the best families are identified.

The third season is used to recombine the best families using remnant seed from the first season. Thus, the second selection cycle begins in the fourth season, with paired crosses between individuals.

To streamline procedures, paired crosses could be done during the third season, combining recombination of the best families with full-sib family development for the next evaluation cycle. Thus, paired crosses between individuals from the selected families can be made. The disadvantage of this is that less recombination occurs between cycles of selection. The obvious advantage is that it allows a cycle to be completed each year, with the use of a single off-season nursery.

Recurrent Selection Among Selfed Families

- Evaluation is based on self-pollinated progeny, usually formed by one generation of selfing, although two generations has also been used.
- *Season* 1: S0 plants from the population are selfed to produces S0: 1 line.
- *Season* 2: S0: 1 lines evaluated, best lines selected
- *Season* 3: Remnant S1 seed used to intercross selected lines, producing the Cycle 1 population with new S0 plants.
- *Season* 4: Self-S0 plants as in Season 1.

- Many modifications are possible. Could test S1: 2 lines or S0: 2 lines, depending on the desire of the breeder. The latter (S0: 2) is useful if you cannot produce enough S0:1 seed for testing.

Methods of Interpopulation Improvement

Reciprocal Recurrent Selection

It designed to improve both the *general* and *specific combining ability* of two populations. The general idea is to test and select in each of two populations, using the opposite population as a tester. Alleles do not migrate between populations, however! Two starting populations should be genetically divergent (although selection will probably drive them apart anyway)–the point is that they should combine well together.

Reciprocal half-sib Selection

- *Season* 1: Individual plants (*e.g.*, 100) from each population are (1) selfed and (2) outcrossed to several (*e.g.*, 6) randomly chosen plants from the opposite population to generate ½ sib families.
- *Season* 2: 100 ½ sib families from each population are evaluated in replicated tests; top 10 are identified.
- *Season* 3: Go back to selfed seed of superior individuals and intercross to form Cycle 1 population *within each population*. DO NOT mix the selections of the two populations!
- *Season* 4: Begin next cycle using Cycle 1 seed of each population.
- *Modifications*:
  - Use inbred testers developed from the opposite population. The inbred tester would be replaced by superior ones as the programme progressed.
  - Produce half-sib seed of each population as in ear-to-row selection. Then plant half-sib fam ilies in alternate rows with the opposite tester population, detassel the half-sib lines and use the open pollinated seed for testing. The best lines are then intermated using the remnant half-sib seed of each parent. The disadvantage is that only the female parent is controlled.

Reciprocal full-sib selection

- Used when the desired outcome is commercial hybrid seed.
- *Season* 1: 200 phenotypically desirable S0 plants in Pop A paired with 200 from Pop B. Each plant is both selfed and crossed to produce S1 and full-sib seed. Selfed seed can be stored for intercrossing of desirable parents and also advanced to develop inbred lines.

- *Season* 2: Evaluation and selection.
- *Season* 3: Intercrosss selfed seed of top 20 individuals from each population–>Cycle 1 populations for A and B.
- *Season* 4: Begin again.
- Alternative if can't produce both selfed and crossed seed at the same time: use S0: 1 lines to make crosses. Adds a year.

MARKER-ASSISTED SELECTION

Marker-assisted selection (MAS) is a method of selecting desirable individuals in a breeding scheme based on DNA molecular marker patterns instead of, or in addition to, their trait values. When used in appropriate situations, it is a tool that can help plant breeders select more efficiently for desirable crop traits.

However, MAS is not always advantageous, so careful analysis of the costs and benefits relative to conventional breeding methods is necessary.

Conventional plant breeding is dependent on appropriate environmental conditions in which to identify and select desirable plants. Typically, breeders improve crops by crossing plants with desired traits, such as high yield or disease resistance, and selecting the best offspring over multiple generations of testing.

A new variety could take 8 to 10 years to develop. Breeders are very interested in new technologies to speed up this process or make it more efficient.

Requirement for an Efficient Marker-Assisted Selection Programme

High Throughput DNA Extraction

Most breeding programmes would need to screen hundreds to thousands of plants for desired marker patterns. In many cases, the results will be needed quickly to allow the breeder to make selections in a timely manner.

Both of these considerations demand a DNA extraction system that can handle a large number of samples in a streamlined operation. Many labs conducting MAS extract DNA from small tissue samples in 96- or even 384-well plates.

Genetic Markers

The species of interest should have available a technology that allows detection of markers along all of the plant's chromosomes.

Although DNA markers have received the most attention, other types of markers (protein, morphological, cytological) can also be used in MAS programmes. DNA polymorphisms as markers have the advantages that large numbers are available throughout the genome, their presence or absence is unaffected by environment, and they usually do not directly affect the phenotype.

Commonly used DNA markers are referred to by their acronyms: RFLP, SSR, RAPD, AFLP, SCAR, and SNP. Each type of marker has advantages and disadvantages for specific purposes.

For efficient MAS, important attributes of markers include:

- Ease of use
- Small amount of DNA required
- Low cost
- Repeatability of results
- High rate of polymorphism
- Occurrence throughout the genome
- Codominance

Knowledge of Associations between Molecular Markers and Traits of Interest

The most crucial ingredient for MAS is knowledge of markers that are associated with traits important to a breeding programme. This information might come from gene mapping or QTL studies, bulked segregant analysis, classical mutant analysis, or some other means.

Data Management System

Large numbers of samples are handled in a MAS programme, with each sample potentially evaluated for multiple markers. This situation requires an efficient system for labeling, storing, retrieving, and analysing large data sets, and producing reports useful to the breeder.

Potential Advantages of MAS

MAS can theoretically enhance selection efficiency because:

- It can be Performed on Seedling Material, thus reducing the time required before a plant's genotype is known. In contrast, many important plant traits are observable only when the plant has reached flowering or harvest maturity. Knowing a plant's genotype before flowering can be particularly useful in order to plan the appropriate crosses between selected individuals.
- MAS is not Affected by Environmental Conditions Some crop production constraints (such as disease, insect pests, temperature and moisture stress) occur sporadically or non-uniformly. Therefore, evaluating resistance to those constraints may not be possible in a given year or location. MAS offers the chance to determine a plant's resistance level independent of environment.
- When recessive alleles determine the trait of interest, they cannot be detected through phenotypic evaluation of heterozygous backcross plants, because their presence is masked by the dominant allele. In a

traditional backcross programme, plants with recessive alleles are identified by progeny evaluation after self-pollination or testcrossing to a recessive tester. This time-consuming step can be eliminated in a MAS programme, because recessive alleles are identified by linked markers.

- Similarly, when multiple resistance genes are "pyramided" (combined) together in the same variety or breeding line, the presence of each individual gene is difficult to verify phenotypically. The presence of one resistance gene may conceal the effect of additional genes. This problem can be overcome if markers are available for each of the resistance genes.
- Environmental variation in the field reduces a trait's heritability, the proportion of phenotypic variation that is due to genetics. In a low heritability situation, progress from phenotypic selection will be slow, because so much of the variation for the trait is due to environmental variation, experimental error, or genotype x environment interaction, and will not be passed on to the next generation. If a reliable marker for a trait is available, MAS can result in greater progress than phenotypic selection in such a situation.
- MAS may be cheaper and faster than conventional phenotypic assays, depending on the trait. For example, evaluating nematode resistance is usually an expensive operation because it requires artificial inoculation of plants with nematode eggs, followed by a labour-intensive technique to count the number of nematodes present. Selecting on the basis of a reliable marker would probably be cost-effective in this case. On the other hand, plant height is cheap and easy to measure, so there may not be an economic advantage in using markers for that trait.
- A consideration that may affect cost effectiveness of MAS is that multiple markers can be evaluated using the same DNA sample. Extraction of DNA from plant tissue is one of the bottlenecks of MAS. Once DNA is extracted and purified, it may be used for multiple markers, for the same or different traits, thus reducing the time and cost per marker.

Potential Drawbacks of MAS

MAS is not universally advantageous. Some limitations of the technique are as follows:

- MAS may be more expensive than conventional techniques, especially for startup expenses and labour costs.
- Recombination between the marker and the gene of interest may occur, leading to false positives. For example, if the marker and the

gene of interest are separated by 5 cM and selection is based on the marker pattern, there is an approximately 5 per cent chance of selecting the wrong plant. This is based on the general guideline that across short distances, 1 cM of genetic distance is approximately equal to 1 per cent recombination. The breeder will need to decide the error rate that is acceptable in the MAS programme, keeping in mind that errors are also usually involved in phenotypic evaluation.

- Sometimes markers that were used to detect a locus must be converted to "breeder-friendly" markers that are more reliable and easier to use.

  Examples are:

  - *RFLP markers converted to STS markers* (Ribaut *et al*., 1997).

    Evaluation of RFLP markers requires several steps and a large quantity of highly purified DNA. STS (sequence tagged site) markers can be detected via PCR using primers developed from RFLP probe sequences. Thus the same locus can be detected with the two types of marker, but the STS marker is far more efficient.
  - *RAPD markers converted to SCAR markers* (Hernandez *et al*., 1999).

    Results of RAPD reactions may vary from lab to lab, and therefore, may be considered less reliable for MAS. Variable RAPD results are due largely to short (10-base) PCR primers, resulting in low binding specificity. SCAR markers are developed by sequencing RAPD bands and designing more specific 18-25 base PCR primers to amplify the same DNA segment more reliably.
- Imprecise estimates of QTL locations and effects may result in slower progress than expected. Many QTLs have large confidence intervals of 20 cM or more or their relative importance in explaining trait inheritance has been over-estimated.
- Markers developed for MAS in one population may not be transferrable to other populations, either due to lack of marker polymorphism or the absence of a marker-trait association.

Conducting a MAS Programme

The basic procedure for conducting MAS with DNA markers is as follows:

- Extract DNA from tissue of each individual or family in a population.
- Screen DNA samples via PCR for the molecular marker (SSR, SNP, SCAR, etc.) linked to the trait of interest.
- Separate and score PCR products, using an appropriate separation and detection technique.

- Identify individuals exhibiting the desired marker allele.
- Combine the marker results with other selection criteria (*e.g.*, phenotypic data or other marker results), select the best fraction of the population, and advance those individuals in the breeding programme.

Future Outlook

MAS has already proven valuable for backcrossing of major genes into elite parents, using both foreground and background selection. This use is expected to increase as new genes and associated markers for economically important traits are identified. As an example of current opportunities for MAS in wheat, protocols for over 20 trait-associated markers are posted on the web site MAS Wheat: Bringing Genomics to the Wheat Fields. Knowledge of actual gene sequences and design of PCR primers for specific alleles (*e.g.*, Ellis *et al.*, 2002), will make MAS more powerful and informative across a range of genetic backgrounds.

The use of MAS for multiple QTLs for complex traits is not as clear-cut. This is due to imprecise estimates of QTL location and effect size, as well as interactions of QTLs with environment and genetic backgrounds. Ongoing improve-ments in statistical methodologies and experimental design (*e.g.*, larger population sizes grown in more environments) will help remedy this situation. In many cases, integration of MAS for QTLs with phenotypic selection would seem a reasonable approach, but the optimum strategies for accomplishing this are not well developed. Efficient incorporation of MAS into an applied breeding programme will likely require a rethinking of the structure and implementation of the entire programme (Ribaut and Hoisington, 1998). Finally, economics will be a major driver of the application of MAS. For certain traits that are expensive or logistically difficult to evaluate, MAS is an attractive alternative. Time savings obtained through MAS may be as important as cost savings where there are competitive markets for improved cultivars. Any cost change in DNA extraction or genotyping methods, or on the other hand, in phenotypic evaluation methods, will affect the relative economic benefits of MAS.

In summary, MAS is a methodology that has already proved its value in some plant breeding situations. It is likely to become more valuable as a larger number of genes are identified and their functions and interactions elucidated. Reduced costs and optimized strategies for integrating MAS with phenotypic selection are needed before the technology can reach its full potential.

PHOTOLYSIS AND CARBON FIXATION

PHOTOSYNTHESIS

Photosynthesis is the means that primary producers (mostly plants) can obtain energy via light energy. The energy gained from light can be used in various processes mentioned below for the creation of energy that the plant will need to survive and grow. Photosynthesis is a reduction process, where hydrogen is reduced by a coenzyme. This is in contrast to respiration where glucose is oxidised. The process is split into two distinct areas, *photolysis* (the photochemical stage) and the *Calvin Cycle*. The diagram below gives a summary of the reaction, where light energy is used to initiate the reaction in its presence:

$$CO_2 + H_2O \rightarrow \text{glucose} + \text{oxygen}$$

PHOTOLYSIS

This part of photosynthesis occurs in the *granum* of *a chloroplast* where light is absorbed by *chlorophyll*; a type of photosynthetic pigment that converts the light to chemical energy. This reacts with water (H_2O) and splits the oxygen and hydrogen molecules apart. From this dissection of water, the oxygen is released as a by-product while the reduced hydrogen acceptor makes its way to the second stage of photosynthesis, the Calvin cycle. Overall, since the water is oxidised (hydrogen is removed) and energy is gained in photolysis which is required in the Calvin cycle

The Calvin Cycle

Also known as the carbon fixation stage, this part of the photosynthetic process occurs in the *stroma* of chloroplasts. The carbon made available from breathing in carbon dioxide enters this cycle, which is illustrated below:

Just like the *Kreb's Cycle* in respiration, a substrate is manipulated into various carbon compounds to produce energy. In the case of photosynthesis, the following steps occur, which create glucose for respiration from the carbon dioxide introduced into the cycle;

- Carbon from CO_2 enters the cycle combining with Ribulose Biphosphate (RuBP)
- A compound formed is unstable and breaks down FROM its 6 carbon nature to a 3 carbon compound called glycerate phosphate (GP)
- Energy is used to break down GP into triose phosphate, while a hydrogen acceptor reduces the compound therefore requiring energy
- Triose Phosphate is the end product of this, a 3 carbon compound which can double up to form glucose, which can be used in respiration.
- The cycle is completed when the leftover GP molecules are met with a carbon acceptor and then turned into RuBP, which is to be joined with the carbon dioxide molecules to re-begin the process.

The energy that is used up in the Calvin cycle is the energy that is made available during photolysis. The glucose that is made via GP can be used in respiration or a building block in forming *starch* and *cellulose*, materials that are commonly in demand in plants.

Limiting Factors in Photosynthesis

Some factors affect the rate of photosynthesis in plants, as follows

- *Temperature* plays a role in affecting the rate of photosynthesis. Enzymes involved in the photosynthetic process are directly affected by the temperature of the organism and its environment
- Light Intensity is also a *limiting factor*, if there is no sunlight, then the photolysis of water cannot occur without the light energy required.
- Carbon Dioxide concentration also plays a factor, due to the supplies of carbon dioxide required in the Calvin cycle stage.

Overall, this is how a plant produces energy which supplies a rich source of glucose for respiration and the building blocks for more complex materials. While animals get their energy FROM food, plants get their energy FROM the sun.

CONTROL OF BIOTECHNOLOGY DEVELOPMENT

The role of the public sector to undertake research and development of agricultural technology as a public service to firms which might profit from translating that R & D into a commercial product or process, to farmers who might profit from adoption of the technology, and perhaps most importantly to consumers at home and abroad who benefited from the lower prices resulting from greater abundance. This was particularly true of the phase of change dominated by improvements in plant and animal breeding, the 'old' biotechnology. While the public sector had a role in basic research for chemical and mechanical technologies, the benefits of research expenditure in these areas were easier to capture by private companies investing in R & D, so that increasingly the public sector has taken a smaller role in these areas although maintaining a strong regulatory role in regard to agricultural chemicals in particular. Where it is impossible to prevent others from escaping payment for the research costs, either because the product is easily copied (seeds which can be regenerated by farmers or other firms) or because proposals for new methods can be readily implemented, private firms are understandably unwilling to invest.

Machines, insecticides, fungicides and other manufactured inputs do lend themselves more readily to private exploitation. Nevertheless the returns to R & D in these products do depend upon the degree of difficulty potential competitors would have in copying the product. In some cases there are inherent technical difficulties in copying the process, or it would be prohibitively expensive, but in other cases it is the ability to obtain patent protection which creates legal barriers to potential competitors' ability to become 'free-riders' and which protects incentives to private investment in R & D. Traditionally, however, it has been impossible for plant and animal breeders to obtain patent rights for their products, which is one reason why public-sector R & D has remained so important in this area.

For, as under the European Patent Convention (EPC) of 1973, it has been judged impossible for plant varieties to satisfy one of the key criteria to qualify for a patent, namely proof of 'an inventive step'. The application of standard breeding practices to generate new varieties by crossing existing plants or animal strains has not been deemed to be invention. Thus under the EPC one set of exclusions from patentability is 'Plant or animal varieties or essentially biological processes for microbiological processes or the products thereof.' In the absence of patent rights plant-breeding firms in particular have worked hard to obtain other means of protection and royalties for their products.

The history of the development of Plant Breeders' Rights (PER) is presented by Mooney, and is of particular interest because of the concerns he expresses about the consequences of allowing the basic genetic stock, which underpins agriculture and hence our whole society, from becoming private rather than public property. While Mooney's concerns are expressed in relation to 'old' biotechnology varieties of plants, they are of particular importance with respect to the products of the 'new' biotechnology. For bio-engineered products are capable of meeting the inventive-step criterion of patentability, and both micro-organisms and transgenic animals have now been patented in the USA. Before examining developments in the seed industry it is worth touching upon the implications of changes in research policy which stress increasing reliance upon private R & D to develop and exploit the new biotechnology and other technologies. In the UK government has decided that it should withdraw from funding what it terms 'near-market research', that is, research beyond the basic phase and which is preparation for commercial exploitation. This policy is based in part upon the arguments that industry should be investing more heavily in R & D, and that research strategies at the near-market stage should be driven by assessments of likely commercial success which can best be made by the firms involved, and will therefore lead to greater efficiency in allocating research funds.

This has led to the closure of a number of government-financed agricultural research institutes, the scaling down of others and the sale of the National Seeds Organization by auction to Unilever; Unilever won against competition with BP and other major public companies. This deliberate attempt to switch an increasing proportion of agricultural R & D expenditure from the public to the private sector carries with it a number of risks. In the first place there is a controversy about whether there is under-investment in R & D so that the returns to extra expenditure are high, or whether the converse is the case. If there is under-investment, then withdrawal of public support for applied research will exacerbate this, as the private sector will only undertake R & D expenditure on those products and processes from which exclusive benefits can be captured by the investor.

Thus in relation to biotechnology in the USA, where again public support for applied biotechnology research is relatively small, Stallman and Schmid argue that there will be emphasis on technologies which are applied in factory conditions where secrecy and control can be maintained. For technologies which cannot be confined to factories, such as seeds, Stallman and Schmid state that 'Firms are also considering mechanisms which "scramble" the genome of a plant in the second generation', in order to prevent farmers from reproducing seed with the enhanced, engineered characteristics. Clearly such actions are designed to frustrate the maximum spread of benefits from the technology and to maximize private profit for companies investing in research.

Another facet of commercially orientated biotechnology research is that it will aim at the most important crops and livestock products, those produced by the largest farming units, and those which are most heavily subsidized. In the latter case this will worsen the budgetary problems of adjusting agricultural policy in OECD countries. The probable neglect of minor crops, difficult habitats and small farmers means that the public sector will have a defined role in agricultural technology research, but one in which it will be relegated to the second division and where it is unlikely to prove successful in terms of the commercial yardstick of rates of return which is increasingly emphasized by public-research policy.

AGRICULTURAL BIOTECHNOLOGY REGULATION

For traditionally bred plants, regulators rely on plant breeders to conduct appropriate safety testing and to be the first line of defence against genetic alterations that might prove dangerous. The same should be true for agricultural products developed using biotechnology. There is no reason for regulators to treat biotech plants any differently than traditionally bred plants, particularly given the fact that biotechnology provides greater control over gene manipulation.

Despite fears about gene manipulation, traditional crossbreeding has altered plant genes and improved the human diet for the past 30 years. In fact, people worldwide safely consume these plants every day. Examples include wheat, corn, potatoes, tomatoes, and countless other staples of the American diet. In 1984, the biotechnology industry began experimenting with "gene-spliced plants" — the more advanced approach to gene manipulation that we now call "biotechnology."

DEPARTMENT OF AGRICULTURE

USDA regulates the release of all genetically engineered agricultural plants under statutes giving USDA's Animal and Plant Health Inspection Service (APHIS) the authority to regulate plants that may be or may become weeds or other nuisances — what the statutes call "plant pests." Although the rules apply in a general sense to novel or exotic varieties of both gene-spliced and conventional plants, APHIS typically only requires field testing of conventional plants that are new to a particular U.S. ecosystem (transplanted from another continent, for example).

However, all genetically engineered agricultural plants face a higher regulatory bar — making it more difficult to expand the use of biotechnology. Genetically engineered crops must be field tested under APHIS regulations prior to commercialization. For example, a new variety of corn produced with conventional hybridization requires no government-mandated field testing, but all new varieties of genetically engineered corn do, even though there is no logical reason for the regulatory disparity. For most genetically engineered plants, APHIS requires the company producing the plants to submit notice detailing the gene or genes that have been inserted, where the plant will be tested, and other relevant characteristics of the plant prior to receiving permission to conduct the field trials. Once the company completes field testing, APHIS reviews the results and makes a determination on whether or not the product should be "deregulated" and can be released into the market.

REGULATORY SCHEME

At the time, the White House Office of Science and Technology Policy began crafting a framework wherein existing federal agencies would regulate genetically engineered organisms on the basis of their characteristics, not the method of production — thus wisely deferring to the scientific community's judgement that regulation ought to address the products, not the process of biotechnology.

Engineered organisms would not require extra scrutiny simply because genetic engineering produced them (the process). Instead, they would be subject to heightened scrutiny only if the individual organisms expressed characteristics that posed some conceptually heightened risk (the product).

The federal government divided regulatory jurisdiction among agencies already involved in agricultural, food, and environmental regulation. These include the United States Department of Agriculture (USDA), the Environmental Protection Agency (EPA), and the Food and Drug Administration (FDA). While each of these agencies considers the characteristics of individual products in their regulation, only FDA followed the general scientific thinking that genetically engineered and non-genetically engineered products should be regulated similarly. Both USDA and EPA automatically subject all genetically engineered plants as a class to premarket approval requirements not ordinarily applied to conventionally bred plants.

ENVIRONMENTAL PROTECTION AGENCY

EPA regulates plants that are modified to produce substances that act like pesticides: that is, substances used by a plant to protect itself from pests, such as insects, viruses, and fungi. EPA's proposed rule for these "plant-incorporated protectants" is not yet finalized. In the interim, however, EPA has regulated such plants by applying its proposed guidelines, which are functionally similar to rules for the registration of synthetic chemical pesticides. Again, biotech products face higher regulatory hurdles, even though plant-incorporated protectants developed through conventional breeding are exempt from these requirements.

HEALTH AND HUMAN SERVICES: FOOD AND DRUG ADMINISTRATION

FDA is responsible for ensuring that food items, including foods derived from genetically modified plants, are safe to eat. Following the general regulatory framework that emphasizes product regulation rather than process regulation, FDA rightly does not treat foods derived from genetically engineered plants to be inherently unsafe. Food producers are not required to seek premarket approval from FDA unless there is a substantive reason to believe that the novel trait (or traits) in the food pose a safety question. The initial determination of safety is left to the producer, but FDA has encouraged producers to consult with agency scientists prior to marketing a food produced with biotechnology to ensure that the appropriate determination is made. Recently, FDA published a proposed rule that would require producers to notify the agency at least 120 days before marketing, which may be a signal that FDA could abandon its more reasonable approach in the future.

In addition, FDA does not require labelling of foods derived from biotechnology unless the genetic insertions so alter the food that the common name no longer describes it adequately. Examples of this phenomenon would include such alterations of the food product that would raise a safety issue or change the product's nutritional composition or its storage or preparation characteristics.

INTERNATIONAL TRADE

While U.S. consumers do not appear to be strongly opposed to biotech foods (they, in fact, seem rather indifferent), a strong anti-biotechnology movement has arisen in several European countries over the last few years. The European Union (EU) has established strong restrictions on the commercial planting of genetically engineered plants, and European food processors and retailers are reluctant to import harvested agricultural products derived from biotechnology — thus jeopardizing the marketability of U.S. commodity grain exports. And the EU is now negotiating for strong restrictions on agricultural biotech products in important international agreements governing food safety and environmental protection. Very strong restrictions were included in the Biosafety Protocol, finalized in January 2000. Perhaps, more important, though, are the Codex Alimentarius Commission standards for food safety. Without convincing scientific evidence that genetically engineered crop plants pose a heightened environmental or human health risk, restrictions on agricultural biotech imports could be challenged under the General Agreement on Tariffs and Trade and the World Trade Organization. Thus the EU has resorted to justifying its decisions on the basis of a risk-management philosophy known as the "precautionary principle," and the EU advocates inclusion of the precautionary principle in international environmental and food safety agreements (such as the Codex Alimentarius), as well as within GATT itself. Its inclusion would give nations greater latitude in restricting imports of U.S. agricultural products.

No single definition exists for the precautionary principle, but its general meaning is that when an activity raises threats of harm to human health or the environment, regulatory measures should be taken to prevent or restrict the activity even if the risks have not been demonstrated scientifically. Although the EU asserts that the precautionary principle is an unbiased risk-management philosophy, critics have noted that its lack of definition and evidentiary standards makes it all too easy to abuse for the purpose of masking trade protectionism, and that its very approach to risk management is inherently flawed and may, in fact, increase net risk.. Nevertheless, the precautionary principle has already been included in several international environmental treaties, including the Convention on Biological Diversity and the Biosafety Protocol.

LABELLING ISSUES

Some activists, however, argue that the government should mandate the labelling of all genetically engineered foods. They assert that consumers have a "right to know" how their foods have been altered, and that a mandatory label would best allow consumers to choose between genetically engineered and conventional foods. Biotechnology advocates have argued against mandatory labelling because such requirements raise food costs— something that mostly harms lower-income Americans and people on fixed budgets. Perhaps more important, while biotech products are not substantially different from other products, special labels would likely make consumers think these products were more dangerous.

Hence, rather than serving educational or "right to know" purposes, such labels promise to simply confuse consumers. A government-mandated label on all genetically engineered foods also would raise important First Amendment free speech issues. In 1996, the U.S. Court of Appeals, in *International Dairy Foods Association et al. v. Amestoy*, ruled unconstitutional a Vermont statute requiring the labelling of dairy products derived from cows treated with a genetically engineered growth hormone, noting that food labelling cannot be mandated simply because some people would like to have the information. "Absent ... some indication that this information bears on a reasonable concern for human health or safety or some other sufficiently substantial governmental concern, the manufacturers cannot be compelled to disclose it."

In other words, to be constitutional, labelling mandates must be based in science and confined to requiring disclosure of information that is relevant to health or nutrition. Furthermore, consumers need not rely on mandatory labelling of biotech foods to truly have a choice. Real-world examples show that market forces are fully capable of supplying information about process attributes (including kosher and organic production standards) that consumers truly demand. The same can be said about non-biotech foods, and the FDA recently published proposed guidelines to assist producers in voluntarily labelling both genetically engineered and non-genetically engineered foods. Additionally, the USDA's newly published rule for organic certification necessarily excludes biotech products from organic food production. Consequently, consumers wishing to purchase non-biotech foods need only look for certified organic products.

Even as underdeveloped nations clamor for biotechnology applications, and as countries like China continue to experiment with and use agricultural biotechnology, opponents of agricultural biotechnology in the West, particularly Europe, attack it as an unnatural process that will destroy the world, not better it. They argue that biotechnology should be heavily regulated, if not banned. Already, however, genetically engineered plants are subject to strict regulatory oversight that is equal to or greater than that advocated by the vast majority of scientific specialists.

Additional regulation will slow down research and development of genetically engineered crops, keep beneficial products off the market, and raise the cost of products that do make it to consumers. Furthermore, the inclusion of similar restrictions — or inclusion of the precautionary principle— in international agreements will greatly impact the international trade of agricultural goods and delay their introduction into the marketplace. Each of these problems could prevent this technology's benefits from being introduced to industrialized nations and, more important, the developing world.

PAST SUCCESSES IN REDUCING POVERTY THROUGH AGRICULTURAL SCIENCE

Science and technology underpinned the economic and social gains in Asia over the past 30 years. In agriculture, these gains came to be known as the Green Revolution. Between 1970 and 1995, cereal production in Asia doubled, calorie availability per person increased by 24 per cent, and real food prices halved. Although the region's population grew by 1 billion people, overall food production more than kept pace with population growth.

These food production increases were achieved largely by the cultivation of high-yielding varieties (HYVs) of rice and wheat, accompanied by expansion of irrigated areas, increases in fertilizer and pesticide use, and greater availability of credit. The scientific basis for the Green Revolution stemmed from national and international research programmes that led to the development and distribution of new HYVs, particularly of rice and wheat. The first generation of these new varieties were based on the introduction of new genes for dwarfing that made the HYVs shorter, more responsive to fertilizers, and less prone to falling over or lodging when fertilized and irrigated.

Subsequent varieties also carried genes that gave increased pest and disease resistance and improved taste and grain quality. The key elements in improving food security in Asia from 1970-95 were government policies reflecting a belief that investments in increasing agricultural productivity were a prerequisite to economic development. These national policies were supported by political leaders in Asia and by both the public and private sectors of the international community.

This mix of supportive public policies, scientific discoveries, and public and private investments in rural Asia, particularly in irrigation, credit, and inputs, led to substantial reductions in poverty and improved food security throughout Asia over the past 30 years. Increased agricultural productivity, rapid industrial growth, and expansion of the nonfarm rural economy have all contributed to almost a tripling of per capita gross domestic product across Asia since 1970.

PRESENT PROBLEMS

Despite these successes, problems remain. The intensification of agriculture and the reliance on irrigation and chemical inputs has led to environmental degradation, increased salinity, and pesticide misuse. Deforestation, overgrazing, and over-fishing also threaten the sustainable use of natural resources. Green Revolution technologies had little impact on the millions of smallholders living in rain-fed and marginal areas, where poverty is concentrated. Furthermore, the Green Revolution has already run its course in much of Asia. Wheat and rice yields in the major growing areas of Asia have been stagnant or declining for the past decade, while population continues to increase.

The key lessons learned from the Green Revolution are:

- It has benefited farmers in irrigated areas much more than farmers in rain fed areas thus worsening the income disparity between the two groups,
- It overlooked the rights of women to also benefit from the technological advances,
- It promoted an excessive of use of pesticides that are harmful to the environment.

As countries became self sufficient in food, government investments declined in the agricultural sector and in science and technology across the region. This reflects a worldwide trend towards declining public investments in the rural sector and in agricultural research and development (R&D), nationally and internationally. In Asia, private sector investments in the rural sector and related R&D have concentrated on export commodities.

The downward trends in public investments by governments and development agencies in smallholder agriculture over the past decade have not been matched by a concomitant rise in private investments. Similarly, there is little (and few incentives for) private R&D on the food crops, livestock, fisheries, and aquaculture systems important for food security and poverty reduction in rural Asia.

FUTURE CHALLENGES

The population of Asia is projected to increase from 3.0 billion to 4.5 billion in the next 25 years. During the same period, the urban population will nearly double from 1.2 billion to 2.0 billion, as rural people move to the cities in search of employment. These increases will place massive pressure on developing member countries (DMCs) of ADB to increase food production. Food demand is influenced by population growth, urbanization, income, and associated changes in dietary preferences.

Urbanization and income growth frequently lead to shifts from a diet based on root crops (cassava, yam, and sweet-potato), sorghum, millets, and maize to rice and wheat, which require less preparation time, and to more meat, milk, fruits, vegetables, and processed foods. This dietary transition has already happened in much of the region. Meeting the food needs of Asia's growing and increasingly urbanized population requires increases in agricultural productivity and matching these increases to dietary changes and rising incomes. To meet this demand, cereal production will need to be increased by at least 40 per cent from the present level of about 650 million tons annually, most of which will have to come from yield increases. In addition, meat demand will double during the period. Production increases will have to be achieved by increasing yields in a sustainable way to conserve diminishing and degraded natural resources. Nearly all of these production increases will need to take place in DMCs themselves because on average 90 per cent of the world's food is consumed in the country where it is produced. Food imports are not only expensive but discourage the creation of employment, which is badly needed in the rural areas.

In this millennium, we face a food, feed, and fibre production challenge in highly complex farming systems for several reasons:

- Water will become the most important limiting factor in agricultural production because the quality and quantity of water will decline as a result of pollution, forest degradation, and increased agricultural, domestic, and industrial use.
- Urbanization will mean the loss of agricultural land to residential and industrial development, and a decline in the number of farm workers.

- Most farmers are poor with small landholdings.
- Farming systems are commonly heterogeneous with mixes of food crops, livestock, and trees.
- About 70 per cent of the cultivated land is rain fed with unreliable distribution and intensity of rainfall.

Thus, the increase in food production during the next 25 years will have to be achieved using less labour, water, and cultivated land. This can be done only if scientists can develop new crop varieties with high yield potential and high water use efficiency.

New understanding of plant and animal genes may offer ways to increase crop yields to the levels required to adequately and sustainably feed the growing population in Asia. Thus, developments in modern biotechnology could make extremely important contributions to future agricultural growth, food security, and poverty reduction. Increasing smallholder agriculture productivity will not only increase food supplies, but will reduce poverty and malnutrition, increase food access, and, improve living standards of the poor.

Rationale for ADB's Involvement in Biotechnology

Biotechnology has the potential to:

- Increase crop and animal productivity;
- Improve nutritional quality;
- Broaden tolerance of crops for drought, salinity, and other abiotic stresses;
- Increase resistance of crops to pests and diseases.

These potential benefits will have significant impact in increasing food production and reducing poverty in DMCs if they can be applied to problems of the poor farmers in the tropics. As with any technology, biotechnology brings with it potential risks. To maximize the benefits and minimize the risks, the introduction and use of biotechnology in DMCs or elsewhere must be thoughtfully managed by the public and private sectors. The key risks that relate to the application of new developments in biotechnology for the public good are food and environmental safety, economic concentration, and intellectual property (IP) management.

To explore the opportunities and risks of biotechnology, ADB has decided to undertake a comprehensive study on the use of biotechnology to reduce poverty and achieve food security in cooperation with the Australian Centre for International Agricultural Research (ACIAR)/Australian Agency for Agricultural Development (AusAID).

There are many reasons to support this initiative:

- As a premiere development institution, ADB should be well informed on the development of biotechnology and the issues surrounding modern biotechnology.

- Biotechnology has been identified by the Rural Asia Study as one of the emerging challenges in Asia.
- Biotechnology has been identified by the Consultative Group on International Agricultural Research (CGIAR) as a powerful tool to fight poverty when used in conjunction with other agricultural research.
- ADB is responsible for assisting the DMCs in dealing with potential risks of biotechnology and providing information on various biotechnology issues.
- ADB's Policy on Agriculture and Natural Resources Research recommends that ADB support research to increase farm productivity through modern biotechnology approaches, while giving appropriate consideration to indigenous and traditional knowledge.

ADB's Poverty Reduction Strategy

While the need for further intensification of agricultural production in Asia is clear, intensification strategies must change to avoid adverse environmental impact and to reverse the effects of earlier practices. Enhanced, but inefficient use of irrigation and mineral fertilizers over the past three decades has had negative side effects such as soil salinity and nutrient leaching. With crop intensification, incidences of pests and diseases also increased.

The following strategies are needed to meet the food demand in Asia over the next 25 years:

- Sustainable productivity increases in food, feed, and fibre crops.
- Reducing chemical inputs of fertilizers and pesticides and replacing these with biologically based products.
- Integrating soil, water, and nutrient management.
- Improving the nutrition and productivity of livestock and controlling livestock diseases.
- Sustainable increases in livestock, fisheries, and aquaculture production.
- Increasing trade and competitiveness in global markets.

As rural poverty persists in Asia, agriculture will play a prominent role in achieving equitable and sustainable rural growth in the twenty-first century. Even when rural people do not work directly in agriculture, they rely on nonfarm employment and income closely related to agriculture. Where there are large numbers of rural poor, agricultural growth is a catalyst for broad-based economic growth and development.

Agriculture's linkages to the nonfarm economy generate employment, income, and growth in the rest of the economy. A healthy agricultural economy also offers incentives for natural resource conservation. The ADB Poverty Reduction Strategy sees three factors, pro-poor, sustainable economic growth; good governance; and social development as key elements for reducing poverty.

Biotechnology may contribute towards achieving the poverty reduction goals in several components of the strategy:

- Increasing priority for agriculture and rural development.
- Addressing environmental considerations.
- Increasing the public and private sector roles in poverty reduction.
- Encouraging regional and subregional cooperation.
- Coordinating efforts of funding agencies.

The challenge is how to use new scientific developments such as biotechnology, together with information and communications technology, to make the complex agricultural systems of Asia more productive and sustainable. Good governance is also crucial to ensure that new agricultural biotechnology reaches the poor.

PROMETHEAN SCIENCE

The pace of change in modern science has led some to term it Promethean science, acknowledging both its risks and benefits. Modern science encompasses new developments in the biological, physical, and social sciences. In biology, discoveries over the past 20 years allow the better understanding of the structure and function of human, animal, and plant genes.

At the same time, new discoveries in the physical sciences underpin the revolution in information and communications technologies. Geographic information systems enable characterization of agro-ecosystems and offer means by which new technologies can be customised to the needs of particular agro-ecosystems.

The biological and physical sciences also interact in new ways. For example, the ability to analyse large volumes of data is a critical component of various genome projects that are mapping all the genes in an organism, as in the Human Genome Project.

New developments in the social sciences underpin community participation in technology development and evaluation (sometimes termed agro-ecological methods). Participatory methods developed in the social sciences can help in understanding problems and the researchable issues, particularly those of small farmers operating in marginal environments. They may also be used to clarify the concerns of rural and urban dwellers in regard to the deployment of new technologies, including the products of biotechnology. Integration of all branches of modern science and traditional knowledge is required to develop knowledge-intensive solutions to the problems of rural Asia.

These solutions need not be only technically feasible but also socially acceptable. Indeed, the potential value of modern science to agriculture and the environment in Asia will require the efforts of all stakeholders, including civil society, farmer cooperatives, producers, consumers, governments, and development agencies.

SCOPE OF BIOTECHNOLOGY

Biotechnology, broadly defined, includes any technique that uses living organisms, or parts of such organisms, to make or modify products, to improve plants or animals, or to develop microorganisms for specific use. It ranges from traditional biotechnology to the most advanced modern biotechnology.

Biotechnology is not a separate science but rather a mix of disciplines (genetics, molecular biology, biochemistry, embryology, and cell biology) converted into productive processes by linking them with such practical disciplines as chemical engineering, information technology, and robotics.

Modern biotechnology should be seen as an integration of new techniques with the well-established approaches of traditional biotechnology such as plant and animal breeding, food production, fermentation products and processes, and production of pharmaceuticals and fertilizers.

The key components of modern biotechnology are listed below:

- *Genomics:* The molecular characterization of all genes in a species.
- *Bioinformatics:* The assembly of data from genomic analysis into accessible forms, involving the application of information technology to analyse and manage large data sets resulting from gene sequencing or related techniques.
- *Transformation:* The introduction of one or more genes conferring potentially useful traits into plants, livestock, fish and tree species.
- Genetically improved organism.
- Genetically modified organism (GMO).
- Living modified organism (LMO).
- *Molecular breeding:* Identification and evaluation of useful traits in breeding programmes by the use of marker-assisted selection (MAS);
- *Diagnostics:* The use of molecular characterization to provide more accurate and quicker identification of pathogens;
- *Vaccine technology:* The use of modern immunology to develop recombinant deoxyribonucleic acid (rDNA) vaccines for improved control of livestock and fish diseases.

Biotechnology consists of a gradient of technologies, ranging from the long-established and widely used techniques of traditional biotechnology to novel and continuously evolving modem biotechnology techniques.

During the 1970s scientists developed new methods for precise recombination of portions of deoxyribonucleic acid (DNA), the biochemical material in all living cells that governs inherited characteristics, and for transferring portions of DNA from one organism to another. This set of enabling techniques is referred to as rDNA technology or genetic engineering.

Modern biotechnology presently includes the various uses of new techniques of rDNA technology, monoclonal and polyclonal antibodies, and new cell and tissue culture methods. A chronology of the development of modern biotechnology.

Over the past two decades the number of significant advances in modern biotechnology for understanding and modifying the genetics of living organisms has increased dramatically.

That has led to greatly increased interest and investment in biotechnology, and increasing concerns as to the power of the new technologies and their safety.

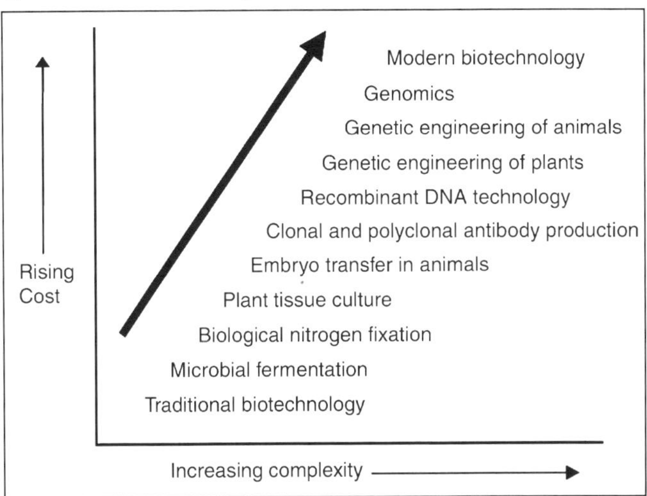

Fig. Gradient of Biotechnologies

ECONOMIC CONCENTRATION IN AGRICULTURAL BIOTECHNOLOGY

Modern biotechnology R&D has been conducted in an institutional and economic environment that differs significantly from the development of the earlier Green Revolution technologies. While the latter were essentially the prerogative of public research institutions and philanthropic foundations, the application of modern biotechnology to agriculture is essentially a competitive, commercial endeavor in which powerful private sector interests compete.

Similarly, while the Green Revolution technologies were essentially dedicated to the public, the strengthening and extension of IP protection, particularly since the conclusion of the Uruguay Round of trade negotiations has increased the private character of biotechnologies. Multinational companies in the seed, agricultural chemical, pharmaceutical, and food-processing industries play a major role in biotechnology research.

They have invested heavily in in-house research facilities, commissioned research, taken equity positions in new biotechnology firms, and entered into contractual arrangements with public research institutions or universities. As a result of mergers and acquisitions in the past few years, the development of new biotechnology applications in agriculture has become increasingly concentrated in the hands of a decreasing number of companies.

AGRICULTURAL BIOTECHNOLOGIES IN DEVELOPING COUNTRIES

Here we provide a brief overview of the main kinds of agricultural biotechnologies that have been used in developing countries over the past 20 years and that should be covered in the e-mail conference. They are described separately, although in practice more than one may be used together in certain situations. Note, new biotechnologies that are still at the research level, be it in the laboratory or at the field trial stage, but which have not yet been applied (*i.e.*, used for commercial production by farmers) in developing countries are not included.

A short description of the different biotechnologies is provided below, indicating also what they are used for, the food and agricultural sectors involved and giving some examples of their applications in specific developing countries. Regarding the examples, their inclusion in the document does not imply that these applications have been a partial or complete success (or, conversely, that they have been any kind of a failure). Indeed, these are the kind of issues to be addressed by participants during this e-mail conference. Also, it should be kept in mind that, although not the subject of this e-mail conference, the pathway from a research development in the laboratory to its eventual application in the field (*e.g.*, farmers cultivating a new genetically improved plant variety or using a new vaccine against an animal disease) can be long, resource-demanding and unsuccessful, so many biotechnologies of seemingly high promise at the experimental stage have had limited applications in developing countries so far.

As many of the biotechnologies are related to molecular biology and genetic material, some basic terminology is introduced here. Living things are made up of cells that are programmed by genetic material called DNA. A DNA molecule is made up of a long chain of nitrogen-containing bases. Only a small fraction of this DNA sequence typically makes up genes *i.e.*, that code for proteins, which are molecules essential for the functioning of living cells, made up of chains of amino acids. The remaining and major share of the DNA represents non-coding sequences whose role is not yet clearly understood.

The genetic material is organized into sets of chromosomes (*e.g.*, 5 pairs in Arabidopsis thaliana-a model plant species; 30 pairs in cattle), and the entire set is called the genome. In a diploid individual (*i.e.*, where chromosomes are organized in pairs), there are two alleles of every gene-one from each parent-transmitted by gametes (reproductive cells) that are normally haploid (having just one of each of the pairs of chromosomes). A typical genome contains several thousand genes *e.g.*, about 30,000 genes in grasses like rice and sorghum.

MOLECULAR MARKERS

Molecular markers are identifiable DNA sequences, found at specific locations of the genome, transmitted by standard Mendelian laws of inheritance from one generation to the next. They rely on a DNA assay and a range of different kinds of molecular marker systems exist, such as restriction fragment length polymorphisms (RFLPs), random amplified polymorphic DNAs (RAPDs), amplified fragment length polymorphisms (AFLPs) and microsatellites. The technology has improved in the past decade and faster, cheaper systems like single nucleotide polymorphisms (SNPs) are increasingly being used. The different marker systems may vary in aspects such as their technical requirements, the amount of time, money and labour needed and the number of genetic markers that can be detected throughout the genome.

Molecular markers have been used in laboratories since the late 1970s and they are applied across all the food and agricultural sectors. They are very versatile and can be used for a variety of purposes.

Thus, they are used in genetic improvement, through so-called marker-assisted selection (MAS), where markers physically located beside (or, even, within) genes of interest (such as those affecting yield in maize) are used to select favourable variants of the genes. MAS is made possible by the development of molecular marker maps, where many markers of known location are interspersed at relatively short intervals throughout the genome, and the subsequent testing for statistical associations between marker variants and the traits of interest. Marker maps are now available for a wide range of economically important agricultural species. Progress in the field of genomics (the study of an organism's entire genome) has also provided much useful information for MAS, enabling in some cases markers to be used that are located within the genes of interest.

Molecular markers are also used to characterize and conserve genetic resources, where some of the approaches can be applied in each of the crop, forestry, livestock and fishery sectors (*e.g.*, estimating the genetic relationships between populations within a species). Other uses again are more sector-specific, such as their utilization to identify duplicate accessions in crop genebanks; monitor effective population sizes (Ne) in capture fish populations or carry out biological studies (*e.g.*, of mating systems, pollen movement and seed dispersal) in forest tree populations. They are also used in disease diagnosis, to characterize and detect pathogens in livestock, crops, forest trees, fish and food.

Molecular markers have been used in a number of developing countries. In livestock, for example, they have been used in four African countries for characterization of genetic resources and in eight Asian countries, where six used them for genetic distance studies and two for MAS. In Latin America and the Caribbean, most countries have used molecular techniques, primarily for characterization purposes, while their use has been limited in the Near and Middle East. In crops, several examples of new hybrids and varieties developed through MAS are available, and in progress, in different crops, such as pearl millet, rice and maize, and in several developing countries like Bangladesh, India and Thailand. Different centres of the Consultative Group on International Agricultural Research (CGIAR) have been working with partners in developing countries to accelerate plant breeding practices through MAS.

GENETIC MODIFICATION

A genetically modified organism (GMO) is an organism in which one or more genes (called transgenes) have been introduced into its genetic material from another organism. The genes may be from a different kingdom (*e.g.*, a bacterial gene introduced into plant genetic material), a different species within the same kingdom or even from the same species. For example, so-called 'Bt crops' are crops containing genes derived from the soil bacterium Bacillus thuriengensis coding for proteins that are toxic to insect pests that feed on the crops. The issue of GMOs has been highly controversial over the past decade. Many countries have introduced specific frameworks to regulate their release and commercialization.

GM crops were first grown commercially in the mid 1990s. While the majority continues to be grown in developed countries, an increasing number of developing countries are reported to be cultivating them.

Recent estimates indicate that 10 developing countries planted over 50,000 hectares of GM crops in 2008 *i.e.*, Argentina (21.0 million hectares), Brazil (15.8), India (7.6), China (3.8), Paraguay (2.7), South Africa (1.8), Uruguay (0.7), Bolivia (0.6), Philippines (0.4) and Mexico (0.1). For comparison, in 1997 the only developing countries reported were Argentina (1.4 million hectares), China (1.8) and Mexico (less than 0.1). Almost all GM crops grown commercially are genetically modified for one or both of two main traits: herbicide tolerance (63 per cent of GM crops planted in 2008) or insect resistance (15 per cent), *i.e.* Bt crops, while 22 per cent have both traits. Commercial release of GM forest trees has been reported in one country, China. In 2002, approval was granted for the environmental release of two kinds of Bt trees, the European black poplar (Populus nigra) and the hybrid white poplar clone GM 741, together representing about 1.4 million plants on 300-500 hectares (FAO, 2004). Regarding GM livestock or fish, there has been no commercial release for food and agriculture purposes in any developing or developed country.

Although documentation is generally quite poor, use of genetically modified microorganisms (GMMs) in the agro-industry and animal feed sector is routine in developed countries and is also a reality in many developing countries. In the agro-industry sector, use of enzymes, *i.e.*, proteins that catalyse specific chemical reactions, is important. Many of the enzymes used in the food industry are commonly produced using GMMs. For example, since the early 1990s, preparations containing chymosin (an enzyme used to curdle milk in the preliminary steps of cheese manufacture) derived from GM bacteria have been available commercially. Similarly, many colours, vitamins and essential amino acids used in the food industry are also from GMMs.

In animal nutrition, feed additives such as amino acids and enzymes are widely used in developing countries. The greatest use is in pig and poultry production where, over the last decade, intensification has increased, further accelerating the demand for feed additives. For example, most grain-based livestock feeds are deficient in essential amino acids such as lysine, methionine and tryptophan and for high producing monogastric animals (pigs and poultry) these amino acids are added to diets to increase productivity. The amino acids in feed, L-lysine, DL-methionine, L-threonine and L-tryptophan, constitute over half of the total amino acid market. In India alone, the amino acid market amounts to about 5 million US dollars. The essential amino acids are produced in some cases by GM strains of Escherichia coli.

In the dairy industry, recombinant bovine somatotropin (rBST), a protein hormone from an Escherichia coli K-12 bacterium containing the cow somatotropin gene, has been used to increase milk production in a number of developing countries. Chauvet and Ochoa (1996) report that rBST was first used in Mexico in 1990 and has been sold in a number of other developing countries, including Brazil, Malaysia, South Africa and Zimbabwe.

CHROMOSOME NUMBER MANIPULATION

As mentioned earlier, genetic material is organized into sets of chromosomes and each plant and animal species has a characteristic number of chromosomes. Manipulation of whole sets of chromosomes is possible and is used for a range of different purposes in agriculture. For example, fish and shellfish have been extensively studied for manipulation of their chromosomes during early stages of development. Using relatively simple techniques such as cold and hot shocks it is possible to induce triploidy (*i.e.*, with three sets of chromosomes), leading to the production of nearly completely sterile populations. Sterility may be desirable in conservation programmes, where it can prevent introgression of escaped individuals from commercial stocks into natural populations. It may also be of interest in commercial fish operations, *e.g.*, when developing hybrid stocks or to prevent the side effects of sexual maturation on carcass quality. As in fish, induction of sterility in crops may be desirable in certain breeding programmes, *e.g.*, to produce seedless fruits, and one of the most rapid and cost-effective approaches is to create polyploids (*i.e.*, with more than 2 complete sets of chromosomes), especially triploids. Triploid varieties have been produced in numerous fruit crops including most of the citrus fruits, acacias and the kiwifruit.

Another example of chromosomal number modification in fish is the production of haploid individuals after eggs are fertilized by sperm which do not contribute genetic material (a process called gynogenesis) or when normal sperm fertilize eggs whose DNA has been deactivated (a process called androgenesis). In both cases the haploid chromosomes can then be duplicated using shocks. The importance of gynogenesis/androgenesis is that it is possible to develop inbred individuals, which may be useful in fish breeding experiments aimed at producing clonal lines for detecting genomic regions affecting quantitative traits.

In crops, chromosome doubling is one of the most important technologies for the creation of fertile inter-specific hybrids (wide-crosses). Wide crossing involves hybridizing a crop variety with a distantly related plant from outside its normal sexually compatible gene pool. Its usual purpose is to obtain a plant that is virtually identical to the original crop, except for a few genes contributed by the distant relative. The technique has enabled breeders to access genetic variation beyond the normal reproductive barriers of their crops. For example, the New Rice for Africa (NERICA) hybrids are derived from crossing two species of cultivated rice, the African rice and the Asian rice, combining the high yields from the Asian rice with the ability of the African rice to thrive in harsh environments. Wide-hybrid plants are often sterile so their seed cannot be propagated, due to differences between the sets of chromosomes inherited from genetically divergent parental species that prevent stable chromosome pairing during meiosis. However, if chromosome number is artificially doubled, the hybrid may be able to produce functional pollen and eggs and be fertile. Colchicine has been used for chromosome doubling in plants since the 1940s and has been applied to more than 50 plant species, including most important annual crops. More recently, several additional chromosome doubling agents, all of which act as inhibitors of mitotic cell division, have been used in plant breeding programmes. To date, with the help of chromosome doubling technology, hundreds of new varieties have been produced worldwide.

In crops and forest trees, chromosome doubling has also been used, as for fish, to generate 'doubled haploids'. The haploid plants can be produced using anther culture which involves the in vitro culture of immature anthers. As the pollen grains are haploid, the resulting pollen-derived plants are also haploid. Doubled haploid plants were first produced in the 1960s using colchicine and today, thermal shock or mannitol incubation can be used. They may also be produced from ovule culture. Breeders value doubled haploid plants because they are 100 per cent homozygous so any recessive genes are readily apparent. The time required after a conventional hybridization to select pure lines carrying the required recombination of characters is thus drastically reduced. Since the 1970s, doubled haploid methods have been used to create new varieties of barley, wheat, rice, melon, pepper, tobacco, and several Brassicas. In the developing world, a major centre of such breeding work is China, where numerous haploid crops have been released and many more are being developed. By 2003, China was cultivating over 2 million hectares of doubled haploid varieties, the most important of which are rice, wheat, tobacco and peppers.

FUTURE OF AGRICULTURAL BIOTECHNOLOGY

There are divergent views about the rate of uptake of new agricultural biotechnologies and their impact. Kalter and Tauer and the US Office of Technology Assessment anticipate comparatively rapid take-up, while others (Buckwell and Moxey and Farrington do not foresee much impact until the next century, that is until ten to fifteen years have elapsed. The caution of the latter commentators is probably justified, for there are various hurdles to be jumped before biotechnologies can gain acceptance. One hurdle to be overcome is the economic one.

Self-evidently there must be some economic advantage to farmers or agro-industry from adopting the technology. Frequently what captures the scientific imagination turns out to be economically unviable in use, although it is only with use that efficiency can be improved and costs brought down. A more demanding test still for biotechnologies will be to obtain legislative approval and public acceptance. These are intimately connected.

In the case of machinery developments, provided safety standards are observed there are no legal impediments to developing new and improved machines, since there are no obvious problems of hazard to the public and therefore public acceptance. Chemical insecticide, fungicides and herbicides do, however, pose much greater problems because of concerns with toxicity, and elabourate testing and licensing procedures have evolved after a relatively haphazard set of procedures in the 1950s and 1960s, when widespread use was made of DDT and organo-phosphorous compounds without adequate recognition of their toxicity.

In significant part, because of the problems which have been experienced with agro-chemicals, the licensing of biotechnologies will inevitably be based on testing at least as stringent as that which now exists for chemicals. Indeed the licensing procedures are likely to be more stringent because of public and scientific concerns. Public acceptance of biotechnology in the food chain will be made more difficult by confusion about differences between biotechnologies. In the early 1980s there was a crescendo of concern about the use of steroids to promote faster liveweight gain in calves and cattle.

The public outcry eventually led to the banning of such hormones for meat animals; but the legacy is a profound distrust of and hostility to new products such as synthetic Bovine Somatrophin (BST), which has the capacity to increase the milk yield of those dairy cows which are 'relatively' deficient in what is a naturally occurring hormone. Although initial scientific results are favourable, and some minor doubts remain, the major influence leading the European Commission to impose a moratorium on the use of BST until the end of 1990 at least is concern about public perception; the Commission has stated 'It would be a serious setback to producers and to the Community's milk policy were present [positive] trends in consumption to be reversed as a result of adverse consumer reaction. With newly bioengineered plants and animals scientific and public concerns are emerging which are of a different order from those associated with previous biotechnology. One relates to the possibility that crop failures may become more frequent if biotechnology leads to a reduction in genetic diversity in crops being grown; such a tendency has already been observed in relation to the Green Revolution technology for wheat. There are also concerns about upsetting the balance of nature in unforseen ways, akin to the unanticipated consequence of introducing rabbits into Australia and then trying to control these by introducing myxomatosis.

Thus questions arise such as what happens if herbicide resistance introduced into a commercial crop plant transfers itself by cross pollination into a closely related weed species, or indeed if the herbicide-resistant crop should colonize wild habitats. There are, of course, more lurid and absurd notions bandied about in the popular press which are similar in nature to the attacks made on Darwinism. The questions, both absurd and real, raised in relation to agricultural biotechnology will undoubtedly slow the rate at which it is adopted.

Nevertheless instances of adoption are increasing: BST in the USA, a bioengineered baker's yeast in the UK, bioengineered sheep producing insulin, etc. As these become more widespread and increasingly affect lower-valued, bulk agricultural products, so the impacts of biotechnology on structural change will increase. The balance of economic power will switch increasingly to industry, to high-technology large farms and against smaller farmers in disadvantaged regions and countries. That, unfortunately, is a seemingly inevitable consequence of what we consider to be economic progress.

2

Genetically Engineered Plants

Technologies, by definition, are neither acts of God nor nature; they are the embodiment of specific human purposes and intentionality. Far from being inevitable, they are researched and developed by those entities with sufficient power to mobilize social institutions to bring the technology into being, according to their own goals and normative considerations. 'Biotechnology is not neutral. It shares the propensity of modern materialistic science to desacralise, dominate and manipulate life. It reduces all living things to a mechanism which it can manipulate according to engineering standards.' Genetic engineering is a technological process for undertaking activities which do not, and cannot, occur in nature. In this sense it is perfectly legitimate, therefore, to label genetic engineering as 'unnatural'. The European Community has incorporated this concept in its definition of genetically modified organisms, and it has been adopted by the UN Environment Programm (in the working of its Fourth Expert Panel set up under the Biodiversity Convention as part of the follow through to the United Nations Conference on the Environment and Development held in Rio de Janeiro in June 1992): 'organisms in which the genetic material has been altered in a way that does not occur naturally by mating and/or natural recombination'.

Thus, hybrids and other modified organisms obtained by traditional breeding techniques are excluded from the definition of genetically modified organisms. It is important to note that the process as well as the products are novel. The promoters of the new technology believe that a concern for the process itself is specious, downplaying its novelty whenever confronted by discussions of regulatory oversight. However, one of the bases of the decision of the US Supreme Court in the *Chakrabarti* case (which, 5 to 4, upheld the patentability of genetically engineered microbes, and which has been unquestionably accepted by foreign governments as disposative of the issue of patentability of genetically engineered life forms) was that the element of 'novelty' required by the patent law was to be found in the process by which the microbe had been created. One reason why we cannot ignore the powerful novel aspects of the processes of genetic manipulation is because we are not omniscient as to what will, in fact, happen when we alter genomes. Genetic manipulation is not like a child's game of Legos or Tinkertoy, in which parts can be rearranged or linked up with only simple mechanical and relatively predictable consequences. In calculating any risk from a transgenic organism, one should consider four elements: the host organism, the foreign genes, the interaction between the foreign genes and the rest of the genome, and the environment in which the organisms will be used.... [in regard to the last two elements] the literature contains many examples of genetic manipulations where inserted genes did not respond in their new environments the way they did in their old ones or where alterations with one part of the genome caused surprising activity in other parts of the genome. There is an element of arrogance to scientific assertions which assure us that the process itself poses no risks, since scientists know so little about actual ecosystems.

For example, I have been told by agronomists that over 80 per cent of the organisms which can be identified in a soil sample from my garden are completely unknown to the scientific literature. Indeed, the Ecological Society of America itself has specifically warned about the problems presented in our lack of ecological knowledge and the resulting consequences from releasing (intentionally or accidentally) genetically altered organisms into ecological systems.

Foods

The Food and Drug Administration, responding to corporate pressures to remove the prospect of regulation of genetically altered foods, issued a set of rules which have largely left responsibility for protecting the public health and safety in the hands of the industry, permitting products to be marketed without scrutiny unless the industry indicated to the agency that it believed governmental oversight was justifiable. However, the proposal as published by the FDA in the Federal Register perversely noted several important problem areas: creating allergens, additions of genes from sources which might violate religious and cultural norms (of vegetarians, Jews or Muslims, etc.), and implications for animal welfare and well-being (for example, the incorporation of human growth hormone-producing gene into a pig's genome produced a highly arthritic animal).

Indeed, the commissioner of the FDA, David A. Kessler, and his colleagues noted the possibility that genetically engineered food might 'contain high levels of unexpected, acutely toxic substances'. In order to begin to address ethical issues presented by the genetic modification of foodstuffs, we need to have some ideas of the purposes for which these modifications are performed. (For example, if we wish to approach an analysis on utilitarian cost-benefit grounds.) Despite a great number of general statements, human nutrition and hunger do not appear to be the actual driving forces behind the development of genetically engineered foods. Other than claiming increased shelf life (which can be indirectly related to nutrition in the sense of reducing spoilage or the consumption of foods that have begun to spoil), there is hardly any indication at all that genetic engineering is being directed to create nutritious substances out of non-nutritious ones.

There are few real instances of genetic engineering reducing the cost of foodstuffs, or increasing the quantities of foods actually available to populations that are hungry because of the non-existence of consumables (as opposed to hungry because they lack money to buy sufficient foods, no matter how produced). Indeed, early genetic manipulations seemed perversely designed to minimize the achievement of such goals: the creation of 'herbicide tolerant plants' which do not reduce the applications of dangerous chemicals to agricultural fields (which might occur if genetic manipulations of food crops were designed to make them resistant to insects, fungi, or disease) but instead permit higher levels of chemical application, or the introduction of recombinant bovine growth hormone to produce more milk at a time when developed countries suffer from milk gluts and have instituted programmes to kill cows and physically dump milk (milk as a commodity is price inelastic; quantity increases do not lead to price decreases).

The goals of genetically engineering foods seem to be more closely tied to increasing the economic gain and power of the corporations involved in food production (for example, the development of herbicide-tolerant farm crops has been led by corporations which manufacture herbicides), making production easier for corporations, reducing or altering packaging and transportation costs for the corporations, etc.

Environment

Industrial societies around the world are faced by massive pollution legacies from the previous emphases on chemical and nuclear industrial activities. The inability of the US Department of Energy to find a suitable site for a longterm nuclear waste repository illustrates our very poor track record for dealing with the scientific, technical, economic, and socio-political aspects of prior technological 'revolutions'.

The environmental problems posed by genetically engineered organisms are likely to be substantially more intractable than those posed by these earlier instances of pollution because genetic wastes multiply, migrate, and mutate. A genetically engineered organism once free in the environment is impossible to recall. (Impacts which are irreversible must be considered with higher scrutiny than those which can be undone.) Examples of environmental problems which need to be assessed include the risks of transgenic crops themselves becoming weeds; the risk of gene flow to wild relatives which might become weeds or pests; the growth of antibiotic resistance in species (particularly animals), since resistance genes are used as markers for the genetic engineering sites; the problem of exotic or non-native species taking over ecological niches (such as gypsy moth, starling, kudzu, rabbits in Australia, etc.) leading to extraordinary economic losses as well as ecological ones.

The restriction (rather than increase) of biodiversity by selective advantageous breeding of transgenic organisms as opposed to natural ones, or the predatory results of expanding exotic species (such as the Dutch Elm micro-organism destroying the American Elm and severely restricting the biodiversity of certain areas in the Northeast United States); health issues (both of plant and animal species, as well as humans — worker health and safety as well as community security); and the occurrence of the completely unexpected, the inability to eliminate uncertainty

For example, the late 1993 floods in the Mississippi valley included the flooding of a field of genetically engineered corn and the dispersal of this plant material to unknown sites within the thousands of square miles of downstream flood plain. Prospective ecological assessment is not being performed, despite the fact that its need has long since been recognized. And economic priorities, particularly those accruing directly to the promoters, as well as the nebulous spur of 'competition' with foreign countries, oftentimes overwhelms any interest in doing environmental impact analysis.

GENETIC ENGINEERING

Genetic engineering is a term used for the directed manipulation of genes, i. e. the transfer of genes between organisms or changes in the sequence of a gene. Closely related to this field are methods which use genes or specific sequences for the identification of traits and other analytical purposes. In plant breeding, the most important and already widely used method of this kind is Restriction Fragment Length Polymorphism (RFLP).

Restriction Fragment Length Polymorphism (RFLP)

The techniques of traditional breeding are very time-consuming. By making crosses, also a large number of undesired genes is introduced into the genome of the plant. The undesired genes have to be "sorted out" by back-crossing. The use of Restriction Fragment Length Polymorphism greatly facilitates conventional plant breeding, because one can progress through a breeding programme much faster, with smaller populations and without relying entirely on testing for the desired phenotype. RFLP makes use of restriction endonucleases.

These are enzymes which recognize and cut specific nucleotide sequences in DNA. The sequence GAATTC is cut by the endonuclease EcoRl. After treatment of a plant genome which restriction endonucleases, the plant DNA is cut into pieces of different length, depending on the number of recognition sites on the DNA. These fragments can be separated according to their size by using gel electrophoresis and are made visible as bands on the gel by hybridizing the plant DNA fragments with radiolabeled or fluorescent DNA probes.

As two genomes are not identical even within a given species due to mutations, the number of restriction sites and therefore the length and numbers of DNA fragments differ, resulting in a different banding pattern on the electrophoresis gel. This variability has been termed restriction fragment length polymorphism (RFLP). The closer two organisms are related, the more the pattern of bands overlap. If a restriction site lies close to or even within an important gene, the existence of a particular band correlates with the particular trait of a plant, *e.g.*, disease resistance.

By looking at the banding pattern, breeders can identify individuals which have inherited resistance genes, and resistant plants can be selected for further breeding. The use of this technique will not only accelerate progress in plant breeding considerably, but will also facilitate the identification of resistance genes, thereby opening new possibilities in plant breeding.

Gene Transfer

In conventional breeding, the pool of available genes and the traits they code for is limited due to sexual incompatibility to other lines of the crop in question and to their wild relatives. This restriction can be overcome by using the methods of genetic engineering, which in principle allow introducing valuable traits coded for by specific genes of any organism (other plants, bacteria, fungi, animals, viruses) into the genome of any plant.

The first gene transfer experiments with plants took place in the early 1980s. Normally, transgenes are inserted into the nuclear genome of a plant cell. Recently it has become possible to introduce genes into the genome of chloroplasts and other plastids (small organelles of plant cells which possess a separate genome). Transgenic plants have been obtained using Agrobacterium-mediated DNA-transfer and direct DNA-transfer, the latter including methods such as particle bombardment, electroporation and polyethylenglycol permeabilisation. The majority of plants have been transformed using Agrobacterium mediated transformation.

Agrobacterium-mediated Gene Transfer

The Agrobacterium-mediated technique involves the natural gene transfer system resident in the bacterial plant pathogens of the genus Agrobacterium. In nature, Agrobacterium tumefaciens and Agrobacterium rhizogenes are the causative agents of the crown gall and the hairy root diseases, respectively.

The utility of Agrobacterium as a gene transfer system was first recognized when it was demonstrated that these plant diseases were actually produced as a result of the transfer and integration of genes from the bacteria into the genome of the plant. Both Agrobacterium species carry a large plasmid (small circular DNA molecule) called Ti in A. tumefaciens and Ri in A. rhizogenes. A segment of this plasmid, designated T-(for transfer) DNA, is transmitted by this organism into individual plant cells, usually within wounded tissue.

The T-DNA segment penetrates the plant cell nucleus and integrates randomly into the genome where it is stably incorporated and inherited like any other plant gene in a predictable, dominant Mendelian fashion. Expression of the natural genes on the T-DNA results in the synthesis of gene products that direct the observed morphological changes, i. e. tumor or hairy root formation. In genetic engineering, the tumor inducing genes within the T-DNA which cause the plant disease are removed and replaced by foreign genes. These genes are then stably integrated into the genome of the plant after infection with the altered strain of Agrobacterium, just like the natural T-DNA.

Because all tumor-inducing genes are removed, the gene transfer does not induce any disease symptoms. This reliable method of gene transfer is well suited for plants which are susceptible to infection by Agrobacterium. Unfortunately, many species, especially economically important legumes and monocotyledons such as cereals, do not respond positively to Agrobacterium-mediated transformation. For these plants, the following methods of direct DNA uptake must be applied..

Particle Bombardment

This method, also referred to as biolistic transformation (from biological ballistics) involves coating biologically active DNA onto small tungsten or gold particles (1-5 ?m in diameter) and accelerating them into plant tissue at high velocity. The particles penetrate the plant cell wall and lodge themselves within the cell where the DNA is liberated resulting in transformation of the individual plant cell in an explant. This technique is generally less efficient than Agrobacterium-mediated transformation, but has nevertheless been particularly useful in several plant species, most notably in cereal crops.

The introduction of DNA into organized, morphogenic tissues such as seeds, embryos or meristems has enabled the successful transformation and regeneration of rice, wheat, soybean and maize, thus demonstrating the enormous potential of this method.

Electroporation and Direct DNA Entry into Protoplasts

Electroporation is a process whereby very short pulses of electricity are used to reversibly permeabilize lipid bilayers of plant cell membranes. The electrical discharge enables the diffusion of macromolecules such as DNA through an otherwise impermable plasma membrane. Because the plant cell wall will not allow the efficient diffusion of many transgene constructs, protoplasts (cells without cell walls) must be prepared. This requirement presents a major obstacle for many applications as protocols making possible the regeneration of protoplasts into complete plants do not exist for many species. Gene transfer from Agrobacterium tumefaciens to a plant cell. In nature (route A), the transfer DNA (T-DNA) contains tumor inncluding genes which lead to crown gall disease. In plant genetic engineering (route B), the tumor including genes are removed and replaced by other genes, *e.g*., genes which confer insect resistance.

In nature as well as in genetic engineering, the agrobacteria attach to a plant cell. Then, the T-DNA is cut out of a plasmid (a small circular DNA-molecule) and is transfered to the plant cell. The T-DNA migrates to the nucleus of the plant cell and becomes incorporated into a plant chromosome. In nature (route A), the tumor inducing genes cause uncontrolled (tumor-like) growth at the site of infection after expression of the T-DNA. In genetic engineering (route B, tumor inducing genes removed by other genes), transgenic cells are selected and regenerated to whole plants. All cells of the regenerated plant now contain the transgene. Adapted from DNA uptake by plant protoplasts can also be stimulated by phosphate or calcium/polyethylene glycol (PEG) coprecipitation. However, these methods all suffer from the drawback that they use protoplasts as the recipient host which often cannot be regenerated into whole plants.

Transgene Expression

In most cases, the introduction of a gene into the plant genome will only have an effect on the plant if the transgene is expressed, i. e. transcribed into mRNA and translated into a protein. A promoter is a sequence of nucleic acids where the RNA polymerase (a complex enzyme synthesizing the mRNA transcript) attaches to the DNA template. The nature of the promoter defines (together with other expression-regulating elements), under which conditions and with which intensity a gene will be transcribed.

The promoter of the 35S gene of cauliflower mosaic virus is used very frequently in plant genetic engineering. This promoter confers high-level expression of exogenous genes in most cell types from virtually all species tested. As it is often advantageous to express a transgene only in certain tissues or quantities or at certain times, a number of other promoters are available, *e.g*., promoters inducing gene expression after wounding or during fruit ripening only. Methods of gene transfer currently employed result in the random integration of foreign DNA throughout the genome of recipient cells. The site of insertion may have a strong influence on the expression levels of the exogenous gene, resulting in different expression levels of an introduced gene, even if the same promoter/gene construct was used. The exact mechanism of this phenomenon are not yet fully understood.

Selection and Plant Regeneration

In a transformation experiment, the proportion of transformed cells is usually small compared to the number of cells which remain unaltered. In order to select only cells which have actually incorporated the new genes, the genes coding for the desired trait are fused to a gene which allows selection of transformed cells, so-called marker genes. The expression of the marker gene enables the transgenic cells to grow in presence of a selective agent, usually an antibiotic or a herbicide, while cells without the marker gene die. One of the most commonly used marker genes is the bacterial aminoglycoside-3' phosphotransferase gene (APH(3')II), also referred to as neo-mycin phosphotransferase 11 (NPTII). This gene codes for an enzyme which inactivates the antibiotics kanamycin, neo-mycin and G418 through phosphorylation.

In addition to NPTII, a number of other antibiotic resistance genes have been used as selective markers, *e.g.*, hygromycin phosphotransferase gene conferring resistance to hygromycin. Another group of selective markers are herbicide tolerance genes. Herbicide tolerance has been obtained through the incorporation and expression of a gene which either detoxifies the herbicide in a similar manner as the antibiotic resistance gene products or a gene that expresses a product which acts like the herbicide target but is not affected by the herbicide.

Herbicide tolerance may not only serve as a trait useful for selection in the development of transgenic plants, but also has some commercial interest. Herbicide tolerance transgenic plants are therefore among the first crops approaching market introduction. Transformation of plant protoplasts, cells and tissues is usually only useful if the they can be regenerated into whole plants. The rates of regeneration vary greatly not only among different species, but also between cultivars of the same species. In many cases regeneration of whole plants from cells is not possible or very difficult. Besides the ability to introduce a gene into the genome of a plant species, regeneration of intact, fertile plants out of transformed cells or tissues is the most limiting step in developing transgenic plants.

APPLICATIONS OF GENETIC ENGINEERING

"PHARMING" AND "PLANTIBODIES"

An increasingly viable option is the production of highly valuable enzymes by plants and animals. In addition to production of human proteins in these organisms, other valuable proteins that are currently produced by microorganisms could very well be produced by higher organisms instead. Animal and plant "bioreactors" in some respects may be superior to recombinant bacterial systems, because eukaryotes glycosylate proteins. Whereas the glycosylation pattern may be species-specific, appropriate glycosylation is often required for protein function. Production through these organismal systems may also be cheaper than cell fermentation techniques. Two examples of production of human proteins in plants include the production of human serum albumin in transgenic tobacco and potato, and production of human insulin by tobacco. In both cases, the produced protein appears to be fully effective in humans. Unfortunately, however, one cannot raise his/her insulin level by eating transgenic tobacco leaves, as the protein in most cases will be broken down to amino acids before it reaches the blood stream. Therefore, in these cases one cannot escape the practice of protein isolation and purification before transgenic leaves are converted into drugs. Also antibodies are being produced in plants. Initially, the antibody's light and heavy chains were produced in different plants. But a subsequent cross of these two varieties resulted in progeny carrying assembled and functional antibodies.

REVERSIBLE MALE STERILITY IN PLANTS

As has been indicated earlier, heterozygous individuals often are healthier and stronger than homozygous ones. The only way to guarantee heterozygoiscity in plants is to make sure self-pollination cannot occur. For most crop plants it was very tedious or practically impossible to exclude selfing. To exclude self-pollination, it would be good to introduce male sterility in plants: progeny from such plants are then expected to be 100percent heterozygous (assuming they were pollinated with pollen from an unrelated variety).

To introduce male sterility, a promoter was identified that was turned on exclusively in tapetum cells (a tissue around the pollen sac that is essential for pollen production). This promoter then was linked up to a gene coding for a bacterial ribonuclease (named barnase). This ribonuclease selectively chops up ribonucleic acids. The promoter/ribonuclease construct was then introduced into plants (canola, tobacco, you name it).

Because the promoter allows expression only in tapetum cells, the gene construct disrupts only development of the tapetal tissue and its end product, pollen. Plants transformed with this construct were male-sterile but otherwise normal. Although male-sterile plants are valuable for hybrid seed production, they have limited value when it comes to crop production. Fertility must be restored to crops such as wheat, rice, and tomato, in which the seed or fruit is the harvested product. Fortunately, the ribonuclease is inhibited very much by a simple protein, named barstar. One can thus cross the male-sterile plant with a male-fertile variety in which the gene for barstar has been introduced, and the result is progeny with viable pollen and restored fertility. A closely related approach has been criticized as "terminator technology" as it is seen by its critics as a way for companies to protect and enforce their patents.

ANTISENSE RNA

Antisense RNA refers to nucleotide strands that are produced in a cell and that are complementary to a particular mRNA. Antisense RNA can be produced, by inverting the coding region of a gene with respect to its promoter. The antisense RNA can hybridize with its corresponding mRNA, making it double-stranded. The double-stranded mRNA no longer can be recognized by the protein-synthesizing machinery (the ribosomes), and thus expression of this mRNA is suppressed.

Also, in many systems double-stranded mRNA is very unstable and is broken down quickly. Thus, one can inactivate specific genes while not interfering with others. Antisense approaches already are used to protect plants from damage by plant viruses. Reversal of a gene from bean yellow mosaic virus (BYMV), and putting it into tobacco under a reasonably strong promoter, has led to a tobacco variety that is quite resistant to BYMV.

A similar approach is used to transfer viral resistance to other plants. This finding is of significance, in that currently no effective, environmentally friendly methods exist to control many plant viruses. Very related to this approach is the RNAi (RNA interference) approach. This is very useful for both agriculture and medicine, and the first examples of practical applications of this RNAi technology are appearing. As with any new technology, the initial pilot projects are sort of pedantic. However, more exciting application possibilities abound.

Obviously, RNAi technology provides an excellent approach for reverse genetics in eukaryotes. With this method one can turn off genes.

Perhaps indicative of the large potential and relative ease of genetic engineering, developing countries (particularly China) are progressing rapidly in development and application of genetically engineered crops. Some have gone into commercial production well ahead of similar crops in the US. In China, tomatoes that have been engineered for improved virus resistance have been on the market since late 1992. There are two main reasons for the more rapid commercialization of bioengineered crops in the developing world:

- Less tight governmental approval mechanisms
- Hungrier populations.

While in developed countries the main value of biotechnological applications may be to reduce production costs, in the developing world a main factor is the production of more food. Indeed, genetic engineering applications seem to be pretty successful to cut down on pathogen-induced losses. Genetic modification of papaya plants (expression of the ringspot virus coat protein in the plant) protects very well against the very destructive ringspot virus.

GENETICALLY ENGINEERED LIFE-FORMS IN THE ENVIRONMENT

The idea of controlling and improving upon nature has been a persistent theme in the history of technology. It has always been met with its skeptics, who are no less zealous in their opposition to the hubris of science as the proponents of change are to their attack on nature worship. At one end of the continuum are those that argue there is nothing in nature that cannot be improved upon. At the other extreme are those who maintain that evolution, without human intervention, results in the optimality of life processes. Nature knows best, according to the popular aphorism. When molecular genetics gave birth to gentechnics, there were already numerous examples of efforts to control and harness nature that had resulted in mixed outcomes.

The mining and burning of coal and oil has resulted in new forms of pollution and has been a major contributor to the "greenhouse" effect or the warming of the earth. Synthetic, non-biodegradable chemicals, which were responsible for the second American industrial revolution, have become ubiquitous residents in the landscape posing remediation problems of enormous magnitude and economic burden. The agricultural system has become hostage to chemical pesticides that continue to pose risks to human health and disrupt natural ecosystems. The early promise of biotechnology included products and processes designed to "fix" the undesirable outcomes of other technologies. It also incorporated research to improve upon existing plants and animals as food sources. In preparation for reviewing the expectations of biotechnology during the first decade, I have classified the environmental applications into five categories: remediation of ecologically harmful technologies; improvement of existing plants; modification of animals; microorganisms to mine the earth; biological sensors.

REMEDIATION

Pollution Cleanup. Research efforts are directed at genetically engineering microorganisms that could degrade oil spills, detoxify chemical pollutants in soil and water, and improve biodegradation in wastewater treatment plants. The problems of off-shore oil spills, toxic wastes at Love Canal, New York, and Woburn, Massachusetts, polychlorinated biphenyl (PCB) contamination of the Hudson River, dioxin in Times Beach, Missouri, and scores of other cases highlighted the importance of dealing with toxic chemicals in the environment.

With the tools of classical microbiology, plasmid engineering, and recombinant DNA techniques at their disposal, scientists began experimenting with microorganisms that break down complex hydrocarbons. Early studies showed promise of a bacterium that could digest crude oil. In 1980 General Electric won a patent claim for an oil-degrading strain of Pseudomonas, the first such claim issued for a microorganism *sui generis*. Nine years later, the B. F. Goodrich Co. received a patent for a novel plasmid that degrades ethylene dichloride, a compound found in many toxic waste dumps. Other investigations sought microbes that could neutralize the synthetic herbicide 2,4,5,T (an ingredient in Agent Orange) and PCBs, as well as organisms that could degrade explosive substances and soil contaminated with nuclear materials. The disposal of radioactive wastes has been a persistent problem for the nuclear industry and the Department of Energy. Community resistance to the sitting of low-level radioactive waste land fills remains high. The prospect of a biotechnological solution to radionuclides could soften the public's opposition to nuclear power. *Biological Alternatives to Agricultural Chemicals*. Promotional materials emanating from the new biotechnology industry spoke enthusiastically of replacing chemical pesticides with biological alternatives that are species-specific and safe for humans and other living things. Some of the new firms embraced the rhetoric of the environmental movement in describing their research agenda and their concern for things ecological. The following is an excerpt from an annual report of the Mycogen Corporation San Diego, California, circa 1988.

Pesticides are necessary to grow crops economically and preserve the quality of our food supply. Worldwide, a staggering $14 billion are spent annually on crop protection products — $5 billion in the U.S. alone. Despite these huge expenditures, many destructive pests go unchecked and cause major crop losses.

The pesticide industry is currently dominated by manufacturers of synthetic chemicals. However, the interplay of environmental and economic forces is creating tremendous pressure for new products which are effective and economical while being safe to use and being compatible with the environment. Biological products are ideally suited to fill the market needs. Among the major chemical companies, Monsanto was recognized for its strong development effort in agricultural biotechnology. Monsanto's R&D programme included the development of a tomato plant genetically engineered with a protein from the naturally occurring bacterium Bacillus thuringiensis that is supposed to resist insects. Several environmental groups, including the Environmental Defence Fund, the National Audubon Society, and the National Wildlife Federation supported a project of the firm Crop Genetics that involved injecting a microbe called Cxc into the seedlings of corn. This microbe, which ordinarily lives in corn plants, was endowed with a gene from Bacillus thuringiensis, the protein product of which is toxic to certain corn pests.

Another group of agricultural research programmes was directed at slowing down the escalating use of artificial chemical fertilizers, which run off into surface waters and foul lakes and streams by the buildup of organic matter — a process known as eutrophication. The expectation was that plants or symbiont microrganisms could be genetically engineered to fix their own nitrogen from the vast reservoir available in the atmosphere. Agracetus began marketing a microbial soybean seed treatment — a product with genetically modified strains of *Bradyrhizobium japonicum* to improve nitrogen-fixing ability. Another strategy involved transferring expressible nitrogen-fixation genes of organisms like *Klebsiella pneumoniae* into the chloroplasts of plants, endowing the latter with a selffertilizing capacity.

Agricultural Harvesting and Distribution. Fruits and vegetables are frequently harvested in a preripened stage to accommodate harvesting machines and mass distribution networks. However, the quality of the produce is often affected by being cut off from critical enzymatic changes that take place during the natural ripening process: "While today's tomatoes are the most impressive since the dawn of produce in their resilience, disease-resistance, and travel-hardiness, they are also the most inedible."

Some genetic engineering research is directed at modifying crops so that they will be compatible with the harvesting technologies while also retaining the taste and quality of naturally ripened produce. The challenge for biotechnology is to recapture the taste of produce as it was when grower and consumer were in close proximity, and to accomplish this without disturbing the existing agricultural system and by a means that does not compromise the environment.

IMPROVEMENT OF PLANTS

Adaptability to Harsh Environmental Conditions: Agricultural crops are adversely affected by many environmental factors, including frost, wind, mineral toxicity, soil acidity, salinity, and of course insect pests. A number of research programmes were devoted to redesigning plants to be more adaptable to harsh environmental conditions. Kenneth A. Barton and Winston J. Brill saw the possibility of mixing and matching in nature by transferring the stress resistant qualities of weeds to food crops. Among the harvests expected of the technology are disease and drought-resistant strains, self-fertilizing seeds, crops with tolerance to high salinity, and plants with enhanced photosynthetic qualities.

Nutritional Value: Improvements sought for existing crops went beyond their growing characteristics and their resistance to natural enemies. Research was also directed at modifying the protein structure of the crops by altering the amino acid chain in the plant's DNA. Among the expected benefits of this research is the enhancement of the usable protein in staple crops like corn and wheat. In their 1983 article in Science, Barton and Brill outlined the possibilities.

The seeds of legumes and cereal grains provide humans directly with approximately 70 percent of their dietary protein requirement. Throughout seed development, storage proteins are synthesized and accumulated within the seed, apparently to provide a source of amino acid reserves during early seed germination. High levels of such protein in seeds provides an enriched amino acid source for both human and animal consumption. However, various deficiencies of seeds in certain essential amino acids do not allow either cereal grains or legumes to provide a balanced diet without supplementation of the limiting amino acids from other sources.

Thus, one approach towards the problem of malnutrition is the improvement of nature's "nutritionally deficient seeds" by encoding them with useful storage proteins. Molecular genetics is viewed as a tool for creating perfect or nearly perfect food sources — foods that carry a complete nutritional balance of amino acids. For lesser developed countries that must import food products to round out the nutritional complement of indigenous crops, or for nations where there has been a scarcity of certain grains, this could be a great asset. But would this technology empower the lesser developed countries? Or would it create a new form of domination by the industrialized nations — a kind of genetic colonialism?

Herbicide Resistance: Nature did not endow most plants with a resistance to synthetic herbicides. Recently, with the extravagant use of herbicides in agriculture and in war as defoliants, plants have been discovered with natural resistance to herbicides. In 1982, herbicides represented 47 percent of the $3.7 billion a year U.S. pesticide market, eclipsing insecticides and fungicides. Biotechnology offered the herbicide industry a way to expand markets and at the same time create a growing environment that simulated a weed-free green house.

By engineering crops with herbicide-resistant traits, the herbicide could be used at any stage in the growing cycle. This so-called improvement upon nature does not take into consideration the effects of the increased use of the herbicide. Guido Ruivenkamp, a Dutch researcher at the University of Amsterdam, wrote that "big industry may see the early opportunity of developing *pesticide-resistant plants* rather than undertaking the longer term effort of developing *pest-resistant plants.*"

Plants that Grow Faster: A report issued by the National Academy of Sciences titled Biotechnology: *An Industry Comes of Age describes an enzyme* with the acronym Rubisco that is an important catalyst in photosynthesis. If scientists learn how to move the genes for Rubisco it may be possible to amplify the genes' function and accelerate plant growth. An obvious advantage to the farmer would be the addition of another growth cycle, making more efficient use of agricultural land.

MODIFICATION OF ANIMALS

Transgenics is the term used to describe the application of rDNA techniques to genetically modify animal species. In some instances, the current research strategy represents an extension of traditional goals to create animals that meet certain requirements of the consumer (as interpreted by agribusiness) or of the food processing industry. For example, there is an interest in creating livestock with leaner meat, and in producing breastier fowl. Since these animals are used exclusively in food production they are not suited to survive anywhere beyond the feedlot. As such they are unlikely to pose environmental problems from a release. The genetic redesign of animal species has other applications which might affect the environment should the animals be released.

There are research programmes to use transgenic animals as functioning, living bioreactors. For example, an animal may be genetically engineered to secrete a human protein in its milk or blood stream. In 1987 a transgenic mouse was produced that secreted a clot-dissolving drug in its milk. Mice are produced twice their normal size by the endowment of a human growth gene. In yet another case, the gene that codes for the AIDS virus was implanted in the fertilized egg of a mouse and brought to term. As a result, the AIDS gene is found (albeit not necessarily expressed) in every cell of the mouse. Release of such a mouse into the environment could represent a new route for the spread of the disease.

It is for this reason that NIH took special precautions in handling these experimental mice by keeping them in high-containment facilities. There has also been experimentation with transgenic fish to determine if they can be genetically designed to survive in colder water or if a larger species can be introduced to repopulate a river. The lingering question is: what impact would such releases have on the aquatic ecology?

MINING OF THE EARTH

Oil Recovery. Microorganisms can be used to enhance recovery methods for removing crude oil from wells. Initial research suggested that certain organisms can reduce oil viscosity, improve rock porosity and permeability, and decrease water-oil interfacial tension. The technique of Microbiologically Enhanced Oil Recovery (MEOR) was first developed in the 1940s. Through genetic engineering, there was a new impetus to improve MEOR by enhancing the traits of organisms like *Clostridium acetobatylicum,* which has been used in the extraction of oil. The microorganism *Xanthomonas campestris,* which produces xanthum gum, is used for oil recovery by increasing the viscosity of water that is pumped into wells to displace the oil. Another MEOR approach that claims success in recovering oil from marginal wells creates a mixture of microbes and molasses which is injected into oil wells. The gas produced by the microbial growth increases the pressure, forcing out the oil.

Microbial Mining. Microorganisms have also been useful in recovering certain metals from ore-bearing minerals. Copper and uranium are mined and commercially purified by a process that includes leaching by bacteria such as *Thiobacillus ferroxidans.* The leaching process may be done in situ or in a controlled environment. Efforts are underway to improve the leaching properties of the organisms through recombinant DNA techniques. R. H. Zaugg and J. R. Swarz cite some potential hazards in using genetically modified organisms in the mining industry. All foreseeable applications of biological processes . . . involve microbial systems operating in relatively open environments, such as slag heaps or tailing ponds. Consequently, there are risks that microorganisms or their metabolic products will inadvertently contaminate the local ecology. The authors offer three examples: (1) bacterial leaching operations that generate large quantities of sulfuric acid could acidify water supplies; (2) *Thiobacilli* and related species may acquire the traits to infect humans; (3) metals concentrated by bacteria from dilute mine waters can accumulate in the food chain.

BIOSENSORS

Microorganisms may also be used as markers in the environment. The concerns about releasing novel organisms have produced an interest in the methods of monitoring the dispersal of the agents. Monsanto Company has developed a genetically engineered strain of *Pseudomonas* carrying natural resistance to antibiotics and genes from *E. coli* that code for proteins that enable the microbe to metabolize lactose.

These traits make the organism a possible marker for field tests. The first approval by the EPA of a field test of a genetically engineered organism under the Toxic Substances Control Act (TSCA) was for Monsanto's genetically marked microbe. The test was performed in Blacksville, South Carolina. The environmental uses of applied genetics constitute an important part of the industrial agenda. But the prospects of releasing genetically novel plants, microbes, and animals have activated a lively public policy debate. The next generation of environmental experiments and applications of biotechnics brought a new group of scientist-stakeholders into the discourse of risks and benefits of new technologies.

PLANTS, POWER, AND INTELLECTUAL PROPERTY

Biotechnology, along with closely related issues of food, farming, and intellectual property rights, has become a flashpoint in multilateral trade and environmental negotiations between developing nations, the United States, and Europe. Sharp disagreements about trade in genetically engineered products and about the patenting of living things have sparked disputes about the powers and scope of emerging institutions of global governance. Central to these controversies are tensions between the principles andjurisdictions of the World Trade Organization (WTO) and those of the international Convention on Biological Diversity (CBD) and the new Cartagena Protocol on Biosafety.

In addition, there are contradictions within the biodiversity convention itself. The WTO, established in 1994, fosters marketbased regulation of biotechnology and genetic resources. The CBD, which was ratified in 1993, does this as well but also invokes environmental and social criteria for management of biodiversity and biotechnology.

Contrasting understandings of biotechnology's effects are pivotal in these disputes. New agricultural biotechnologies promise control over the traits and reproduction of food crops and livestock. Advocates of crop genetic engineering argue that transgenic crops can increase world food production while limiting environmental damage from agriculture. Their critics contend that claims about the precision and power of genetic engineering are dangerously exaggerated. Widespread adoption of transgenic crops would at best permit only temporary foodproduction increases, they say, and would endanger agricultural genetic diversity, the livelihoods of farmers, and the food security of countries that depend on food imports. Genomic mapping and molecular bioengineering have opened new opportunities for capital accumulation in agricultural research and development, technology licensing, and sales of seed, food, and pesticides. These new profit opportunities have speeded the mergers of agribusiness, pharmaceutical, and chemical corporations and takeovers by these firms of seed and biotechnology research companies.

One result is the consolidation of technological and genetic resources, economic clout, and political influence in a handful of transnational corporations. These firms are in a position to dominate international markets in agrochemicals, germplasm (seeds and varietal breeding lines) of major commercial crops, and biotechnology itself (equipment, expertise, genome databases, and other proprietary information).

Facilitating this trend is the expansion of intellectual property rights (IPRs). In some countries, those who discover or devise new types of plants, animals, microbes, or genes; novel uses for them; or processes for altering them at the molecular level may be granted patents or other private-ownership rights to such inventions.

These "life patents" are controversial in part because they enable patent holders, which are primarily corporations based in the global North, to profit from products they have developed from crop varieties and medicinal materials obtained from the biodiversity-rich global South, sometimes by means of only minor modifications.

Because such patents restrict the rights of farmers to save or exchange seeds, critics contend, they may undermine food security. In addition, many indigenous peoples' groups and some governments object on moral or political grounds to the private ownership of living things.

Meanwhile, public institutions have become dependent on private-sector partnerships for access to privately owned materials and techniques. As a result, for-profit companies are gaining growing influence over the research agendas of universities and the priorities of public agricultural research and extension services around the world. Moreover, the trend towards intellectual enclosure— the use of IPRs to restrict what scholars and research organizations may publish or share with students, colleagues, and the public— has begun to impede access to new technologies and the exchange of scientific knowledge.

Such concerns have raised the stakes in long-simmering international controversies about food, farming, and trade policies and intellectual property rights. These debates have grown more heated as many governments and social movements have begun to question the benefits of economic globalization and the terms on which it is taking place. In the 1990s, concerns about the growing power of the multinational "gene giants" added fuel to this fire.

New global environmental institutions, particularly the Convention on Biological Diversity and its Cartagena Protocol on Biosafety, have become staging grounds for resistance to WTO rules and to the market-based management of genetic resources that the WTO supports. This resistance is rooted in preexisting patterns of inequitable resource flows and the resulting inequality, mainly between the gene-rich global South and the technology-strong North.

During the past two decades, developing countries have fought to include in international environmental accords provisions that offer opportunities to redress—or at least not replicate—the exploitative relations of the past. As a result, the new environmental treaties provide openings for the inclusion of social equity and environmental justice as principles of international governance.

Nevertheless, these new institutions contain their own contradictions. The CBD embodies a deep tension between market-oriented and alternative or pluralist approaches to biotechnology regulation and resource management, a tension reflected in ongoing disputes about the role of intellectual property and biotechnology in the treaty's implementation.

Southern governments and activist nongovernmental organizations (NGOs) have represented the CBD as counterbalancing the more narrowly economic principles of the WTO. However, the United States is determined to make the WTO the primary, overarching regime for regulation of biotechnology trade and food and environmental standards, and to limit the purview of the CBD and its offspring, the biosafety protocol.

This agenda is linked to U. S. agricultural and technology export goals. The continuing repercussions of these disputes are evident in the WTO and other multilateral forums, particularly with regard to intellectual property rights to the raw materials and products of biotechnology. These disputes have contributed to shifts in the pattern of alliances between major grain-exporting industrial countries, especially the United States and its former European allies, as well as the developing countries of South and Southeast Asia, Africa, and much of Latin America.

ENVIRONMENTAL-ECONOMIC GOVERNANCE

At stake in the conflict between the WTO and the CBD is a model of global economic governance that subordinates social, environmental, and ethical concerns to the overriding objectives of economic growth and trade liberalization. U. S. delegates to the biosafety protocol and WTO talks insist that the genetic-resource inputs to and the genetically altered products of biotechnology are ordinary commodities, subject to standard rules of transnational commerce and to the jurisdiction of the WTO and its Agreement on Trade Related Intellectual Property Rights (TRIPS). Biotechnology's raw materials (seeds and plant and animal samples), its technological tools and related knowledge (genetic information, databases, and product formulas), and its products, according to this view, should be managed as private property, freely bought and sold, with minimal labeling requirements or restrictions on their import, export, and use.

Any benefits from the commercial sales of genetically engineered products ought to be allocated strictly under the terms of standard, two-party business contracts. The theoretical rationale implicit in this approach is a neoliberal version of environmental economics. It assesses the values of genetic and other natural resources in terms of their world-market prices or the dollar costs of replacing them, regardless of the fact that lack of hard currency leaves the world's majority with no purchasing power in globalresource markets.

Neoliberal environmental economics takes scant account of nonmonetary and long-term values of ecosystems and their components, or of the place-specific values of nature to local communities. Instead, this economics constructs food, ecosystems, and organisms as stocks of industrial raw materials, fungible units of natural capital and genetic information for sale to the highest bidders.

Yet agro-biotechnology and related food-security and environmental issues are proving difficult to subsume under this economic reductionist paradigm. The conceptualization of biodiversity as an export commodity and a technological input is contested by some governments and by networks of indigenous peoples and other NGOs. In WTO negotiations, Southern-country coalitions have attempted to delay implementation of the TRIPS Agreement and to widen TRIPS loopholes that permit the use of social and moral criteria in national policies on intellectual property.

These TRIPS critics argue that uniform application of WTO rules will foster even greater North-South inequality and that the TRIPS Agreement in particular conflicts with the CBD. In the words of Cameroon's ambassador to the European Union, Philomenon Yang, speaking on behalf of the African Group of CBD delegations, "The TRIPS Agreement creates potential for disastrous conflicts between the technologically advanced and the less technologically advanced countries. It will endanger the traditional rights of farmers and of local communities all over the world ... [and] greatly jeopardize the application of the [Biodiversity] Convention. "

ORIGINS, LIMITATIONS, AND CONFLICTS

The U. S. government wants to strengthen WTO TRIPS rules that make it illegal under most circumstances for local citizens, businesses, or government agencies to duplicate or use proprietary medicines, plant varieties, gene sequences, therapeutic techniques, or research technologies. The U. S. goal of obtaining stronger TRIPS rules for crop varieties is supported by Australia, Canada, a few Latin American food-exporting states, and, with reservations, by some European governments and Japan.

Against these economic powerhouses stands a large group of developing countries that are opposed to the strengthening of TRIPS. In the period leading up to the Seattle Ministerial Conference, more than one hundred developing countries endorsed proposals to roll back the 1993 TRIPS accord. Conflicting efforts to amend TRIPS have continued in the WTO's Council for TRIPS.

The U. S. position is that "unimproved" genetic materials taken from crops developed by informal breeding or from wild organisms belong to whoever would make use of them, as part of humankind's "common heritage." As such, the United States recognizes, these genetic-resource inputs of biotechnology are covered by the CBD, which requires that its member governments make their genetic resources accessible to others.

In contrast, access to and regulation of biotechnology industry outputs—genetically engineered products, their genetic recipes, and the tools and know-how for producing them—fall outside the CBD mandate, in the U. S. view, because they are private, tradable commodities and thus subject to WTO rules and to intellectual property rights.

While the CBD was being negotiated from 1989 to 1992, the United States pressed this position forcefully in the Uruguay Round of the General Agreement on Tariffs and Trade. These negotiations transformed the GATT into the WTO, expanded its purview to include trade in agriculture and services as well as in goods, and added the TRIPS Agreement. The United States, the main force in the WTO, wanted member governments to open their markets to foreign exports and investments, especially in industries where the United States is relatively strong, such as agriculture, financial services, computer electronics, entertainment media, and biotechnology.

The TRIPS accord, initiated and pushed by a coalition of European, Japanese, and U. S. multinational corporations, stipulates that WTO parties must adopt laws to enforce patents "in all fields of technology". It requires WTO member countries to recognize the proprietary rights of local or foreign citizens or enterprises to crop varieties, whether conventional or genetically engineered, and to genetically altered microorganisms and other biotechnological innovations.

During the GATT IV talks, developing countries were able to obtain small but significant exceptions to the blanket requirement for IPR coverage of living organisms and related technologies. Section 5, Article 27, Section 3(b) of the TRIPS Agreement allows WTO member governments to exclude from patentability "plants and animals other than microorganisms, and essentially biological processes for the production of plants or animals other than non-biological and microbiological processes.

However, Members shall provide for the protection of plant varieties either by patents or by an effective *sui generis* system or by any combination thereof. " The TRIPS accord also permits states to adapt their IPR regimes in ways necessary to safeguard the environment, morality, or *ordre publique*. As a further concession to developing countries, TRIPS was slated for review in 2000–2001. As the ramifications of IPRs in crop and medical biotechnology have become more apparent, developing-country WTO members have requested reconsideration of the application of TRIPS rules to food crops, microorganisms, and "essentially biological" processes and have called for the recognition of non-economic factors in environmental and biotechnology regulation.

These ongoing disputes highlight the limitations of "free trade," economic efficiency, and private property as the paradigmatic principles of global governance. The growing perception that trade liberalization favours the economically strong helped to precipitate the WTO ministerial meltdown in Seattle and has contributed to continuing deadlocks in the WTO.

The issues of biotechnology regulation and private property in organisms and their parts are especially inflammatory. They not only raise unprecedented questions about the powers and rights of individuals and companies to own and manipulate life but also are linked to controversies about international economic inequality and environmental justice.

THE CBD BATTLEGROUND

In contrast to the WTO, the Convention on Biological Diversity establishes an arguable basis in international law for taking non-economic criteria into account in biotechnology regulation. At the same time, contradictions built into the CBD make it a source of continuing conflict between market-oriented and equity-oriented approaches.

Because the CBD is a framework treaty that requires further elaboration to be put into practice, its parties meet every two years to adopt guidelines for its implementation. CBD articles addressing access to genetic resources, distribution of the benefits these resources provide, the transfer of biotechnology, biosafety, and related intellectual-property issues have been hotly disputed throughout this process. Some CBD articles commit signatory countries to goals that, at least implicitly, conflict with the privatization and market-based valuation of nature.

The CBD recognizes the sovereignty of states over genetic and other resources within their territories. It calls for the in situ and ex situ conservation of biological diversity, including crop genetic resources, for protection of the "traditional lifestyles relevant for the conservation and sustainable use of biological diversity" of "local and indigenous" communities and of the "customary use of biological resources in accordance with traditional cultural practices".

The CBD also calls for prior informed consent as a precondition for access to local genetic resources, for national policies to promote conservation and sustainability, and for equity and "fairness" in genetic-resource trade and in the distribution of technology and its benefits. Given the disparities in economic power between transnational corporations and most developing countries and communities, and the ubiquitous economic incentives and opportunities for short-term exploitation of natural resources, none of the above objectives is likely to be achieved by means of the market mechanisms and private-property systems fostered by the WTO.

In addition, resolutions by the CBD's Conference of the Parties have recognized alternatives to individual or corporate property rights, such as collective property and indigenous traditions of shared knowledge, concepts not recognized in the WTO. However, there are other CBD provisions that foster a commodity-based framework for resource management. The influence of these market-oriented approaches has been amplified by the "green-developmentalist" bias of the Secretariat of the CBD.

Why—in a treaty initiated as a *conservation* compact, with a focus on wilderness, forests, and wildlife—have biotechnology and IPRs become so pivotal? The answer lies both in patterns from the past (the long history of removal of genetic and other resources from colonized regions) and in the political economy of the present (the importance of IPRs and other private property rights to the global extension of commodity relations and the unequal impacts of economic globalization in different world regions). The controversies about life patents and genetic-resource access are linked to the economic agendas of countries with growing biotechnology industries.

These agendas, in turn, arouse fears among food-import-dependent and genetic-resources-provider countries that the growth of giant biotechnology-agrochemical firms and the extension of IPRs will lead to their further impoverishment and dependency. To see these connections, we need to recall the context in which these CBD conflicts arose.

GOALS FOR A BIODIVERSITY CONVENTION

The Convention on Biological Diversity is one of two international conservation treaties launched at the 1992 Earth Summit in Rio de Janeiro. Its primary objectives are "the conservation of biological diversity, the sustainable use of its components and the fair and equitable sharing of the benefits arising out of the utilization of genetic resources, including by appropriate access to genetic resources and by appropriate transfer of relevant technologies, taking into account all rights over those resources and to technologies, and by appropriate funding".

The original sponsors of the CBD definitely did not intend to investigate the degree to which existing international power relations, dominant development models, or present international distributions of resources contribute to environmental crises. These contentious issues, however, are proving to be inseparable from the more narrowly defined environmental goals of the treaty's original sponsors.

The Northern states (mainly Germany, France, and, later, the United States) that first called for a global conservation treaty in the mid-1980s wanted to limit the expansion of polluting industrialization in the global South, particularly increased emissions of greenhouse gases, and to preserve some critical mass of tropical forests as a carbon sink.

They also hoped to slow the rate of species extinctions and the destruction of "wilderness" areas and to guarantee continued Northern access to the biological resources of Southern regions.

In contrast, the immediate goal of most Southern-government signatories was to obtain additional foreign aid in the context of shrinking overall development assistance. Some diversity-rich countries also hoped that the CBD would help them to establish their own biotechnology enterprises or at least to obtain revenues from the export of their genetic resources under the terms of bioprospecting contracts with Northern pharmaceutical firms and research institutions.

Negotiation of the CBD from 1989 to 1992 involved intense disputes, mainly between the G77-plus-Chinabloc of Southern states and the more powerful industrialized nations of the OECD (Organization for Economic Cooperation and Development). When conflicts over the draft text could not be resolved, the disputed text was either excluded, as in the case of proposals for strong forest-conservation language, or diplomatically finessed.

As a consequence, key provisions of the CBD are ambiguously worded and open to conflicting interpretations. These contradictions might remain moot were it not for the activism of transnational alliances of social movements and NGOs. In contrast to other agencies of global governance (such as the WTO, the World Bank, and the International Monetary Fund) and in contrast to many other United Nations bodies, the CBD has permitted, and even depended on, the contributions of civil-society organizations. Mainstream conservationist NGOs have helped to draft CBD decisions.

Advocacy-oriented networks of NGOs, in alliance with Southern states and some European states, have pushed for the CBD to take on issues of biotechnology and its environmental hazards and social consequences, much to the consternation of the U. S. government.

The paradoxical role of the United States in the CBD sheds light on the treaty's internal tensions and external pressures. The Reagan administration, lobbied by conservation biologists and mainstream environmental organizations, was an early supporter of plans for a single, global conservation accord. Less publicized was the desire of agricultural and biotechnology interests to guarantee continued access to Southern genetic resources.

During the CBD negotiations, the United States and the OECD negotiating bloc worked to ensure that the convention would make it easier for firms and research institutions to continue to survey and select crop varieties and pharmacologically active substances from the territories of other states. This goal took shape as CBD Article 15, "Access to Genetic Resources, " which establishes that "Each Contracting party [to the CBD] shall endeavor to create conditions to facilitate access to genetic resources for environmentally sound uses by other contracting parties and not to impose restrictions that run counter to the objectives of this convention.... Access, where granted, shall be on mutually agreed terms ... subject to prior informed consent".

CROP PATENTS AND PIRACY

The concept of "genetic resources," which emerged in environmental discourse in the 1980s, refers to the information contained in the genomes of plants, animals, and microorganisms, many of them as yet "undiscovered"—that is, unknown to Western science—and in danger of extinction. Organisms containing pharmacologically active compounds, found in greatest abundance in tropical regions, are potential sources of medically and commercially important drugs or models for new synthetic drugs.

As advances in biotechnology and information technology made possible the rapid screening, genetic mapping, and molecular manipulation of natural substances, genetic resources came to be seen as an important raw material for the U. S. economy. By the late 1980s, the perceived value of the medicinal genetic resources of the tropics, especially rainforests, had soared, albeit temporarily.

Another form of genetic resources of great interest to U. S. negotiators was agricultural biodiversity: the myriad unique, locally specialized foodcrop varieties bred and conserved by farmers worldwide. Genetic, varietal, and crop diversities are the cornerstones of the survival strategies of many small-scale cultivators, the sector that still produces most of the world's food).

Locally adapted crops and their wild relatives are also important to large-scale commercial agriculture as sources of traits for crop improvement because crop varieties, especially in industrial, monocrop farming systems, must be altered continually to defeat pests and diseases that have adapted to existing varieties and pesticides.

Most of the crops produced in the world's major agricultural exporting regions—Europe, the United States, Canada, Australia, Brazil, and South America's Southern cone—are derived from plants originally domesticated in Latin America, Southeast Asia, and western Asia.

Samples of some of these varieties are conserved in public and private gene banks, and firms and research agencies from the relatively "gene-poor" global North have tapped these collections most frequently. But while the CBD was being framed, developing countries were already challenging the right to use crop genetic resources without compensation for their owners.

The decade preceding the CBD had seen unresolved North-South disputes over asymmetrical international flows of genetic resources and profits derived from them. Until the 1980s, genetic resources in fields, farms, and gene banks had been treated by seed companies, botanic gardens, and research institutions as a global commons, free for the taking.

In the early 1980s, however, governments of the South began to object to the treatment of genetic resources in their territories as a worldwide open-access resource. They sought international recognition of their rights to produce and market useful plants and needed medicines, including those based on processes or substances patented in other countries.

Under pressure from Mexico and other diversity-rich states, the UN Food and Agricultural Organization's Commission on Plant Genetic Resources for Food and Agriculture adopted the International Undertaking on Plant Genetic Resources for Food and Agriculture (IU) in 1983. The original version of the IU declared *all* plant genetic resources, including varieties covered by intellectual-property claims in their countries of origin, to be part of the "common heritage" of humankind.

The Southern sponsors of this particular reinterpretation of "common heritage" intended it to mean that their farmers or enterprises would not be blocked by patents or other forms of IPRs from reproducing, breeding, and selling hybrid or genetically engineered plants or seeds. Industrial-country governments and transnational firms strenuously resisted this interpretation.

In response to U. S. charges of patent "piracy, " Southern spokespeople raised the issue of "biopiracy, " pointing out that brand-name pharmaceuticals and crop varieties being exported to their countries were in many cases derived from medicines and crops discovered or bred in the global South. At least, they argued, the new biodiversity treaty ought to recognize their right to be compensated for the profitable development by others of their own biological patrimony.

As a condition for promising access to their genetic resources, Southern governments and NGOs insisted that the CBD both recognize the principle of national sovereignty over genetic resources and also include provisions for the sharing of geneticresource benefits with resource providers and for the transfer of new biotechnologies to developing countries. The United States and the OECD negotiating bloc agreed to accept this as a compromise, but declined to endorse any explicit formulas for defining "fair" compensation, "equitable" benefit sharing, or "appropriate" technology transfer.

The U. S. delegation insisted that in addition to providing access to genetic resources, the new treaty should endorse "the adequate and effective protection [of] patents and other intellectual property rights". As a countermeasure, India won inclusion of a sentence to the effect that the IPR reference should not be applied in a way that undermined the CBD's objectives. Disputes over IPRs and biotechnology nearly derailed the CBD negotiations.

Then, the Bush administration refused to sign the treaty when it was launched at the Earth Summit, on the grounds that its IPR references were not sufficiently strong. In short, the U. S. government's commitment to promote the international expansion of its agricultural, pharmaceutical, and technology industries prevailed over U. S. environmentalists' desire for a comprehensive conservation treaty.

U. S. INDUSTRY, THE CBD, AND INTELLECTUAL PROPERTY RIGHTS

The White House alleged that the new convention "threatened to retard biotechnology and undermine the protection of ideas". This defensive, even paranoid, response has its roots in the importance of biotechnology and other high-technology sectors to the United States' international economic position and in the growing importance of intellectual property to U. S. industries.

By the 1980s, protection of U. S. proprietary technology and related intellectual property rights had become a cornerstone of U. S. trade policy. Exports of new technologies—along with agrochemicals and the food surplus produced by state-subsidized agriculture— were seen as counterweighing negative U. S. trade balances in other sectors. Federal policymakers were swayed by industry arguments that genetic engineering was about to produce a bonanza of therapeutic compounds that would be commercialized only if they were covered by IPRs.

Biotechnology lobbyists were in close touch with White House officials during the early CBD talks, and the Japanese, British, and other OECD delegations shared many of their concerns. Industry representatives argued that the treaty could be invoked to pressure them to cede technological secrets to potential competitors and that it might undermine U. S. goals in the GATT IV trade talks.

The United States also tried to prevent the adoption of the CBD language calling for international regulation of genetically engineered organisms and their products. This CBD provision established the grounds for the biosafety protocol and was energetically advocated by a coalition of Southern states and transnational NGOs. It was included in the convention text over strong U. S. and Japanese objections, with the details to be negotiated at a later date.

Later, at the second conference of the CBD parties in 1994, a coalition spearheaded on the NGO side by the Third World Network, the German Green

Party, and Greenpeace International, among others, convinced the CBD parties to begin protocol talks, a process that is still embroiled in controversy.

Six months after the CBD was introduced in 1992, the Clinton administration convened a review of the treaty by representatives of three biotechnology firms and three conservationist groups, who concurred that the CBD's IPR provisions were, after all, adequate. The president then signed the convention, but the Senate has not ratified it. In any case, conditions attached to the White House interpretation of the CBD would reduce U. S. obligations under the treaty to the same sort of market-based determination of genetic-resource and biotechnology distribution that would prevail even without the CBD.

A proposed "statement of understanding" to be attached to the U. S. signature, should the Senate decide to ratify the CBD, declaims that "As such, the [WTO] TRIPs agreement will function as a 'floor' for substantive protection for intellectual property rights by GATT TRIPs parties under the biodiversity convention".

This interpretation of the TRIPS Agreement is the opposite of the interpretation of the CBD by those NGOs and Southern states who see it as a basis for opposition to TRIPS and other WTO free-trade rules and to the uncompensated commoditization of biodiversity. Indeed, the more that the United States has pressed in the WTO and other forums for genetic-resources IPRs and market-based resource valuation, the more that gene-rich countries have insisted on their right to limit access to their biological resources by taxonomists and plant breeders as well as by bioprospectors.

Brazil and other "megadiversity" countries interpret the CBD provisions recognizing national sovereignty over biodiversity as a sort of national-level property right. They hope to sell their genetic resources or use biodiversity access as a bargaining chip to obtain aid, technology, or other benefits. Such a commodity-based approach is unlikely to work to their advantage, however, in a global genetic-resources market where providers have little power and supply exceeds demand.

Differing interpretations of the CBD and its purview continue to provoke controversy as its member governments, with the United States as a powerful, nonmember presence, battle over the treaty's implementation. I have argued that there is an internal tension in the CBD between an approach based on private property and globalized, market-based resource management, and a more pluralist approach that recognizes differences in the development needs, cultures, and property-rights systems of various countries and communities, as well as differences in how they use and value genetic resources.

This tension grows out of global inequalities produced during the colonial era and compounded since then as would-be developing countries attempt to enter—or are pushed to join—an integrated world economy from a position of disadvantage. The tension between market-based and equity-oriented frameworks also reflects disunity within the political-economic project of modern environmentalism. Over the past two decades, a green-developmentalist approach has come to dominate both the discursive practices of mainstream conservationist organizations and the greening policies of the World Bank. This approach has also influenced multilateral environmental institutions, including the CBD. Green developmentalism proposes that environmental problems can be corrected by market solutions. In this worldview, "natural capital" can be assigned property rights and can be traded transnationally.

Forest, mineral, and water resources and ecosystem services, as well as organisms and their parts, are assigned monetary prices based on actual or hypothetical markets. The result is a pan-planetary metric for valuing resources and managing their exchange. This universalizing discourse makes it possible to speak of the "global" management of environmental problems and to act on the assumption of compatibility between capitalist growth and ecological sustainability. The discursive practices of green developmentalism also further the shift from the direct appropriation, or "primitive accumulation," of genetic resources—the mode that prevailed for more than five hundred years—to the market exchange of genetic raw materials.

In theory, green developmentalism provides nature with the means to earn its own right to survive in a world-market economy. Conservation projects are to be financed by exports of environmental assets—access to ecotourism sites, rights to use ecological services (*e.g.*, carbon emission credits), and intellectual property rights to medicinal plants, shamans' recipes, traditional crop varieties, and the genetic information they contain.

Other greendevelopmentalist policies include "green conditionalities" attached to development aid; "capacity-building" projects to re-educate Southern inhabitants, train environmental managers, and construct environmental regulatory agencies within Southern states; and biodiversity surveys and assessments. These discursive practices revalue the South's natural resources from a "global" (read Northern) perspective. However, this revaluation of Southern resources according to methods of Western taxonomy and neoclassical economics constitutes a devaluation of those resources.

The denomination of biodiversity values in dollars discounts the greater part of the values of Southern natural resources to the people who live in direct interdependence with those resources: their tangible use-values, their symbolic values, and their exchange values in local and domestic markets. To obtain their "fair share" of the "benefits of biodiversity" as promised by the CBD, diversity-rich countries and indigenous communities are encouraged to assert their own intellectual property rights to genetic resources in their territories and then sell those rights.

Prominent conservation biologists have argued that selling rights of access to living pharmacies will provide resources and incentives to preserve natural areas. As the Smithsonian's leading conservation scientist told a World Bank workshop, prospects for saving biodiversity are now linked to biotechnology's new ability to "generate wealth at the level of the molecule."

Bioprospecting agreements have proliferated between Southern suppliers (governments, parastatals, and NGOs and Northern buyers (large and small pharmaceutical firms, universities, and agencies such as the U. S. National Institutes of Health). NGO critics of bioprospecting as a strategy for benefit sharing contend that it will facilitate the mining of Southern genetic raw material by bioprofiteers and reinforce the idea that, having paid a "fair market price" for this property, drug and seed firms are justified in selling products derived from collected materials back to their countries of origin at vastly higher prices.

Meanwhile, some of the early proponents of bioprospecting have since concluded that "regrettably, ... genetic prospecting may not help much in the struggle to preserve habitats rich in biological diversity" and that the prospects of substantial bioprospecting resource transfers from North to South are negligible.

These analysts acknowledge that, given the economic and legal resources of transnational biotechnology firms and the fact that genetic-resource supplies already exceed industry's demand, most gene-source countries are in a weak position to bargain for "fair" compensation. Most local, direct providers of organic samples and knowledge are in a worse position, especially when national governments do not acknowledge their rights.

Furthermore, the United States holds that if there is to be any sharing of the benefits of genetic resources, it should be at this initial, bioprospecting stage—when the commercial value of the resources sold is unproved and the sale price low—rather than after higher-priced products have been made by altering the natural materials or "discovering" new uses for them.

This early valuation would offer little to providers of medicinal and crop samples but is consistent with the U. S. position that genetic resources "enhanced" by biotechnology are ordinary commodities subject to free-market rules.

CBD SUPPORT FOR THE COMMODIFICATION OF GENETIC RESOURCES

The administrative body of the CBD has been markedly influenced by the green-developmentalist perspective of Western governments and environmental organizations such as the World Conservation Union (IUCN). Interpretations of the CBD mandate prepared by the Secretariat of the CBD reflect this influence. One such document's proclamation that "modern biotechnology offers the potential to invent sustainable systems of the future, to be accompanied by a new paradigm for industry" reveals how central biotechnology has been to the interpretation of the treaty.

Indeed, the word *biotechnology* appears six times in the CBD text and merits an entire section, Article 19, the only CBD article heading that addresses the distribution of the benefits of biodiversity. "The private sector is the key player in benefit-sharing arrangements," the aforementioned secretariat document states flatly. Another secretariat note identifies a "policy setting" conducive to benefit sharing as one that encourages "access legislation, incentives, partnerships and contracts".

This emphasis on market transactions and business partnerships illustrates how CBD administrators have promoted the marketbased management of genetic resources. Case studies compiled by the secretariat interpret biodiversity benefit sharing almost exclusively in terms of North-South bioprospecting deals. In effect, access to "biodiversity benefits" under the CBD is being made contingent on the participation of diversity-rich countries in a global genetic-resource market.

But when CBD documents define biodiversity benefits as benefits to be derived from the commercialization of genetic resources by biotechnology industries, they fail to recognize that biodiversity benefits already exist—that is, that the benefits of natural resources are known and valued by people who depend on them directly for sustenance, shelter, aesthetic pleasure, and spiritual significance. This equation of biodiversity benefits with genetic resources reduces biological variety to its purported essence as a commodity, separable from its complex interrelationships with the rest of nature and society. It is as if the values of biological variety come into existence only when its genetic information is "developed"—codified, counted, and commercialized—by biotechnology.

A broader conceptualization would centre on the incalculable present and future benefits of healthy ecosystems and diversity-based farming systems both to people locally and—insofar as such eco-social systems may conserve and generate biodiversity—to people in all countries.

Measuring the value of a country's biological resources on the basis of their commercial potential, or in terms of the market costs of replacing them, is compatible with the dominant discourse of environmental managerialism. However, this green-developmentalist approach privileges those aspects of nature that can be removed from their local contexts, transformed by investment, and sold.

The approach fosters a view of ecosystems as warehouses of potential commodities to meet the demands of foreign consumers rather than as the bases of local and national life, as the sources of material necessities and meanings, and as the biophysical contexts of cultures. In this way, green developmentalism divorces the problems of biotechnology and genetic-resource management from the development needs of gene-rich but hard-currency-poor countries. The globalized, market-based management of biodiversity requires clear property rights to natural resources, and the CBD secretariat has devoted what might seem, in a conservation treaty, inordinate attention to intellectual property. Conference decisions and secretariat documents display an almost schizophrenic ambivalence. The predominant assumptions in these documents support Northern intellectual-property models and their international extension, but references to IPRs are counterbalanced in nearly every statement by references to "alternative systems" or to the "concerns of indigenous and local peoples" about the impacts of IPRs.

This controversy has been kept alive by the active influence in the CBD and related international forums of social movements that oppose the privatization of genes and associated knowledge. On the basis of ethical, equity, and ecological concerns, the NGOs and indigenous peoples' and farmers' organizations that constitute this transnational movement have denounced the patenting and commoditization of life.

With the support of Sweden and a number of Southern states, these organizations have energetically resisted Northern proposals that the CBD recognize the WTO as the appropriate body for settling international disputes over property rights to organisms and knowledge about nature. In light of fears among many Southern delegations of losing sovereignty to multilateral trade bodies; growing doubts about the benefits of industrial agriculture and the safety of genetic engineering; and questions about the fairness of intellectual-property regimes, these NGOs have found a growing audience among developing-country delegations.

Benefits and Risks of Biotechnology

Plants developed through genetic engineering were first grown on 1.7 million hectares in 1996 in the United States, but by 2002 they were grown on 58.7 million hectares in 16 countries. By far the major use of the present plants is to manage pests—weeds, insects and diseases. Weed management with genetically engineered plants is accomplished because the plants have a modified enzyme (a protein) that allows them to survive an application of a specific herbicide that normally acts on that enzyme. Growers can plant the herbicide-tolerant seeds, allow the plants to emerge along with any weeds in the field and then treat the field with an herbicide.

The result is that the weeds, but not the crops, die. The advantage to growers is that they spend less time on weed management, have enhanced weed control, use safer herbicides, and in many cases use less herbicides. Additionally, this technology allows growers to use soil conservation practices such as reduced or no-tillage, thus helping to retain soil structure and moisture and reduce erosion. Herbicide tolerant crops (soybean, canola, cotton and maize) were grown on 48.6 million hectares in 2002. Insect-resistant crops developed through genetic engineering utilize the common soil bacterium, Bacillus thuringiensis (Bt), which has been commercially used for more than 50 years, as an insecticide spray.

Although safe to humans and the environment, when a susceptible insect ingests Bt, the Bt protein binds to specific molecular receptors in the gut and creates pores causing the insect to starve to death. Insecticidal products containing Bt were first commercialized in France in the late 1930s, but even in 1999 the total sales of Bt products constituted less than 2 per cent of the total value of all insecticides. Bt, which had limited use as a foliar insecticide, became a major insecticide only when genes that produce Bt toxins were engineered into major crops. The Bt crops available at present are maize and cotton. These were grown on a total of 14.5 million hectares in 2002. Virus-resistant crops were created by inserting a non-infective part of a plant virus into a plant, essentially "vaccinating" the plant to protect it from the virus.

This technique is called "pathogen-derived resistance." Squash and papaya have been engineered to resist infection by some common viruses and are approved for sale in the United States. There are fewer than 1 million hectares of these crops. The bioengineered plants available at present provide growers with better tools to manage pest problems. As with any technology, there are risks and benefits to currently available genetically engineered plants, but the present body of information indicates their use has enhanced pest management, substantially reduced the amounts of pesticides used in some crops, enabled growers to use safer pesticides, and contributed to enhanced safety for humans and the environment.

The regulatory process for managing these plants and their effects on the environment and humans has evolved with the technology and the scientific community's knowledge of these tools. Many of the more controversial issues surrounding genetic engineering of plants—such as pesticide resistance, gene flow and intellectual property issues—are not unique to this new technology but pertain to all types of agriculture. Some species of insects have developed resistance to sprays of Bt, indicating the potential for some species to become resistant to Bt plants. However, despite Bt plants being grown on more than 62 million hectares worldwide from 1996 to 2002, there have been no documented cases of resistance development. The reasons for this lack of resistance appear to involve not only biological factors of the insects and the Bt plant, but also the fact that the regulatory agency (the Environmental Protection Agency) in the United States requires a resistance management plan for growing Bt plants. No other insecticide has such strict regulations. Still, growers, companies and federal regulatory agencies must be vigilant about resistance developing for biotech crops used to manage insects, weeds and viruses as they also must with non-biotech pest management tactics.

It will be important to consider the accrued environmental and health benefits of these biotech crops prior to the development of any resistance and how resistance can be managed if and where it occurs. In addition to pesticide resistance, gene flow from biotech to non-biotech crops may also be a concern. However, the risk of gene flow varies with each crop and each gene. Pollen flow in soybeans is very limited so the risk of a biotech soybean crop crossing with a non-biotech soybean is minimal, but this may be different for another crop. Likewise, if the gene in the biotech crop that provided a pest management trait, such as insect resistance, moved into a non-biotech plant, such as a weed, any selective advantage of the insect-protected weed in the ecosystem should be assessed. These same questions should also be answered with non-biotech crops, but these have not received the same level of attention as biotech crops because of the latter's higher profile.

Horizon of Biotechnology

In the future, the potential uses of plant biotechnology are far more wide-ranging than the pest-management biotech crops of today. Plants are being developed that serve as production "factories" for medically important drugs, sources of alternative energy, tools for cleaning toxic waste sites, and biomaterials including dyes, inks, detergents, adhesives, lubricants, plastics and the like. Consumers may see these products as more directly enhancing their quality of life than the pest-management biotech crops of today. Perhaps an even more dramatic advantage to consumers will be seen when plants are genetically engineered to have enhanced health benefits such as disease-fighting chemicals or increased amounts of essential vitamins and minerals. A healthy and well-informed discussion of the risks and benefits involved in agricultural biotechnology is needed to ensure a proper role for this new technology in our future food and health systems. No one should believe that any technology, including biotechnology, will completely solve the world's agricultural problems. Many people familiar with biotechnology, however, believe it to be an important component of the solution.

GENETICS AND EVOLUTION

MEIOSIS: THE GENETICS OF REPRODUCTION

The genetic information found in *DNA* is essential in creating all the characteristics of an organism. This remains the case when passing genetic information to offspring, that can occur via a process called *meiosis* where four *haploid* cells are created from their *diploid* parent cell. For a species to survive, and genetic information to be preserved and passed on, reproduction must occur. This can be done by passing on the information found in the chromosomes via the gametes that are created in meiosis.

CHROMOSOME COMPLEMENT

Humans are diploid creatures, meaning that each of the *chromosomes* in our body are paired up with another. Haploid cells possess only one set of a chromosome. A diploid human cell possesses 46 chromosomes and a gamete created by a human is haploid possesses 23 chromosomes. Tetraploid organisms possess more than 3 sets of a particular chromosome.

REPRODUCTION

Reproduction occurs in humans with the fusion of two haploid cells (*gametes*) that create a *zygote*. The nuclei of both these cells fuse, bringing together half the genetic information from the parents into one new cell, that is now genetically different from both its parents. This increases *genetic diversity*, as half of the genetic content from each of the parents brings about unique offspring, which possesses a unique *genome* presenting unique characteristics. Meiosis as a process can increase genetic variation in many ways, explained soon.

THE PROCESS OF MEIOSIS

The process of meiosis essentially involves two cycles of division, involving a gamete mother cell (diploid cell) dividing and then dividing again to form 4 haploid cells. These can be subdivided into four distinct phases which are a continuous process

1st Division

- *Prophase: Homologous chromosomes* in the nucleus begin to pair up with one another and then split into chromatids (one half of a chromosome) where crossing over can occur. Crossing offer can increase genetic variation.
- *Metaphase:* Chromosomes line up at the equator of the cell, where the sequence of the chromosomes lined up is at random, through *chance*, increasing genetic variation via independent assortment.
- *Anaphase:* The homologous chromosomes move to opposing poles from the equator
- *Telophase:* A new nuclei forms near each pole alongside its new chromosome compliment.

At this stage two haploid cells have been created from the original diploid cell of the parent.

2nd Division

- *Prophase II*: The nuclear membrane disappears and the second meiotic division is initiated.
- *Metaphase II*: Pairs of chromatids line up at the equator

- *Anaphase II*: Each of these chromatid pairs move away from the equator to the poles via *spindle fibres*
- *Telophase II*: Four new haploid gametes are created that will fuse with the gametes of the opposite sex to create a zygote.

Overall, this process of meiosis creates gametes to pass genetic information from parents to offspring, continuing the family tree and the species as a whole.

INDEPENDENT ASSORTMENT AND CROSSING OVER

The previous page investigates the process of meiosis, where 4 *haploid* gametes are created from the parent cell. Half the genetic information from a parent is present in these haploids, which fuse with gametes of the opposite sex to create a *zygote*, with a complete chromosome compliment that will create offspring after prolonged growth.

The process of *meiosis* increases *genetic diversity* in a species. The sex organs which produce the haploid gametes are the site of many occurrences where genetic information is exchanged or manipulated.

INDEPENDENT ASSORTMENT OF CHROMOSOMES

Alleles for a particular *phenotype* determine what characteristic an organism will express, as with the following example where

- Chromosome 1 contains an allele for blonde hair
- Chromosome 2 contains an allele for brown hair
- Chromosome 3 contains an allele for blue eyes
- Chromosome 4 contains an allele for brown eyes

The top assortment to the left produces 2 blonde hair/blue eyes gametes while the below produces 2 brown hair/brown eyes gametes.

The top assortment on the right produces 2 blonde hair/brown eyes gametes while the below produces 2 brown hair/blue eyes gametes.

The above indicates that even though the two homologous chromosomes contain the same genetic information, the assortment of the chromosomes (the order they lie in) can determine what genetic information is present in each of the 4 gametes produced. With 23 chromosomes in a human gamete, their are 2^{23} combinations (8388608 combinations)

CROSSING OVER

During meiosis, when *homologous chromosomes* are paired together, there are points along the chromosomes that make contact with the other pair. This point of contact is deemed the *chiasmata*, and can allow the exchange of genetic information between chromosomes. This further increases *genetic variation*.

CROSSING OVER AND GENETIC DIVERSITY

Gregor Mendel, an Austrian monk, is most famous in this field for his study of the *phenotype* of pea plants, including the shape of the peas on the pea plants.

GREGOR MENDEL'S WORK

Mendel's goal was to have a firm scientific basis on the relationship of genetic information passed on from parents to offspring. In light of this he focused on how plant offspring acquired the phenotype of their seeds. The plants that were used in the experiment had to be *true breeding*, *i.e.*, those plants with round seeds must have had parents with round seeds, who in turn had parents producing round seeds etc. This is done to increase the accuracy of results.

After successfully producing two generations from these true breeding plants, the following was evident:

- The first generation of plants produced all had a round seed phenotype.
- When these first generation plants were crossed, a ratio of 3 round seeds averaged every 1 wrinkled seed.
- The ratio of 3:1 was not exact, though this is because of the randomness of the processes that are executed to produce these plants. Independent assortment is completely random, as are mutations, therefore variable results occur producing a sampling error.
- Due to the scale of the experiment done by Gregor Mendel, the sampling error was smaller than that of a smaller scale experiment.

Mendel successfully hypothesised that the reason for this trend in phenotypes from generation to generation was down to the fact that genetic information was being passed on from their parents. The fact that round seeds appeared more frequently than wrinkled seeds is due to round seeds being the dominant phenotype, which when present effectively 'masks' the phenotype of the recessive (wrinkled seed) gene.

DOMINANT AND RECESSIVE ALLELES

- All plant seeds produced in the first generation were round
- 3 out of every 4 plant seeds produced in the second generation were round.

The parents, one possessing wrinkled seeds the other possessing round were crossed together, for some reason in the first and second generation the presence of the round seed gene in offspring superceded the presence of the wrinkled seed. This is called dominance.

MENDEL'S LAW AND MENDELIAN GENETICS

Mendel's First Law

"The alleles of a gene exist in pairs but when gametes are formed, the members of each pair pass into different gametes. Thus each gamete contains only one allele of each gene."

INCOMPLETE DOMINANCE

When a particular gene possesses both *dominant* and recessive alleles, it is possible for *incomplete dominance* to occur, where the organism at hand expresses a phenotype morphed by the expression of both the dominant and recessive alleles. In essence, *heterozygous* (possessing opposing alleles Rr) organisms derived from *homozygous* (possessing the same alleles RR or rr) are created, they possess a phenotype different to that of both their parents.

MULTIPLE ALLELES

Diploid organisms naturally have a maximum of 2 alleles for each gene expressing a particular characteristic, one deriving from each parent.

In some cases, however, more than two types of allele can code for a particular characteristic, as is the case of genetic coding for blood type in humans. Their are up to 6 possible genotypes that code for the four blood groups, A, B, AB and O.

EXAMPLE OF A CROSS

The following dihybrid cross involves two true breeding pea plants, where two factors are looked at, the shape of the seed and the colour of the seed.

MENDELIAN GENETICS

The past few pages have elaborated on the work of Gregor Mendel and how his work has paved the way to predicting the characteristics of offspring. However, a degree of randomness is involved, when involving factors such as independent assortment during meiosis and the possibility of genetic mutations.

CHROMOSOME MUTATIONS

It is natures intention that the exact genetic information from both parents will be seen in the offspring's *DNA* in the the critical stages of fertilisation. However, it is possible for this genetic information to mutate, which in most cases, can result in fatal or negative consequencies in the outcome of the new ogranism.

Non-Disjunction and Down's Syndrome

One well known example of mutation is *non-disjunction*. Non-disjunction is when the spindle fibres fail to seperate during meiosis, resulting in gametes with one extra chromosome and other gametes lacking a chromosome.

If this non-disjunction occurs in chromosome 21 of a human egg cell, a condition called *Down's syndrome* occurs. This is because their cells possess 47 chromosomes as opposed to the normal chromosome compliment in humans of 46. The fundamental structure of a chromosome is subject to mutation, which will most likely occur during crossing over at meiosis. There are a number of ways in which the chromosome structure can change, as indicated below, which will detrimentally change the genotype and phenotype of the organism. However, if the chromosome mutation effects an essential part of DNA, it is possible that the mutation will abort the offspring before it has the chance of being born. The following indicates types of chromosome mutation where whole genes are moved:

Deletion of a Gene

As the name implies, genes of a chromosome are permanently lost as they become unattached to the centromere and are lost forever

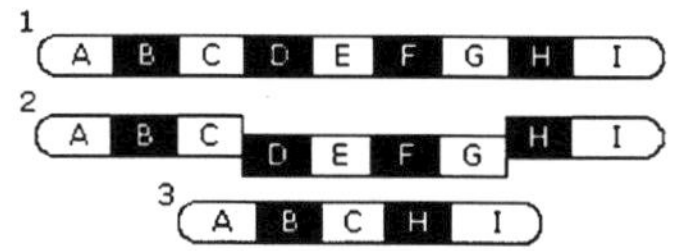

- Normal chromosome before mutation.
- Genes not attached to centromere become loose and lost forever.
- New chromosome lacks certain genes which may prove fatal depending.

Duplication of Genes

In this mutation, the mutants genes are displayed twice on the same chromosome due to duplication of these genes. This can prove to be an advantageous mutation as no genetic information is lost or altered and new genes are gained

- Normal chromosome before mutation
- Genes from the homologous chromosome are copied and inserted into the genetic sequence
- New chromosome possesses all its initial genes plus a duplicated one, which is usually harmless

GENETIC MUTATIONS

This page continues from the previous page investigating genetic mutations.

Inversion of Genes

This is where the order of a particular order of genes are reversed as seen below:

- Normal chromosome un-altered
- The connection between genes break and the sequence of these genes are reversed
- The new sequence may not be viable to produce an organism, depending on which genes are reversed. Advantageous characteristics from this mutation are also possible

Translocation of Genes

This is where information from one of two homologous chromosomes breaks and binds to the other. Usually this sort of mutation is lethal

- An un-altered pair of homologous chromosomes
- Translocation of genes has resulted in some genes from one of the chromosomes attaching to the opposing chromosome

Alteration of a DNA Sequence

Mutation have investigated changes at the chromosome level. The sequence of nucleotides on a DNA sequence are also susceptible to mutation.

- Deletion: Here, certain nucleotides are deleted, which affects the coding of proteins that use this DNA sequence. A gene coded for alanine, with a genetic sequence of C-G-G, and the cytosine nucleotide was deleted, then the alanine amino acid would not be able to be created, and any other amino acids that are supposed to be coded from this DNA sequence will also be unable to be produced because each successive nucleotide after the deleted nucleotide will be out of place.
- Insertion: Similar to the effects of deletion, where a nucleotide is inserted into a genetic sequence and therefore alters the chain thereafter. This alteration of a nucleotide sequence is known as frameshift
- Inversion: Where a particular nucleotide sequence is reversed, and is not as serious as the above mutations. This is because the nucleotides that have been reversed in order only affect a small portion of the sequence at large
- Substitution: A certain nucleotide is replaced with another, which will affect any amino acid to be synthesised from this sequence due to this change. If the gene is essential, *i.e.*, for the coding of haemoglobin then the effects are serious, and organisms in this instance suffer from a condition called sickle cell anaemia.

All of the genetic mutations looked at through the last 2 pages more or less have a negative impact and are undesired, however, in some cases they can prove advantageous.

Genetic mutations increase genetic diversity and therefore have an important part to play. They are also the reason many people inherit diseases.

MUTATION FREQUENCY AND POLYPLOIDY

The previous two pages have investigated mutations, and this page continues with more information related to genetic mutations.

Polyploidy

Humans are diploid creatures, meaning for every chromosome in our body, there is another one to match it. Read the following:

- Haploid creatures have one of each chromosome
- Diploid creatures have two of each chromosome
- Triploid creatures have three of each chromosome
- Polyploid creatures have three or more of each chromosome
- They can be represented by n where n equals haploid, 2n equals diploid and so on.

It is possible for a species, particularly plant species, to produce offspring that contains more chromosomes than its parent. This can be a result of non-disjunction, where normally a diploid parent would produce diploid offspring, but in the case of non-disjunction in one of the parents, produces a polyploid. In the case of *triploids*, although the creation of particular triploids in species is possible, they cannot reproduce themselves because of the inability to pair homologous chromosomes at meiosis, therefore preventing the formation of gametes.

Polyploidy is responsible for the creation of thousands of species in today's planet, and will continue to do so. It is also responsible for increasing genetic diversity and producing species showing an increase in size, vigour and an increased resistance to disease.

Mutation Frequency

This page and the previous two have investigated the different ways that mutations arise, and the following elaborates on the ways in which mutations are instigated.

Barring all external *factors*, mutations occur very rarely, and are rarely expressed because many forms of mutation are expressed by a recessive allele.

However there are many mutagenic agents that artificially increase the rate of mutations in an organism. The following are some factors that increase genetic mutations in organisms

- Members of species in a particular geographic area or ethnic origin are more susceptible to mutations
- High dosages of X-Rays or ultraviolet light can increase the likeliness of a mutation
- Radioactive substances increase the rate of mutations exponentially

As mentioned previously, genetic mutations are a source of new variation in a species because it physically alters the sequence of nucleotides in a given sequence, therefore altering the genome in a unique way.

THEORY OF NATURAL SELECTION

In the 19th century, a man called *Charles Darwin*, a biologist from England, set off on the ship HMS Beagle to investigate species of the island.

After spending time on the islands, he soon developed a theory that would contradict the creation of man and imply that all species derived from common ancestors through a process called natural selection. Natural selection is considered to be the biggest factor resulting in the diversity of species and their genomes.

The principles of Darwin's work and his theory are stated below:

- One of the prime motives for all species is to reproduce and survive, passing on the genetic information of the species from *generation* to generation. When species do this they tend to produce more offspring than the *environment* can support.
- The lack of resources to nourish these individuals places pressure on the size of the species *population*, and the lack of *resources* means increased competition and as a consequence, some organisms will not survive.
- The organisms who die as a consequence of this competition were not totally random, Darwin found that those organisms more suited to their environment were more likely to survive.
- This resulted in the well known phrase *survival of the fittest*, where the organisms most suited to their environment had more chance of survival if the species falls upon hard times. (This phrase if often associated with Darwin, though on closer inspection Herbert Spencer puts the phrase in a more accurate historical context.)
- Those organisms who are better suited to their environment exhibit desirable characteristics, which is a consequence of their genome being more suitable to begin with.

This 'weeding out' of less suited organisms and the reward of survival to those better suited led Darwin to deduce that organisms had evolved over time, where the most desirable characteristics of a species are favoured and those organisms who exhibit them survive to pass their *genes* on.

As a consequence of this, a changing environment would mean different characteristics would be favourable in a changing environment. Darwin believed that organisms had 'evolved' to suit their environments, and occupy an *ecological niche* where they would be best suited to their environment and therefore have the best chance of survival. As the above indicates, those alleles of a species that are favoured in the environment will become more frequent in the genomes of the species, due to the organisms higher likeliness of surviving as part of the species at large

DARWIN'S FINCHES AND NATURAL SELECTION

Darwin's finches are an excellent example of the way in which species' gene pools have adapted in order for long term survival via their offspring. The Darwin's Finches diagram below illustrates the way the finch has adapted to take advantage of feeding in different ecological niche's.

Their beaks have evolved over time to be best suited to their function. The finches who eat grubs have a thin extended beak to poke into holes in the ground and extract the grubs. Finches who eat buds and fruit would be less successful at doing this, while their claw like beaks can grind down their food and thus give them a selective advantage in circumstances where buds are the only real food source for finches.

Industrial Melanism

Polymorphism pertains to the existence of two distinctly different groups of a species that still belong to the same species. Alleles for these organisms over time are governed by the theory of natural selection, and over this time the genetic differences between groups in different environments soon become apparent, as in the case of industrial melanism.

Industrial melanism occurs in a species called the peppered moth, where the occurrence has become of more frequent occurrence since the beginning of the industrial age. The following argument elaborates the basis of principles involved in natural selection as far as industrial melanism is concerned.

- *Pollution*, which is more common in today's world since the industrial age causes a change in environment, particularly in the 1800's when soot would collect on the sides of buildings from chimneys and industries and make them a darker colour.
- The resultant effect was that the peppered moth, which had a light appearance was more visible against the darker backgrounds of sooty buildings.
- This meant that *predators* of the peppered moth could find them more easily as they are more visible against a dark background.
- Due to mutations, a new strain of peppered moth came to existence, where their phenotype was darker than that of the white peppered moth.
- This meant that these new, darker peppered moths were once again harder to track down by their prey in environments where industry has taken its toll.
- In this instance, natural selection would favour the darker moths in polluted environments and the whiter moths in the lesser polluted environments due to their ability to merge in with their environmental colours and lessen the chances of them being prone to a predator.

Sickle Cell Trait

Consider this argument of natural selection in the case of sickle cell trait, a genetic defect common in Africa:

- Sickle cell trait is a situation that occurs in the presence of a recessive *allele* coding for *haemoglobin*, a substance in the blood responsible for the transport of gases like oxygen. The presence of the allele is either partially expressed recessively (sickle cell), or fully expressed by a

complete recessive expression which results in full blown *anaemia*. If this particular allele is dominant, no sickle cell trait is expressed in the phenotype.
- The above occurrences in the case of a recessive allele result in structural defects of red blood cells, severely reducing the organisms capacity to uptake oxygen.
- It was pointed out that in Africa, there is a high frequency of this mutation, where cases of *malaria* were high.
- A substantiated link was made noting those who suffer sickle cell trait or anaemia were immune to the effects of malaria.
- This is yet again natural selection at work. Although sickle cell trait or anaemia are not advantageous characteristics on their own, they prove to be advantageous in areas where malaria proves to be a greater threat to preserving the genome (*i.e.*, surviving).
- The *incomplete dominance* of this genetic expression proves favourable either way.

This is how science has understood natural selection since the first studies involving Darwin. In the 21st century, humans selectively breed species to create hybrid species possessing the best genes of both parents via a process known as selective breeding.

SELECTIVE BREEDING

As mentioned in previous pages, scientists from the past harnessing the knowledge of genetics has resulted in many scientific breakthroughs and uses of this knowledge. Most notably, Gregor Mendel's studies into Monohybrid and Dihybrid crossing and Charles Darwin's study of evolution and natural selection has meant that humans have learnt to actively manipulate the *phenotype* of offspring by selective breeding in animals and plants.

Breeders of animals and plants in today's world are looking to produce organisms that will possess desirable characteristics, such as high crop yields, resistance to disease, high growth rate and many other phenotypical characteristics that will benefit the organism and species in the long term.

This is usually done by crossing two members of the same species which possess *dominant* alleles for particular genes, such as long life and quick *metabolism* in one organism crossed with another organism possessing genes for fast growth and high yield. Since both these organisms have dominant genes for these desirable characteristics, when they are crossed they will produce at least some offspring that will show ALL of these desirable characteristics. When such a cross occurs, the offspring is termed a *hybrid*, produced from two genetically dissimilar parents which usually produces offspring with more desirable qualities. Breeders continuously track which characteristics are possessed by each organism so when the breeding season comes once again, they can selectively breed the organisms to produce more favourable qualities in the offspring.

The offspring will become *heterozygous*, meaning the allele for each characteristic will possess one dominant and one recessive gene. Most professional breeders have a true breeding cross (ie AAbb with AAbb) so that they will produce a *gene bank* of these qualities that can be crossed with aaBB to produce heterozygous offspring. This way the dominant features are retained in the first breeding group and can be passed on to offspring in the second instance.

This process of selecting parents is called artificial selection or selective breeding, and poses no threat to nature from man manipulating the the course of nature. It has allowed our species to increase the efficiency of the animals and plants we breed, such as increasing milk yield from cows by continuously breeding selected cows with one another to produce a hybrid.

Inbreeding Depression

However, while it is an advantage both to the species and to humans to produce these desirable qualities that may benefit the organisms in question, continuous in-breeding and selective breeding of particular genes runs the risk of losing some of the other genes from the gene pool altogether, which is irreversible. This is called in-breeding depression, where the exclusivity of the advantageous genes mean that some other less desirable genes are phased out. In the long term, it is more advantageous for organisms to remain heterozygous;

Genetic diversity means the *gene pool* of a species is prepared for a wide range of scenarios such as food shortage or an epidemic of disease. Some genes in some organisms may provide the organism with *immunity* against the disease or an ability to go long periods of time without food. If continuous in-breeding has occurred in a species, some of these genes may have been phased out due to the breeder wanting other more desirable genes to be present in their crop.... Genetic diversity in the long term is reduced, because many organisms end up with similar *genomes* due breeding with each other constantly. In normal circumstances, this process would be random, and would produce more variable offspring.

With the above facts in hand, breeders need to produce more heterozygous offspring to ensure the long term welfare of the species they are breeding and their livelihood. The most important thing here is to preserve the genetic diversity of a species, and preferably keep the gene pool of a species as diverse as possible.

Humans have realised the above dangers, and instead of harnessing and exhausting natures reserves, we have learned to preserve their genetic information for their long term survival and our own well-being. One species becoming extinct can knock the balance of an ecosystem and have a detrimental knock on effect.

With this in mind, humans have gene banks to preserve the genetic information in the case of extinction, and nurture species that are at dangerously low population levels.

Ironically, the human interference that has disrupted so many species can now provide a means of placing genes into organisms, therefore preparing them for the above hypothetical scenarios such as an epidemic of disease. Genetic engineering would provide the means of allowing organisms to suit their environment without the trial and error over time that comes from natural selection.

DEVELOPMENTS IN THE SEED INDUSTRY

Many major food crops upon which we depend originated and were first cultivated in Third World countries rather than in the industrialized countries where they are most productively exploited today and in which they have been improved by traditional plant breeding. The potato originated in South America, wheat in Ethiopia and the Near East, important maize varieties in South America, and rice in Asia. Moreover the so-called Vavilov centres, which contain the greatest variety of living species and which have been the source of much important gene material, are mainly in Third World countries. The wealth of the industrial countries owes much to their past ability to exploit agriculturally crops and animals originating in other countries.

In the colonial era the success of British scientists in smuggling rubber plants from Brazil to Sri Lanka, Singapore and Malaysia and the comparable more complex route by which coffee was introduced to Latin America from Ethiopia are instances of colonial powers obtaining significant wealth by exploiting plant material from the Third World. In one way or another the process by which institutions and firms in developed countries have continued to collect new plant and animal varieties from less-developed countries has continued. But what is still a major source of friction is the procedures by which legislative protection is developed to give companies in the richer countries commercial rights to exclusive exploitation of varieties developed from plants freely collected from other countries, and to extract high rates of profit from the sale of those varieties. Most extreme is the situation whereby under the USA Plant Patent Act of 1930, which covered asexually propagated plants such as certain fruits, flowers, ornamental shrubs and trees, it has been possible for plants discovered and smuggled out of other countries to be given patent protection in the USA. The key to this is that the Act accepted the unfeasibility of requiring proof of an 'inventive-step' for the granting of a patent, and employed the criterion that as far as can be determined the plant variety is new.

Since some novel variety found in the rain forests of South America may satisfy this criterion, and the source of the plant can be concealed from the authorities, the theft and smuggling of plants can be said to have been encouraged. Seed companies in developed countries have successfully pressed their case for protection, and have progressively extended Plant Breeders' Rights (PBR) through both national legislation and the Union for the Protection of New Varieties of Plant (UPOV), which is an international agreement among signatory countries to honour a system of varietal rights.

In the case of food crops, in the absence of patentability, PBR operates by registering distinct varieties. Pressure has been exerted upon less-developed countries to become signatories of UPOV and thus acknowledge the rights of breeders, most of whom are from developed countries. However, the majority have stalled or rejected membership, and have been fundamentally unhappy about accepting the principle of creating monopoly rights for others over plant material which in some cases originated in their own country. The whole issue of intellectual property rights has been raised to a prominent political level by the USA's insistence that it be included as one of the thirteen areas for negotiation in the Uruguay Round of multilateral bargaining of the General Agreement on Trade and Tariffs.

While the issue is a much broader one than that of plant and animal breeders' rights, it reflects a basic political conflict between the richer countries, determined to obtain secure returns on R & D expenditures by companies, and LDCs struggling to catch up and anxious to avoid having to pay royalties on products protected by PBR and patents. This conflict has broader moral and ethical dimensions when it relates to the issue of charging royalties for the seeds of food crops to countries where malnutrition is widespread. Against the previous argument has to be balanced the need of those incurring R&D expenditure to capture enough of the returns to provide an incentive for R & D. At least that is certainly the case where R & D is to be undertaken by the private sector, which is politically the increasingly preferred solution in the UK, the USA and elsewhere. This, however, leads to another area of concern, which is that a small number of large multinational chemical and pharmaceutical firms have increasingly come to control the seed industry. Companies such as Shell, Sandoz, Dekallb-Pfizer, Ciba-Geigy and BP have become owners of most of what were small independent seed companies. This linking between chemicals and seeds within firms has a strong commercial logic. It facilitates the marketing of packages consisting of seed varieties plus the appropri-ate fertilizers, insecticides, etc. It also allows the companies concerned to integrate their plant breeding, biotechnology and chemical research. One particular synergy for these companies will be to bioengineer varieties of major food crops to resist their own herbicides and weedkillers.

This would permit chemical weed control to be extended to large-scale crops where this is not now possible, and is just one way in which the direction of biotechnolo-gical research may be influenced (biased) by the combination of forces giving a small number of companies extensive control over the basic means of agricultural production. There are many other issues which could be touched on regarding corporate influence over the development of agricultural biotechnology.

It is, however, understandable that the spokesmen for LDC countries, which are profoundly concerned about their dependence upon western technology and exercised by what is seen as colonial and post-colonial exploitation by the West, should be alarmed by the prospect of new rights being created over seeds and other biotechnological products.

Certainly there are widespread fears that they will only gain access to the fruits of biotechnology on unfavourable terms and that their relative dependence on and subservience to western industry will be increased so that benefits to them will be small. There are many western critics who would agree with this, and that the system by which agricultural biotechnology is delivered will favour the 'haves' rather than the 'have-nots'.

HYBRID SEEDS

In agriculture and gardening, hybrid seed is seed produced by cross-pollinated plants. In hybrid seed production, the crosses are specific and controlled. The advantage of growing hybrid seed compared to inbred lines comes from heterosis. To produce hybrid seed, elite inbred varieties with well documented and consistent phenotypes (such as yield) are crossed and the resulting hybrid seed is collected. Another factor that is important in hybrid seed production is the combining ability of the parent plants. Although two elite inbred parent plant varieties may produce the highest yields of their crop, it does not necessarily mean that crossing these inbreds will result in the highest yielding hybrid. Combining ability is the term used to describe the level of heterosis that the parents will generate in the resultant seed. Higher combining ability between the parents results in increased performance in the resulting hybrid seed. Hybrids are bred to improve the characteristics of the resulting plants, such as better yield, greater uniformity, improved colour, disease resistance, and so forth.

Today, hybrid seed production is predominant in agriculture and home gardening, and is one of the main contributing factors to the dramatic rise in agricultural output during the last half of the 20th century. In the US, the commercial market was launched in the 1920s, with the first hybrid maize. All of the hybrid seeds planted by the farmer will be the same hybrid, which causes the first generation of seed from the hybrids planted to be inbred. This is why hybrid seed is generally not saved from subsequent generations and is purchased for each planting. Hybrid seeds are much dearer than normal seeds, due to the technology, time and effort put in to produce them.

Hybrid seed began with maize in the 1920s, and became extended to vegetables and flowers; and more recently, rice and some forage crops. Hybrid seeds are produced from naturally out-breeding crops, from which inbred lines are produced by repeated self-pollination. The established inbred lines are crossed to produce first generations (F_1) hybrid seeds. The hybrid seeds are prized because they produce uniform plants benefiting from the effect called heterosis (hybrid vigour). Heterosis can result in a large increase in yield over the inbred lines or comparable lines that are out-crossing. The precise basis of heterosis is still unclear, but epistasis and over-dominance are thought to be involved.

Epistasis is the interaction between different genes, and over-dominance is a condition where the heterozygotes (genes represented by two different versions) are superior to either homozygotes (gene represented by the same versions). The F_1 hybrid seed is heterozygous in many genes.

Hybrid seed is planted to produce a crop that is harvested for use. Saving seed from the crop and planting it is undesirable because the two different versions of the genes in the F_1 hybrid segregate out in the offspring, producing an extremely variable progeny. In other words, the superior qualities of the F_1 hybrid will have all disappeared. The hybrid is obtained by crossing the inbred lines, which therefore, have to be separately maintained. Thus, only the seed companies produce hybrid seeds, and farmers must buy those seeds from the company every year.

Hybrid maize arose through the advocacy of a few influential Americans. Foremost among the advocates was Henry A. Wallace, who became vice-president of the United States. Wallace graduated from University with an agriculture degree, and studied statistics thereafter on his own. He later taught the subject at Iowa State University and used his knowledge to develop the first commercial hybrid maize.

In 1926, he founded the Hi-Bred Corn company (now Pioneer Hi-Bred Seed Company, a subsidiary of Dupont Chemical Company), and later entered politics. He was made Secretary of Agriculture before being elected vice-president of the United States. Wallace was noted for his concern for the common man and envisioned hybrid corn as a means of providing bountiful food at low prices for the masses. The detailed history of hybrid corn and Wallace makes fascinating reading.

The first corn hybrids were made by detasseling the plants of the maternal inbred-line by removing the male flowers so that the female flowers on the plants can only be fertilized by pollen produced from plants of another, male line. The detasseling operation used to be performed mainly by young girls employed during the summer months. Later on, male-sterile lines were developed that did not produce fertile male flowers or pollen. The male-sterile maternal lines were fertilized with paternal lines that allowed the hybrid seed to produce both male and female flowers. The male-sterile lines are most frequently altered in the mitochondrial genome, leading to the inhibition of male flower development. A number of such lines are now available.

DISASTER STRUCK

The early development of male-sterile lines led to disaster, however. The primary line used in the 1960s contained the T (Texas) cytoplasm male-sterility gene; and by 1970, over 85 per cent of the commercial maize planted contained that gene. The gene also caused a pleiotropic (multiple effects due to a single gene) susceptibility to a fungus disease. During a damp 1970 summer, the disease spread widely particularly in the summer corn belt. The impact on maize production was disastrous, leading to a return to hand-detasseling for a number of years until alternate male-sterility lines could be developed. The lesson that should have been learned was that the absence of diversity is bound to lead to disastrous epidemics; but that lesson tends to get ignored in favour of risky but profitable genetic manipulations.

HYBRIDS GALORE

Rice hybrids have been produced using cytoplasmic male-sterility. Over-dominance and epistatic genes were implicated as the basis for heterosis (and inbreeding depression, a phenomenon in which inbred lines suffer decreased yield). Alfalfa interspecies hybrids showed heterosis, interspecies hybrids are a little different from those originating from inbred lines, but in general they act similarly to inbred lines. A large number of vegetable crops have been hybridized. Hybrid cucumbers have been produced by hand pollination, removal of male flowers, or gynoecy (property of producing only female flowers). There does not seem to be an available male sterility gene. Hot and sweet peppers have been hybridized. Both nuclear and cytoplasmic sterility are used in some cases. Most hybrid-pepper seed production is carried out in China, India or Thailand. About two-thirds of commercial onions are hybrids. These are produced using male sterility lines. Hybrid cabbage shows strong heterosis, and the use of such hybrids is expanding. The seed is produced using male sterile lines.

Most of the male sterile lines used commercially contain mitochondrial genes, but such genes are not readily available in a number of crops. Genetic engineers have developed a system of male-sterility based on transformation of the chloroplast with a gene for beta-ketothiolase that interferes with fatty acid synthesis, leading to disrupted anther tissue and a failure to produce pollen. The beta-ketothiolase gene is controlled by a light sensitive promoter, so that male-fertility can be restored in hybrids using several days of continual illumination. The system was developed in tobacco but may be extended to food crops, barring unforeseen complications.

GENETICALLY MODIFIED MALE-STERILITY

A number of genetically modified (GM) male-sterile crops have been developed and tested in the field. In Canada, a male-sterile transgene was introduced into the nuclear gnome of canola, and that construction was approved for, and has been in commercial production. The transgenic construct included a barnase ribonuclease gene controlled by a tapetum promoter. Barnase kills pollen cells thus rendering the plant male-sterile. In the hybrid male fertility is restored using the barstar inhibitor of barnase, although barnase is well known to be toxic to animal cells. Development continued, and the technology came to be used to protect GM traits patented by agribusiness corporations such as herbicide tolerance under the general rubric of genetic use restriction technology (GURT). Such crops were extensively field tested in Europe; and we have warned that the F_1 hybrid grown in the field will actually spread the barnase transgene as well as the herbicide tolerance gene in pollen with potentially harmful ecological impacts. Furthermore, the toxin may well be carried over into the canola press cake used both for both food and feed.

The development of hybrid seed had left seed production to seed companies for the practical reason that it is the most economical way to maintain appropriate inbred lines, and seed production can be isolated from the food production areas of open pollinating crops. But it had also prevented farmers from saving and replanting seeds, making it necessary to purchase seeds every season. Biotechnology has gone a step further and demanded that seed production be restricted to companies even when there is no rational basis for the restriction, other than corporate greed. Goeshl and Swanson addressed the question of GURT based on the hybrid-crop experience. They argued that developed countries could benefit from the additional production supposedly to be gained by the technology, while developing countries will suffer from their inability to afford the high extra cost. They predict net deterioration in the developing countries due to the widening gap in productivity. These predictions must be taken with a very large grain of salt. There is at present no evidence that genetic use restriction technologies, or indeed, any genetic modification technology have led to increase in crop yield. Furthermore, both hybrids and GM crops lack the diversity required for sustainability in the complex ecosystems of the developing world. What is needed is seed production that takes into account the unique requirements of developing countries, where the farmers' rights to save, replant and exchange seeds are integral to food sovereignty and food security.

Hybrid varieties have been evolved in those high valued vegetable crops which exhibit marked heterosis such as solanaceous vegetables (tomato, eggplant, sweet pepper), cucurbits (melons, watermelon, cucumber, squash, pumpkin and gourds), cole crops (cabbage and cauliflower), root and bulb crops (onion, radish, carrot) and fruit vegetable like okra.

GENETICAL MODIFIED ORGANISM-DERIVED FOODS

The concept of substantial equivalence was originally developed through discussions at the Organisation for Economic Co-operation and Development, though, to a large extent, these discussions built on previous work done by the World Health Organization (WHO) and the Food and Agriculture Organization. In 1991, a Joint FAO/WHO Consultation had concluded that the evaluation of a food derived through modern biotechnology should consider both food safety and nutritional value using similar conventional food products as a standard and taking into account the processing of the food and its intended use. In the 1980s, the OECD established a Group of National Experts on Safety in Biotechnology (GNE), which continued to work through to 1993.

As an intergovernmental organization, OECD's GNE was attended by delegates nominated by the governments of the OECD member countries, who were, for the most part, representatives of those agencies and ministries with a responsibility for safety in biotechnology. As a result of the work of the GNE, the OECD published a number of important documents relevant to the safety assessment of biotechnology-derived products during this period, which deal with a range of issues, including safety considerations for industrial, agricultural as well as environmental applications of organisms derived by recombinant DNA techniques.

It was recognized at this stage that the safety assessment of an organism derived through recombinant DNA techniques would rely heavily on the knowledge of its parental organism, as well as on an analysis of how the new organism appears to differ from the parent. OECD recommended that the considerable data on environmental and human health effects of living organisms that exists should be used to guide the risk assessment.

By the early 1990s, there had already been large numbers of small-scale field trials of new crop varieties derived through recombinant DNA techniques, and it became clear that new foods derived from these varieties would be marketed during the 1990s. In order to proactively address food safety issues related to these novel foods, in 1990, the GNE established a Working Group on Food Safety and Biotechnology comprising experts, mainly from the ministries and agencies responsible for food safety issues in OECD member countries.

The main objective of the Working Group was to elabourate scientific principles and concepts to be used when evaluating the safety of new foods and food components of terrestrial microbial, plant or animal origin. The Working Group did not consider the safety assessment of food additives, contaminants, processing aids or packaging materials. Nor did it consider environmental safety issues which had been (or were being) addressed by other groups of the GNE. Early in the life of the Working Group, an important recognition was that traditionally, the safety of food for human consumption had been based on the reasonable certainty that no harm will result from intended uses under the anticipated conditions of consumption. Foods prepared and used in traditional ways have usually been considered safe on the basis of long-term experience, even though they may have contained natural toxicants or anti-nutritional substances. Normally, new varieties of foods or crops have not been subjected to traditional toxicological testing. In fact, where toxicological testing has been applied to whole foods, the results have often been difficult to interpret. Although the OECD's Working Group recognized that modern biotechnology might extend the scope of genetic changes that can be made - and might even broaden the range of the possible sources of foods - it was recognized that this would not inherently lead to foods that are less safe than those developed through conventional techniques. The Working Group therefore indicated that the evaluation of foods or food components derived through modern biotechnology does not require a fundamental change in established principles, nor does it require a different standard of safety. Realising the importance of using examples of new foods to identify and demonstrate the applicability of the proposed scientific principles, the Working Group organized a number of meetings and intergovernmental consultations which included case study presentations of novel foods such as enzymes, genetically modified bakers' yeast, mycoprotein and a number of genetically modified varieties of crop species.

Although these case studies were not intended to be formal safety evaluations they were crucial in illustrating important concepts in the safety assessment of novel foods. Therefore, it was through their application in real examples that the concepts and principles were identified. The Working Group proposed a scientific approach to the evaluation of foods derived through modern biotechnology, which is based on a comparison with traditional foods that have a safe history of use.

One of the main concepts developed was that of substantial equivalence, and in describing this concept, the Group stressed that this was a principle that had been used in the past (perhaps intuitively) even if it had not been articulated as such. The concept of substantial equivalence is described as embodying the idea that existing organisms used as food, or as a source of food, can be used as the basis for comparison when assessing the safety of human consumption of a food or food component that has been modified or is new. As previously indicated, in elabourating the concept of substantial equivalence, the OECD Working Group noted that food safety is considered as a reasonable certainty, that no harm will result from intended uses under the anticipated conditions of consumption and that the most practical approach to the determination of safety is to consider whether a genetically modified organism (GMO)-derived food is comparable to an analogous conventional food product.

The concept of substantial equivalence therefore relies on the existing history of safe food use of a conventional food product as a useful consideration in framing the safety assessment of a GMO-derived food by permitting the identification of similarities and differences which can be considered in the assessment. In 1996, participants at an expert FAO/WHO consultation recommended that safety assessment based upon the concept of substantial equivalence be applied in establishing the safety of foods and food components derived from genetically modified organisms.

Assessing substantial equivalence was recognized as not being a safety assessment *per se*, but a process which establishes that the characteristics and composition of the new GMO-derived food are comparable to those of a familiar, conventional food which has a history of safe consumption. A Joint FAO/WHO Expert Consultation on Foods Derived from Biotechnology was convened in 2000 to address food safety and nutritional questions regarding foods derived from GM plants, including a review of the scientific basis, application and limitations of the concept of substantial equivalence. It concluded, based on the current application of substantial equivalence and alternative strategies that this concept contributed to a robust safety assessment framework. Moreover, the Consultation noted that substantial equivalence is a concept used to identify similarities and differences between GM food and a comparator with a history of safe food use which subsequently guides the safety assessment process.

While not a safety assessment *per se*, the substantial equivalence approach allows the structuring of the safety assessment through characterizing the similarities and identifying the differences which can then be the focus of further consideration. This approach has been referred to as a useful safety standard. It therefore permits inference that the new food under consideration will be no less safe than the conventional food under conditions of similar exposure, consumption patterns and processing practices.

Substantial equivalence is therefore clearly not intended to be a measure of absolute safety, but instead recognizes that while demonstrating absolute safety is an impractical goal, demonstrating that there is reasonable assurance that the GMO-derived product under consideration is no less safe than a conventional food product is an achievable goal. Since it's development, the use of the substantial equivalence concept in the safety assessment of GMO-derived foods has been subject to misinterpretation and criticism.

The comparative nature of the substantial equivalence concept without prescribing the extent of the phenotypic and compositional comparisons has led some to criticize the concept as not being measurable, and therefore inappropriate in safety assessment. This and other criticisms relate, in part, to the mistaken perception that the determination of substantial equivalence was the end point of a safety assessment rather than the starting point. After several OECD countries had gained experience with safety assessment of GMO-derived foods, an OECD workshop examined the effectiveness of the application of substantial equivalence in safety assessment. This workshop concluded that the substantial equivalence approach provides equal or increased assurance of the safety of foods derived from genetically modified plants, as compared with foods derived through conventional methods. In 2000, the OECD Task Force for the Safety of Novel Foods and Feeds also reviewed the substantial equivalence concept, its interpretation and its application. While noting that the majority of international guidance documents addressing the safety assessment of genetically modified (GM) plants had interpreted substantial equivalence consistently, this Task Force reported that there were differences in how it was applied and that these differences needed to be resolved.

The Codex *Ad Hoc* Intergovernmental Task Force on Foods Derived from Biotechnology pursued the development of international guidance on the safety assessment of GMO-derived foods. To date, this Task Force has developed proposed draft principles and guidelines for the safety assessment of foods derived from modern biotechnology. These guidelines interpreted the concept of substantial equivalence as a way of structuring the safety assessment, consistent with the FAO/WHO interpretation.

3

Status of Agricultural Biotechnology and Food Production

ADVANTAGES AND DISADVANTAGES OF BIOTECHNOLOGY

During the latter stage stages of the 20th century, man harnessed the power of the atom, and not long after, soon realised the power of genes.

Genetic engineering is going to become a very mainstream part of our lives sooner or later, because there are so many possibilities advantages (and disadvantages) involved.

Here are just some of the advantages:

- Disease could be prevented by detecting people/plants/animals that are genetically prone to certain *hereditary* diseases, and preparing for the inevitable. Also, infectious diseases can be treated by implanting genes that code for antiviral proteins specific to each antigen.
- Animals and plants can be 'tailor made' to show desirable characteristics. Genes could also be manipulated in trees to absorb more CO^2 and reduce the threat of global warming.
- Genetic Engineering could increase genetic diversity, and produce more variant alleles which could also be crossed over and implanted into other species. It is possible to alter the genetics of wheat plants to grow insulin.

Of course there are two sides to the coin, here are some possible eventualities and disadvantages:

- Nature is an extremely complex inter-related chain consisting of many species linked in the food chain. Some scientists believe that introducing genetically modified genes may have an irreversible effect with consequences yet unknown.
- Genetic engineering borderlines on many moral issues, particularly involving religion, which questions whether man has the right to manipulate the laws and course of nature.

Genetic engineering may be one of the greatest breakthroughs in recent history alongside the discovery of the atom and space flight, however, with the above eventualities and facts above in hand, governments have produced legislation to control what sort of experiments are done involving genetic engineering. In the UK there are strict laws prohibiting any experiments involving the cloning of humans.

However, over the years here are some of the experimental 'breakthroughs' made possible by genetic engineering:

- At the Roslin Institute in Scotland, scientists successfully cloned an exact copy of a sheep, named 'Dolly'. This was the first successful cloning of an animal, and most likely the first occurrence of two organisms being genetically identical. Note: Recently the sheep's health has deteriorated detrimentally
- Scientists successfully manipulated the genetic sequence of a rat to grow a human ear on its back. (Unusual, but for the purpose of reproducing human organs for medical purposes)
- Most controversially, and maybe due to more liberal laws, an American scientist is currently conducting tests to clone himself.

Genetic engineering has been impossible until recent times due to the complex and microscopic nature of DNA and its component nucleotides. Through progressive studies, more and more in this area is being made possible. For us to understand chromosomes and DNA more clearly, they can be mapped for future reference. More simplistic organisms such as fruit fly (Drosophila) have been chromosome mapped due to their simplistic nature meaning they will require less genes to operate. At present, a task named the Human Genome Project is mapping the human genome, and should be completed in the next ten years.

The process of genetic engineering involves splicing an area of a chromosome, a gene, that controls a certain characteristic of the body. The enzyme endonuclease is used to split a DNA sequence and split the gene from the rest of the chromosome.

This gene may be programmed to produce an antiviral protein. This gene is removed and can be placed into another organism. It can be placed into a bacteria, where it is sealed into the DNA chain using ligase. When the chromosome is once again sealed, the bacteria is now effectively re-programmed to replicate this new antiviral protein. The bacteria can continue to live a healthy life, though genetic engineering and human intervention has actively manipulated what the bacteria actually is. No doubt there are advantages and disadvantages, and this whole subject area will become more prominent over time.

THE GENE POOL AND SPECIATION

We have already discussed some of the reasons why the genetics of a species can change over a long period of time. Charles Darwin's The Origin of

Species went into great depth in this matter, and providing substance into the theory of evolution. The key thing to remember about evolution is that it favours more preferable genes in the gene pool, and over time, these preferable characteristics become more exclusive in the gene pool. This next section rounds up all the factors that can alter the make up of a gene pool

Natural Selection

Natural selection will favour genes that are more suited to their environment and become more exclusive in the gene pool over time in such an environment. Different genes will become more exclusive when the environment changes, or the species migrate.

Mutation

Mutations are random occurrences which change the genome of an organism. They greatly increase genetic diversity, where advantageous mutations are favoured by natural selection and disadvantageous ones are phased out.

Gene Migration

Occurs after *genetic drift*, where two groups of a species become separated and therefore cannot reproduce. The gene pool of these groups differ over time. If these two groups can once again meet up and reproduce, their genetic differences can be merged within the single group and increase *genetic diversity*.

Non-Random Mating

Can cause in-breeding depression by continuous inbreeding, non-random mating is also known as selective breeding, where the breakthroughs of Mendelian genetics have allowed us to predetermine what genes are present in offspring. As advantageous genes are desired by the breeder, some of the less 'popular' genes are lost due to this random mating, therefore decreasing genetic diversity.

It is important for a species to have a large gene pool, because in the event of danger, some alleles will allow the species to survive and reproduce to produce a larger and more variant gene pool. An extremely contagious disease may threaten 99percent of a species, though the remaining 1percent may possess an allele that provides them with resistance to the disease. If this allele was not present in the population, then chances are the entire population would be wiped out

Genetic Drift

Sometimes, species can be split into groups, usually as a result of a geographical factor preventing the two groups from contacting one another.

This means that the two groups are unable to reproduce with one another. As the two groups now live in different environments, natural selection will favour slightly different genes in each of the groups that will favour them in their particular environment. Over time, the difference in gene pool between the two groups can be quite dramatic.

If these two groups once again become re-united, gene migration occurs. If they remain seperated, their genetic differences become greater which can result in the formation of a new species.

Speciation

When the genetic differences become so great, it can come to the stage where the two groups can no longer reproduce with one another. This results in the formation of a new species (because organisms who are capable of reproducing but cant reproduce with a member of the same species are deemed another species). Long ago, the land of Earth was all on one continent. Over time these continents separated, with members of the same species drifting away on each continent. Over time, speciation has occurred. This is evident by looking at the marsupials of Australia, who have been isolated from other mammals in their ancestral line, and therefore have many differences to that of mammals on other continents.

Adaptive Radiation

Adaptive radiation is the slow change of *genotype* and *phenotype* of a species from its common ancestor, meaning that species with a common ancestor become more diversified over time. The next page investigates adaptive radiation.

ADAPTIVE RADIATION

When *Charles Darwin* was in the Galapagos islands, one of the first things he noticed is the variety of finches that existed on each of the islands. All in all, there were many different species of finch which differed in beak shape and overall size. This is adaptive radiation and natural selection at work.

The Marsupials

The marsupial mammals occupy Australia, and dither from placental mammals because they bear their young inside a pouch. Long ago, the land mass of Earth consisted of one single continent, Pangaea, where all animals existed. When this continent fragmented into smaller continents, the geographical barrier meant the mammals of the time could no longer reproduce with one another. From here on in, the gene pool of these groups of mammals would become increasingly different until they could no longer reproduce with one another. This occurrence applies to the marsupial family of mammals that occupy Australia.

All the marsupials in present day Australia would have evolved from one common ancestor. However, over time and via natural selection, the many marsupial species (*i.e.*, kangaroo and koala) have occupied their own ecological niche and adapted accordingly. Kangaroo's have long powerful legs to cover the wide area of land that they occupy while the koala's smaller structure and more centralised centre of gravity allow them to climb trees and obtain the eucalyptus that they feed on.

Humans, the most evolved species on the planet, have also underwent many changes over time from our ancestors:

- Humans are bipedal, meaning we walk on two of our limbs. The other limbs, our arms and subsequent hands, have adapted to do precision tasks such as typing or tying shoe laces, common tasks that we perform in our own environment.
- The amount of melanin in our skin is representative of the environment we live in, *i.e.*, dark skinned people occupy hotter climates.
- The structure of our body gives an indication of the temperature in our climate also. Humans who have produced offspring that successfully live in a cold environment tend to be broader and smaller in stature while hotter environments are occupied by thinner taller humans.

It is worth noting that if two groups of humans were totally isolated over time, it is entirely possible that a different species could spawn from own, because the two groups would be adapting to their own environment over time. This would mean changes in the gene pool in each of the groups, to the point where the two groups could not reproduce with another and produce fertile offspring due to these genetic differences. At this point, with any species, a new one has begun.

MEIOSIS AND ALTERNATION OF GENERATIONS

MITOSIS

The cell cycle contains the process in which cells are either dividing or in between divisions. Cells that are not actively dividing are said to be in interphase, which has three distinct periods of intense activity that precedes the division of the nucleus, or mitosis.

The division of the rest of the cell occurs as an end result of mitosis and this process occurs in regions of active cell division, called meristems. Meristems.

Mitosis is a process within the cell cycle that is divided into four phases which we will sum up here:

- *Prophase*: The chromosomes and their usual two-stranded nature becomes apparent, the nuclear envelope breaks down.

- *Metaphase*: The chromosomes become aligned at the equator of the cell. A spindle composed of spindle fibres is developed and some attach to the chromosomes at their centromere.
- *Anaphase*: The sister chromatids of each chromosome, that is now called the daughter chromosomes, separate lengthwise and each group of daughter chromosomes migrates to the opposite ends of the cell.
- *Telophase*: The groups of daughter chromosomes are grouped within a developing nuclear envelope which makes them separate nuclei. A wall forms between the two sets of daughter chromosomes thus creating two daughter cells.

In plants, as the cell wall is developing, droplets or vesicles of pectin merge forming a cell plate that eventually will become the middle lamella of the new cell wall. The key feature of mitosis is that the daughter cells have the same chromosome number and are otherwise identical to the parent cell.

MEIOSIS INTRODUCTION

Mitosis, as just reviewed above, is a process by which a cell can reproduce itself and the number of chromosomes and the nature of the DNA will be identical to the original parent cell. Very few species will grow or live indefinitely, so there must be some way to ensure the continuity of the species. Reproduction is the only way a species can be perpetuated, without perpetuation the species will become extinct.

Reproduction can occur in several ways as vegetative propagation, such as in the development of runners in strawberry plants, or by special cells called vegetative spores which are products of mitosis. In these processes, the 'offspring' have identical cells and identical chromosomes to the parent cells and thus the processes are called asexual reproduction—a means without, so without sex reproduction. Most plants, however, will undergo sexual reproduction which involves the production and recombining of sex cells called gametes. In flowering and cone-bearing plants this involves the production of seeds. The gametes produced are male and female, and are called sperm cells and egg cells, correspondingly. When the gametes combine together, the cells fuse and form a single cell called a zygote. It is the zygote that will go on to become the plant embryo and eventually a mature, adult plant.

However, in thinking about this process, what would happen if both gametes had the same number of chromosomes as the rest of the cells in the organism? When they fused to become a zygote, they would have two times the number of chromosomes as the rest of the cells in the organism. The number of chromosomes would increase exponentially through the generations if this occurred. This is where meiosis comes in to play. Meiosis is the process by which gametes, sex cells, are formed. It is unique because gametes have exactly half of the total number of chromosomes as the rest of the cells in the parent organism. When two gametes, each with half the number of chromosomes, get together they are able to restore the chromosome number to the same as the rest of the cells in the parent organism. When the zygote develops into a plant embryo and eventually a mature plant, it will have the exact number of chromosome specific to the species. Note that the processes and steps in meiosis are very similar to mitosis, so make certain you have a good understanding of mitosis so that you will be able to compare the two processes. Before we get into the nitty-gritty of meiosis, keep in mind that all living cells have two sets of chromosomes—one from a male and one set from a female parent. The genes in the chromosomes may control the same characteristics but in contrasting ways: genes for plant height, genes for plant colour, genes for fruit colour, etc—the female gamete might code for short plants, while the male gamete might code for tall plants. That is more of a genetics topic though. But you should know that the chromosomes that code for the same characteristics are called homologous chromosomes.

PHASES OF MEIOSIS

The end result of one round of meiosis will be four cells with half the number of chromosomes as the parent cell. The daughter cells are rarely, if ever, identical to each other or the parent cell depending on the organism involved. There are two successive divisions in meiosis, which in plants occur without a pause. Mitosis takes roughly 24 hours, while meiosis takes up to two weeks. In some organisms, meiosis takes weeks or years depending on the organism.

Division I –Reduction Division

The chromosome number is reduced to half the parent cell chromosome number. End result of division one is two cells.

Prophase I

- Chromosomes coil, becoming shorter and thicker, the two-stranded nature becomes apparent, two strands are called a chromatid and chromosomes are aligned in pairs. Each pair of chromosomes has four chromatids and they have a centromere attached in the centre holding the four strands together.
- Nucleolus disassociates and nuclear envelope dissolves.
- Segments of the closely associated pairs of chromatids may be exchanged with each other (between the pair members) this is called crossing-over. Each chromatid contains the original amount of DNA but now may have "traded" genetic material.
- The chromosomes separate. Some spindle fibres are forming and some are attaching to the centromeres of the chromosomes. The fibres extend from each pole of the cell.

Metaphase I

- In pairs, the chromosomes align at the equator of the cell, with the centromeres and spindle fibres apparent.
- The two chromatids, from each chromosome, function as a single unit.

Anaphase I

- One entire chromosome, consisting of two chromatids, migrates from the equator to a pole. The chromosomes do not separate from each other and retain both chromatids when the reach their pole. At each pole, there will be half the chromosome number. If crossing over occurred in prophase then the chromosomes will consist of original DNA and DNA from a homologous chromosome—now at the opposite pole.
- The centromere remains intact in each pair of chromatids.

Telophase I

- What occurs in this step, depends on the species involved, as they may revert to interphase or proceed directly to division II.
- If they revert to interphase, they will only do so partially and the chromosomes will become longer and thinner.
- nuclear envelopes will not form, but the nucleoli will generally recluster.
- Telophase is over when the original cell becomes two cells or two nuclei.

Division II—Equational Division

The chromosome number stays the same, the cells replicate and result in four cells. The events closely resemble the events in mitosis, except that there is no duplication of DNA during the interphase that may or may not occur between the two divisions.

Prophase II—Main Features

Chromosomes of both nuclei become shorter and thicker. The two-stranded nature becomes apparent once again.

Metaphase II—Main Features

- Chromosomes align their centromeres along the equator.
- Spindle fibres form and attach to each centromere, extending from one pole to the other.

Anaphase II—Main Features

The centromeres and chromatids of each chromosome separate and begin their migration to the opposite poles.

Telophase II—Main Features

- The coils of chromatids—now called chromosomes again—relax and the chromosomes become longer and thinner.
- Nuclear envelopes and nucleoli reform for each group of chromosomes.
- New cell walls form between the four groups of chromosomes.
- Each set of chromosomes in the four new cells, has exactly half of the chromosome number of the original number.

ALTERNATION OF GENERATIONS

One member of every original pair of chromosomes end up in each resulting cell, which means each cell has one set of chromosomes. None of the daughter cells is identical to the parent cell. A cell with one set of chromosomes is called haploid and a cell with two sets is called diploid. In sum, a diploid cell undergoes meiosis and results in four haploid cells.

Sex cells, or gametes, of an organism are haploid and when a zygote is formed, the zygote is diploid. This is always true, no matter how many chromosomes an organism might have. In many places, you will find it stated that an organism has n number of chromosomes.

In this case, an organism will have 2n chromosomes in its diploid cells and 1n chromosomes in its haploid cells. Alternation of generations refers to a plants life cycle including sexual reproduction that is characterized by alternating between a diploid (2n) sporophyte phase and a haploid (1n) gametophyte phase.

BRYOPHYTES

Bryophytes are essentially nonvascular plants, meaning they do not have xylem or phloem. The habitations of bryophytes are widely varied and include bare rocks in the scorching sun to frozen alpine slopes. Bryophytes include mosses, liverworts and hornworts. These groups of plants require external water, usually in the form of dew or rain.

Some bryophytes grow exclusively in dark, damp environments in order to provide moisture. Water is essential for bryophyte reproductive activities. Most mosses have water-conducting cells called hydroids in the centers of their stems, and some even have food conducting cells called leptoids. These cells are not nearly as efficient as xylem and phloem and generally, bryophytes are not very tall plants. The lack of vascular tissue leaves the plant body very soft and pliable. Bryophytes are good nest-making material for birds.

The alternation of generations in bryophytes is quite obvious as the gametophyte generation is the 'leafy' plant that is generally visible. The sporophyte generation that produces spores is located at the tips of the 'leafy' gametophyte generation. The sporophyte generally looks like a slender stalk with a cap on top. While the lifecycles of all bryophytes are similar and even their chromosome number and habituation have similarities; they care divided into three distinct divisions based on the few differences in structure and reproduction.

Division Hepaticophyta—Liverworts

Wort means plant or herb, and in ancient times herbalists thought that some bryophytes—specifically the ones that look like liver lobes—were useful in treating liver ailments. Although, the belief was discounted the name stuck.

Structure—About 20percent of the liverworts have a flattened, somewhat leaf like body called a thallus (plural thalli). The other 80percent, which aren't as common in nature, are 'leafy' and look more like mosses. Mosses are more complex than liverworts. They have one-celled rhizoids on their lower surfaces. The rhizoids look like tiny roots and anchor the plants to surfaces and soil particles. The thalli are the gametophyte generation and develop almost directly from spores. They have smoother upper surfaces and the cell wall corners are thickened. Thalloid liverworts—*Marchantia* is the best-known species of thalloid liverwort. It can usually be found on damp soil after a fire. *Marchantia* can reproduce asexually and sexually. Asexual reproduction is accomplished through gemmae, which are tiny lens-shaped pieces of tissue that detach from the thallus.

Gemmae cups are produced along the upper surface of the gametophyte. Raindrops splash onto the cups and the cups detach and may splash up to 3 feet away from the 'parent' gametophyte. Lunularic acid inhibits the Gemmae from growing, as soon as it is out of the cup, though, the inhibition is removed and each may develop into a new thallus. Sexual reproduction in the *Marchantia* involves the interaction between spores on separate male and female gametophytes. The gametangia are formed on gametophores, or umbrella like structure on long stalks.

The stalks are positioned between the grooves of the thallus. The male gametophore is shaped like a disc with a scalloped edge, while the female gametophore looks like the hub and spokes of a wheel. the male gametangia are called antheridia, contain numerous sperms and are produced in rows just beneath the upper surface of the antheridiophore. The female gametangia are flask-like and each one contains a single egg.

They are produced in rows and hang with their neck downward from the archegoniophore. Rain will splash and release the flagellated sperm cells. The stalks of the archegoniophores may not be finished growing at the time of fertilization. The zygote, fertilized egg, will develop into a multicellular embryo (a.k.a. immature sporophyte) which is anchored from the tissues of the archegoniophore by a knoblike foot.

The foot is connected to the sporophyte (also referred to as the capsule, out of which develops the various type of tissues) by a short, thick stalk called the seta. Inside the capsule, spore mother cells undergo meiosis which results in haploid spores. Some capsule cells do not undergo meiosis, but remain diploid and develop into elaters, which are long and pointy and responsive to changes in humidity. The elaters will twists and untwist rapidly in order to disperse the spores.

The young sporophyte will be protected until maturity by a caplike tissue called the calyptra. There are other variations of this cycle in the other thalloid forms. In some floating or amphibious liverworts, the spores are freed only as the thallus decays.

"Leafy" liverworts—These liverworts are usually found in tropical jungles of fog belts. They have two rows of partially overlapping 'leaves' in which the cells contain distinctive oil bodies. They are of course, not true leaves.

They do have folds and lobes that collect water and usually house tiny animals. The male and female gametangia produce antheridia and archegonia in cuplike structures. When the sporophytes mature they will germinate and produce a protonema which has photosynthetic cells and will develop into a new gametophyte plant.

Division Anthocertophyta

- *Hornworts*: The mature sporophytes of a hornwort look like miniature green cattle horns. Their gametophyte generation looks similar to filmy versions of thalloid liverworts. Hornworts are rare in artic regions and are usually found in moist, shady areas, although some do grow on trees. They have between one and eight chloroplasts—usually just one—and the chloroplasts have pyrenoids similar to green algae.
- *Asexual reproduction*: They can reproduce asexually by fragmentation or by developing lobes that are separate from the main part of the thallus.
- *Sexual reproduction*: Hornworts also reproduce sexually and the plants can be unisexual, like mosses and liverworts, or they can be bisexual, meaning the archegonia and antheridia are on the same plant. The sporophytes are distinctive in form and contain numerous stomata. They do not have stalks, setae, and look instead like tiny broom handles rising out of the gametophyte generation.

Meristematic tissue at the base of the sporophyte continually increases the length of the sporophyte until conditions are favourable and the axis (central core) undergoes meiosis to produce spores. The sporophyte tip will split and the spores will be dispersed as the horn peels into ribbon like segments.

Division Bryophyta—Mosses

- *Subclasses*: There are three subclasses of mosses: peat mosses, true mosses and rock mosses. They share similar reproductive and life cycles and they are all distinct from other plant organisms. Moss 'leaves' do not have mesophyll, stomata or veins and all the 'leaf' cells are haploid. The cells in the 'leaf' do contain numerous lens-shaped chloroplasts, except in the midrib. Mosses have transparent water-storage cells that chloroplasts, yet aid in absorbing water and thus providing moisture to the avascular plant. Mosses also have root like rhizoids which do provide for some water absorption. However, most of the water for the plant travels up the moss surface by way of capillarity.
- *Asexual reproduction*: This reproductive process does not rely on a sexual cycle and its alternation of generations, instead it has been demonstrated that moss fragments can produce protonemata which will 'bud' and develop into gametophyte mosses.
- *Sexual reproduction*: Mosses produce gametangia on the same plant or on separate plants. Commonly, the gametangia are produced on the same plant. The multicellular antheridia and archegonia are produced at the tips of 'leafy' shoots. The individual archegonium has a cavity, the venter with a single egg, and a neck through which the sperm gains access to the egg. Sperm cells are produced in the antheridia.

Upon fertilization, the zygote develops into an embryo that remains attached to the gametophyte by an embedded foot. The embryo develops into a sporophyte with a capsule and a seta, a stalk.

The gametophyte produces a calyptra which partially covers the capsule. Inside the capsule, spore mother cells undergo meiosis and produce spores which are then released through the teeth of the peristome located at the tip of the capsule. Until spore maturity, the peristome is protected by an operculum, which will fall off at maturity. After the spores germinate, protonemata will 'bud' and develop into gametophyte mosses.

VASCULAR PLANTS

Ferns and Relatives

These plants are seedless plants, but unlike the bryophytes, they do have vascular tissue (xylem and phloem). Because of the presence of vascular tissue, the leaves of ferns are their relatives are better organized than the mosses and liverworts.

Division Psilotophyta

The sporophytes in this division have neither true leaves nor roots, their stems and rhizomes fork evenly.

Whisk ferns—Whisk ferns are the simplest of the vascular plants. They consist of evenly forking stems with small protuberances called enations. They lack leaves and roots. These ferns have a central vascular cylinder composed of xylem and phloem.

Reproduction—Whisk ferns reproduce via gametangia. Spores will germinate into tiny saprophytic gametophytes on the surfaces which antheridia and archegonia are scattered. The resulting zygote will develop a foot and a rhizome. An upright stem is produced when the foot separates from the rhizome.

Division Lycophyta

These plants have stems that are covered with photosynthetic microphylls. Microphylls are leaves with a single vein and a trace that is not associated with a leaf gap.

Club mosses –there are two types of club mosses that still have living representatives. Ground pines will develop sporangia in the axils of sporophylls. Each gametophyte may be capable of producing several sporophytes. The second type of club moss include the spike mosses. These plants are heterosporous and have a ligule, or little tongue, on each microphyll. The microspores will develop into male gametophytes with antheridia (sperm producing); while the megaspores will develop into female gametophytes with archegonia (egg producing). The biggest difference between the ground pines and the spike mosses is the presence of the ligule (spike mosses) and the spike mosses produce two types of spores and gametophytes (heterospory).

Quillworts

Quillworts are found partially submerged in water for at least part of the year. Their microphylls (leaves) look somewhat like porcupine quills although they lack the rigidity of an actual porcupine quills. The microphylls arise from a corm like base. The corm base has a cambium that will remain active for many years. Club moss spores have many uses including flash powder, medicine, talcum powder, as well as ornamental uses and novelty items.

Division Sphenophyta

The plants in this division have ribbed stems that contain silica deposits in the epidermal cells. Their scale-like microphylls lack chlorophyll. However, the silica in the stems makes these plants useful for scouring. Horsetails and scouring rushes occur in both the branched and unbranched forms. Either way, they are jointed stems with small whorls of scale-like leaves at the base of the plant.

Horsetails, or Equisetum, are hollow in the centre of their stem which contains cylinders of carinal and vallecular canals. The hollow stems arise from extensively branching rhizomes just beneath the surface of the soil. Some types of horsetails have non-photosynthetic stems. They reproduce via strobili, cones, that are produced in the spring in all species.

Horsetail spores have ribbon like elaters that are sensitive to humidity. An interesting note about horsetails; usually equal numbers of male and female gametophytes are produced by these plants, however, the female gametophytes may become bisexual and the development of more that one sporophyte from a gametophyte is common.

As mentioned, horsetails have been used for scouring. They can be eaten if the silica is removed, however, they really don't make for a feast. Plants from this division have also been used for medicinal purposes including as a diuretic, treatment for tuberculosis and treating venereal diseases. Horsetails have been used as an ingredient in shampoos and metal polish.

Division Pterophyta

Ferns are the most common and best recognized examples of the vascular seedless plants. The leaves of ferns have megaphylls—or leaves with more than one vein and a leaf trace/leaf gap association—which are large and usually subdivided into many lobes. Hence, the 'finger' look to a fern plant. Fern fronds, as the leaves are called, typically are dissected and feathery in appearance.

However, realise that they do vary quite a bit as far as external structure and form goes. Fern fronds start as croziers or fiddleheads which unfurl into the main leaf form. On the underside of the frond, patches of sporangia can be found.

These sporangia are usually in sori, which are clusters; and are sometimes covered by a clear flap called the indusium. This setup allows for the maximum protection of the sporangia, but allows for distribution. Individual sporangia have an 'spring-loaded' annulus which allows for mature spores to be catapulted out of the sporangium. The gametophytes of ferns are called prothalli and develop after the spores germinate.

Prothalli contain both archegonia and antheridia and only one zygote develops into a sporophyte. Ferns are used for ornamental purposes as well as stuffing materials for bedding in tropical regions. Fern fronds are used in weaving baskets and hats, brewing some types of ale and numerous folk medicine practices.

SEED PLANTS

There are two main subdivisions of seed plants—the ones without covered seeds, the gymnosperms, and the ones with covered seeds, the angiosperms.

GYMNOSPERMS

Gymnosperms are plants that do not flower and do not bear their seeds in an enclosure such as a fruit. The seeds are produced on the surface of the sporophylls or similar structures until they are dispersed. The sporophylls are usually arranged in a spiral on the female strobili (cones) which develop at the same time as the smaller male strobili. The male strobili produce the pollen which will fertilize the ovules in the female cones. The ovule contains a nutritious nucellus which is itself enclosed in several layers of integument. The integument layers will eventually become the seed coat, after fertilization and further development of the embryo takes place.

Gymnosperms are classified into one division and three subdivisions: division *Pinophyta* with subdivision *Cycadicae* which includes the palm like cycads; subdivision *Pinicae* which includes conifers and class *Ginkgoatae* the Ginko trees; and subdivision *Gneticae* which includes the gnetophytes.

Subdivision Pinicae: Conifers—pine trees and evergreens to the layman.

Structure and form—For the sake of discussion we will look at Pines, which are the largest genus of conifers. Pine needles are their leaf structures. They are usually arranged in clusters or bundles of two to five leaves (needles), although some species have as few as one or as many as eight leaves in a cluster, the clusters are sometimes referred to as fascicles. Each needle is covered with a thick cuticle over the epidermal layer and a layer of thick-walled cells just beneath the epidermis called the hypodermis.

The stomata on the epidermal surface are sunken and are surrounded by an endodermis. The mesophyll cells do not have the wide air spaces as broadleaf and flowering plant leaves. Resin, and resin canals develop noticeably throughout the mesophyll cells.

The canals are tubes in which resin is secreted. Resin is both aromatic and antiseptic and helps to prevent fungal infections and deter insect attacks. Some conifers produce resin in response to injury. The fascicles, needle clusters, will fall off every two to five years after maturing. They do not, however, fall off all at once and unless diseased, will not look bare like other flowering trees. The secondary xylem, wood, in conifers varies in hardness. Most gymnosperm wood consists of tracheids and has no vessel members or fibres as do flowering trees.

Therefore the wood lacks thick walled cells. Conifer wood is considered to be softwood, while the wood of broadleaf trees is considered to be hardwood. The xylem rings in conifers are often fairly wide as a result of rapid growth. Both vertical and horizontal resin canals can be found throughout the wood. Pine phloem lacks companion cells, but has albuminous cells that perform similar function for the phloem. The roots of pine trees are always found in association with mycorrhizal fungi. The fungi perform functions for the roots, which enable normal growth. Pine trees can be found in all types of environments and ones of opposite extremes. *Reproduction*—There are two kinds of spores produced by pine trees. The microspores are produced in the smaller male strobili which develop at the tips of lower branches. At the base of the male cone, the microsporangia develop in pairs and give rise to four-celled pollen grains. Millions of pollen grains are produced per cone. The female cones are larger and produce the megaspores. They are formed in ovules at the bases of female cone scales.

A pore, called the micropyle, allows for access by the sperm (pollen grain) to the ovule. The pore is formed by the overlapped layers of integument protecting the ovule. Each megaspore develops into a female gametophyte. The archegonia is contained in the mature female gametophyte. Prior to the maturity of the archegonia, pollen grains will become lodged in the cone scales in sticky pollination drops. The pollen grains develop a long pollen tube that digests its way down to the developing archegonia.

Upon arrival at the archegonia, two of the original four cells in the pollen grain will migrate into the tube. One sperm will unite with the egg cell to form a zygote. The zygote will continue developing and will become a seed embryo with a membranous wing formed from part of the cone scale. The seed embryo is ready for distribution and upon landing will germinate and become a new tree. Class *Ginkoatae:* Ginkgo trees have small fan-shaped leaves with veins that evenly fork. They have similar reproductive cycles to that of the conifers with the exception that the edible seeds are encased in a fleshy covering. The covering smells like rancid butter at seed maturity.

Subdivision Cycadicae

Cycads—These plants look like little palm trees with unbranched trunks and large crowns of pinnately divided leaves. Their strobili and cones are quite similar to those of conifers, however, their sperms have numerous flagella—much unlike conifers.

Subdivision Gneticae

Gnetophytes—These plants have vessels in their xylem. Most of the species are in the genus Ephedra and have jointed stems and leaves that are nothing more than scales. Sometimes the plants in this genus are called joint firs, as they look like jointed sticks. The plants in this subdivision are adapted to unusually dry environments. They produce tiny leaves in groups of twos and threes, which turn brown as soon as they appear. Male and female strobili may occur on the same plant.

RELEVANCE TO HUMANS

Conifers are sources for paper products and lumber materials. The resin from conifers has historically been used as sealing pitch, turpentine, floor waxes, printer's ink, perfumes, menthol manufacture and rosin for musical instruments. Ginko leaves are used medicinally as are plants from the genus Ephedra. Arrowroot starch was once purified from a cycad species. Teas have been made from conifers.

Angiosperms—Flowering Seed Plants (Covered Seed Plants)

Angiosperms are plants that have seeds encased in a protective covering. That covering is the ovary which is part of the flower structure and distinguishes angiosperms from gymnosperms, the other seed plants. So it can be said that angiosperms are also flowering plants. There is one division of angiosperms, Magnoliophyta, which is divided into two classes: monocots and dicots.

Angiosperms, like gymnosperms, are heterosporus, which means they produce two types of spores and their sporophytes are more dominant than those of gymnosperms. At maturity, the female gametophytes are reduced to a few cells and are completely enclosed within sporophyte tissue; while the male gametophytes consist of a binucleate cell with a tube nucleus which forms a pollen tube much like the one formed in gymnosperm pollination.

DEVELOPMENT OF GAMETOPHYTES

Angiosperm gametophytes develop in separate structures, sometimes on the same plant. The embryo sac, or female gametophyte, develops in the ovule which is surrounded by integuments. The integuments will later become the seed coat after fertilization has occurred. The male gametophytes, pollen grains, develop in the anthers.

Pollination—The process of transferring pollen grains from the anther to a stigma is called pollination. This may occur via wind, insects, birds or other agents. Flowers may be geared towards certain pollinators. Flowers pollinated by bees are usually sweet and fragrant. Their colours are usually blue and yellow. Beetles are attracted to flowers with strong odours and dull colours or white flowers.

There are fly-pollinated flowers that smell like rotting meat. Moths are drawn to pale yellow or white flowers. Birds, akin to bee preferences, like sweet nectar and fragrant flowers with bright colours. Orchids have unusual pollination mechanisms, and in some they literally grab an insect and manually attach two pollen packets to the hind end of the insect before releasing it.

Fertilization and seed development—The pollen grain contains two sperm nuclei (formed from a generative nucleus that splits) and one tube nucleus. The tube nucleus forms a pollen tube that grows through the stigma, the style and into the ovary via the micropyle. Upon arrival at the ovary, one sperm nuclei will fertilize the egg and form a zygote. The other sperm nuclei will fuse with the polar nuclei of the egg in order to form a 3n endosperm nucleus. This will be conserved as food for the plant embryo or may become part of the seed.

Some species will end up with 5n, 9n, or 15n endosperm tissue, depending on how the embryo sac develops. Parthenocarpy—Some fruits can develop without the development or fusion of gametes, but with the otherwise standard structures being involved. These fruits are called parthenocarpic fruits. Not all 'seedless' fruits are parthenocarpic.

CLASSIFICATION TRENDS IN FLOWERING PLANTS

Specializations in flowering plants include an overall efficiency in plant growth, by reducing the number of parts; fusion of parts; the appearance of compound pistils that are composed of several carpels; various positioning of the ovary with respect to the receptacle including inferior and superior ovaries; irregular flowers and unisexual flowers. A monoecious species of flower has both male and female flowers on the same plant, while a dioecious flower would have separate plants for male and female flowers. All of these are characteristics that can be looked at in the classification and categorization of flowering plants.

BIOTECHNOLOGY IN FOOD AND AGRICULTURE

Biotechnology in food and agriculture, particularly genetic engineering, has become the focus of a "global war of rhetoric". Supporters hail genetic engineering as essential to addressing food insecurity and malnutrition in developing countries and accuse opponents of "crimes against humanity" for delaying the regulatory approval of potentially life-saving innovations. Opponents claim that genetic engineering will wreak environmental catastrophe, worsen poverty and hunger, and lead to a corporate takeover of traditional agriculture and the global food supply. They accuse biotechnology supporters of "fooling the world" of *The State of Food and Agriculture* surveys the current state of scientific and economic evidence regarding the potential of agricultural biotechnology, particularly genetic engineering, to meet the needs of the poor.

Agriculture in the twenty-first century is facing unprecedented challenges. An additional 2 billion people will have to be fed over the next 30 years from an increasingly fragile natural resource base. More than 842 million people are chronically hungry, most of them in rural areas of poor countries, and billions suffer from micronutrient deficiencies, an insidious form of malnutrition caused by the poor quality of, and lack of diversity in, their habitual diet. The Green Revolution taught us that technological innovation - higher-yielding seeds and the inputs required to make them grow - can bring enormous benefits to poor people through enhanced efficiency, higher incomes and lower food prices. This virtuous cycle of rising productivity, improving living standards and sustainable economic growth has lifted millions of people out of poverty. But many remain trapped in subsistence agriculture. Can the Gene Revolution reach those left behind?

At the same time, a rapidly urbanizing global population is demanding a wider range of quality attributes from agriculture, not just of the products themselves but of the methods used in their production. The agriculture sector will need to respond in ways beyond the traditional focus on higher yields, addressing the protection of environmental common goods, consumer concerns for food safety and quality, and the enhancement of rural livelihoods both in the South and in the North. Is the rhetoric of war deafening us to a more reasoned debate regarding the hazards and opportunities posed by biotechnology?

There is clear promise that biotechnology can contribute to meeting these challenges. Biotechnology can overcome production constraints that are more difficult or intractable with conventional breeding. It can speed up conventional breeding programmes and provide farmers with disease-free planting materials. It can create crops that resist pests and diseases, replacing toxic chemicals that harm the environment and human health, and it can provide diagnostic tools and vaccines that help control devastating animal diseases. It can improve the nutritional quality of staple foods such as rice and cassava and create new products for health and industrial uses.

But biotechnology is not a panacea. It cannot overcome the gaps in infrastructure, markets, breeding capacity, input delivery systems and extension services that hinder all efforts to promote agricultural growth in poor, remote areas. Some of these challenges may be more difficult for biotechnology than for other agricultural technologies, but others may be less difficult. Technologies that are embodied in a seed, such as transgenic insect resistance, may be easier for small-scale, resource-poor farmers to use than more complicated crop technologies that require other inputs or complex management strategies. On the other hand, some biotechnology packages, particularly in the livestock and fisheries areas, require a certain institutional and managerial environment to function properly and thus may not be effective for resource-poor smallholders.

The safety and regulatory concerns associated with transgenic crops constitute a major hurdle for developing countries, because many lack the regulatory frameworks and technical capacity necessary to evaluate these crops and the conflicting claims surrounding them. Although the international scientific community has determined that foods derived from the transgenic crops currently on the market are safe to eat, it also acknowledges that some of the emerging transformations involving multiple transgenes may require additional food-safety risk-analysis procedures. There is less scientific consensus on the environmental hazards associated with transgenic crops, although there is general agreement that these products should be evaluated against the hazards associated with conventional agriculture. There is also wide consensus that transgenic crops should be evaluated on a case-by-case basis, as is the case with pharmaceuticals, taking into consideration the specific crop, trait and agro-ecological system. Because very few transgenic crops have been evaluated for their ecological impacts in tropical regions, a major research effort is required in this area. Public- and private-sector transgenic crop research and development are being carried out on more than 40 crops worldwide and dozens of innovations are being studied, but there is clear evidence that the problems of the poor are being neglected. Barring a few initiatives here and there, there are no major public- or private-sector programmes to tackle the critical problems of the poor or targeting crops and animals that they rely on. Concerted international efforts are required to ensure that the technology needs of the poor are addressed and that barriers to access are overcome.

PRODUCT QUALITY AND BIOTECHNOLOGY

We are also using biotechnology to change the characteristics of the raw material inputs so that they are more attractive to consumers and more amenable to processing. Biotechnology researchers are increasing the shelf life of fresh fruits and vegetables; improving the crispness of carrots, peppers and celery; creating seedless varieties of grapes and melons; extending the seasonal geographic availability of tomatoes, strawberries and raspberries; improving the flavour of tomatoes, lettuce, peppers, peas and potatoes; and creating caffeine-free coffee and tea.

Japanese scientists have now identified the enzyme that produces the chemical that makes us cry when we slice an onion. Knowing the identity of the enzyme is the first step in finding a way to block the gene to create "tearless" onions. Much of the work on improving how well crops endure food processing involves changing the ratio of water to starch. Potatoes with higher starch content are healthier because they absorb less oil when they are fried, for example. Another important benefit is that starchier potatoes require less energy to process and therefore cost less to handle. Many tomato processors now use tomatoes derived from a biotechnology technique, somaclonal variant selection. The new tomatoes, used in soup, ketchup and tomato paste, contain 30 per cent less water and are processed with greater efficiency. A 1?2 per cent increase in the solid content is worth $35 million to the U.S. processed-tomato industry.

Another food processing sector that will benefit economically from better quality raw materials is the dairy products industry. Scientists in New Zealand have now used biotechnology to increase the amount of the protein casein, which is essential to cheese making, in milk by 13 per cent. Biotechnology also allows the economically viable production of valuable, naturally occurring compounds that cannot be manufactured by other means. For example, commercial-scale production of the natural and highly marketable sweetener known as fructans has long eluded food-processing engineers. Fructans, which are short chains of the sugar molecule fructose, taste like sugar but have no calories. Scientists found a gene that converts 90 per cent of the sugar found in beets to fructans. Because 40 per cent of the transgenic beet dry weight is fructans, this crop can serve as a manufacturing facility for fructans.

SAFETY OF THE RAW MATERIALS

The most significant food-safety issue food producers face is microbial contamination, which can occur at any point from farm to table. Any biotechnology product that decreases microbes found on animal products and crop plants will significantly improve the safety of raw materials entering the food supply. Improved food safety through decreased microbial contamination begins on the farm. Transgenic disease-resistant and insect-resistant crops have less microbial contamination. Biotechnology is improving the safety of raw materials by helping food scientists discover the exact identity of the allergenic protein in foods such as peanuts, soybeans and milk, so they can then remove them. Although 95 per cent of food allergies can be traced to a group of eight foods, in most cases we do not know which of the thousands of proteins in a food triggered the reaction. With biotechnology techniques, we are making great progress in identifying these allergens. More importantly, scientists have succeeded in using biotechnology to block or remove allergenicity genes in peanuts, soybeans and shrimp.Finally, biotechnology is helping us improve the safety of raw agricultural products by decreasing the amount of natural plant toxins found in foods such as potato and cassava.

Food Processing

Microorganisms have been essential to the food-processing industry for decades. They play a role in the production of the fermented foods. They also serve as a rich source of food additives, enzymes and other substances used in food processing.

IMPROVING FOOD FERMENTORS

Because of the importance of fermented foods to so many cultures, scientists are conducting a lot of work to improve the microorganisms that carry out food fermentations. The bacterium responsible for many of our fermented dairy products, such as cheese and yogurt, is susceptible to infection by a virus that causes substantial economic losses to the food industry. Through recombinant technology, researchers have made some strains of this bacterium and other important fermentors resistant to viral infection.

We have known for years that some bacteria used in food fermentation produce compounds that kill other, contaminating bacteria that cause food poisoning and food spoilage. Using biotechnology we are equipping many of our microbial fermentors with this self-defence mechanism to decrease microbial contamination of fermented foods.

FOOD ADDITIVES AND PROCESSING AIDS

Microorganisms have been essential to the food industry not only for their importance as fermentors, but also because they are the source of many of the additives and processing aids used in food processing. Biotechnology advances will enhance their value to the food industry even further. Food additives are substances used to increase nutritional value, retard spoilage, change consistency and enhance flavour. The compounds food processors use as food additives are substances nature has provided and are usually of plant or microbial origin, such as xanthan gum and guar gum, which are produced by microbes. Many of the amino acid supplements, flavours, flavour enhancers and vitamins added to breakfast cereals are produced by microbial fermentation. Through biotechnology, food processors will be able to produce many compounds that could serve as food additives but that now are in scant supply or that are found in microorganisms or plants difficult to maintain in fermentation systems.

Food processors use plant starch as a thickener and fat substitute in low-fat products. Currently, the starch is extracted from plants and modified using chemicals or energy-consuming mechanical processes. Scientists are using biotechnology to change the starch in crop plants so that it no longer requires special handling before it can be used. Enzymes produced by microbial fermentation play essential roles as processing aids in the food industry. The first commercial food product produced by biotechnology was an enzyme used in cheese making. Prior to biotech techniques, this enzyme had to be extracted from the stomach of calves, lambs and baby goats, but it is now produced by microorganisms that were given the gene for this enzyme. The production of high-fructose corn syrup from cornstarch requires three enzymes, and those same enzymes are important in making baked goods and beer. Other enzymes are essential to the production of fruit juices, candies with soft centers, and cheeses. The food industry uses more than 55 different enzyme products in food processing. This number will increase as we discover how to capitalize on the extraordinary diversity of the microbial world and obtain new enzymes that will prove important in food processing.

FOOD SAFETY TESTING

In addition to the many ways biotechnology is helping us enhance the safety of the food supply, biotechnology is providing us with many tools to detect microorganisms and the toxins they produce. Monoclonal antibody tests, biosensors, polymerase chain reaction (PCR) methods and DNA probes are being developed that will be used to determine the presence of harmful bacteria that cause food poisoning and food spoilage, such as *Listeria* and *Clostridium botulinum*.

We can now distinguish *E. coli* 0157:H7, the strain of *E. coli* responsible for several deaths in recent years, from the many other harmless E. coli strains.

These tests are portable, quicker and more sensitive to low levels of microbial contamination than previous tests because of the increased specificity of molecular technique. For example, the new diagnostic tests for *Salmonella* yield results in 36 hours, compared with the three or four days the older detection methods required.

Biotechnology-based diagnostics have also been developed that allow us to detect toxins, such as aflatoxin, produced by fungi and molds that grow on crops, and to determine whether food products have inadvertently been contaminated with peanuts, a potent allergen.

FOOD PRODUCTION AND GREEN REVOLUTION TECHNOLOGY

Mechanization, which often came with the green revolution technologies, was said to be bad because it displaced labour. Again, it often happened that the increase in crops and output per year because of varieties that needed shorter growing seasons and higher yields per crop had labour requirements that more than offset the labour saved by mechanization.

In most areas, the extra crop, thereby demanding greater labour, would have been impossible without mechanization and the green revolution technologies. In the 1970s, saving labour by mechanization was always assumed to be a bad thing, though it might also mean more leisure for the farmer. Increasingly, mechanization means that farmers may earn cash income off the farm, affording school fees for children or consumption of products in the market economy. It was conceded that the demand for agricultural labour was up, but that mechanization, by doing the more menial tasks, displaced women and children from the farm labour force. Many who oppose child labour in factories somehow deem agricultural stoop labour by children in the hot sun to be virtuous. As incomes go up and child labour is displaced, there is a very good chance that they will now be in school. And where incomes go up, other technologies are also acquired that improve the quality of life and save on child labour. The high-yielding varieties of rice have been basically for production in irrigated fields. The upland varieties have barely changed in yields over the last three decades. This has led to the following argument: The Green revolution has given rise to greater regional inequality of income. Careful empirical studies of rice cultivation in Asia have found that laborers from the upland regions migrate to work in the paddies, smoothing out the potential regional differences.

There are also arguments seeking a reduction in genetic variety, soil erosion, or groundwater contamination, which are of serious concern but cannot negate the enormous gain in human nutrition or the human catastrophe that would result from any attempt to reverse the green revolution. The diminution of genetic diversity is not a phenomenon of the green revolution *per se* but of agriculture. The individual farmer will select for yield or other output concerns, not for global genetic diversity. Modern no-tillage (or reduced or minimum tillage) agriculture using pesticides for weed and pest control conserves both soil and water better than its organic competitors.

MODERN FOOD SUPPLY AND SAFETY

Regularized and improved food supply has been a consistent factor throughout history in improving human health and longevity. In more recent times, improved food safety has added another dimension to human health. The evidence on the safety of our modern food supply is quite contrary to the strongly held conventional wisdom. Humans have had to deal with food contamination throughout history. Human foodstuffs have been the carriers of botulism, ergot, and the aflatoxins from the fungus *Aspergillus flavus,* which has caused mass illness, blindness, and largescale death. Even uncontaminated natural foods contain substances that would be considered a threat to human life if they were used as a food additive.

In the United States, during the early twentieth century, contaminated food, milk, and water caused many foodborne infections, including typhoid fever, tuberculosis, botulism, and scarlet fever. Once the sources and characteristics of foodborne diseases were identified, there were technologies that could help to control them before vaccines or antibiotics: "handwashing, sanitation, refrigeration, pasteurization, and pesticide application." Healthier animal care, feeding, and processing also improved food supply safety.

PLANT-PRODUCED CHEMICALS

The buzz word "chemicals," used by some to label and condemn food additives, could appropriately be used to describe what the plants themselves manufacture. Plants of all kinds, and most other life forms, are chemical factories, manufacturing a variety of chemicals, some of which have very active properties. Humans have extracted some of these chemicals for medicinal uses, some as poisons. Given the truism that dose makes the poison, some chemicals from plants that were once used as poisons are now used for medical purposes.

In other words, chemicals from plants are not inherently good or bad; goodness or badness is determined by a variety of factors and the uses to which they are put. One of the reasons for maintaining as much biodiversity as possible is that the earth's plants produce an array of active chemicals, the vast majority of which have not been screened for possible human use. According to Katharine Milton, plants are unable to run from hungry predators and so have developed a variety of defences to avoid the loss of edible components. "These protections include a vast array of chemicals known as secondary compounds (such as tannins, alkaloids and terpenoids). At best these chemicals taste awful; at worst, they are lethal".

Leopold and Ardrey add that these plant secondary substances are "chemicals that do not participate in the basic metabolism of the plant. Among these are many chemicals that serve to repel or discourage the use of the plant by insects, microorganisms, nematodes, grazing animals, and man." These plant toxins apparently are often used to ward off fungi. "While antifungal chemicals in nightshade were distasteful ... there are probably protective compounds in other wild fruits that prevent rotting without spoiling taste, compounds that should be of particular interest to plant breeders". Many plants produce "phytoestrogens," the so-called estrogen disrupters, which might have the effect of reducing the fertility of their predators, the animals that graze or otherwise consume them. "Essentially, they give their predators an oral contraceptive".

Eating fresh, uncooked produce may have many taste and health benefits, but it can also be dangerous unless proper precautions are taken. As epidemiologist Michael T. Osterholmm stated: "The hearthealthy diet may be kind to our cardiovascular system but it is hell on the digestive tract". Rachel Carson, whose book *Silent Spring* is seen as the inspiration of the modern environmental movement, recognized the role of natural toxins—systemic pesticides—in plants. She maintained that the perception of nature as an "enchanted forest" was a "fairy tale". Even some of the weeds associated with cultivars and often eaten with them are endowed with toxins and other substances deleterious to human health. Garn notes, "Weeds growing along with cultivars often included plants with interesting alkaloids, many of them in the tomato family that includes *Datura, Belladonna,* and the like".

TOXINS IN PLANT EVOLUTION

Through time, plants evolved for their own survival, not to serve human needs. "Plants in nature synthesize toxic chemicals in large amounts, apparently as a primary defence against the hordes of bacterial, fungal, and insects and other animal predators *plants in the human diet are no exception.*" The evolutionary process involved in domestication changed the survival strategy for these plants, as they now depended upon the continued cultivation to survive. The toxicity of many plants was reduced compared to their wild progenitors, but in most cases was not completely bred out of plants. In fact, currently there is considerable effort to breed plants with greater disease resistance so as to reduce the need for pesticides, in some instances by increasing the production of these toxins; in other instances, not.

GREEN REVOLUTION PROS AND CONS

It is hard to imagine anyone opposing the green revolution. The increase in yields from the green revolution in rice alone has produced enough to feed one billion people. Growth in yields accounts for 92 per cent of the increase in world cereal production since 1960. The world average for grain yields per hectare rose from 1.1 tons in 1950 to 2.9 tons in 1992. Without the increase in yields, it would have been necessary to bring another 3.6 billion hectares of land under cultivation, which would have almost doubled world cropland from "34 per cent of the earth's surface to 61 per cent". The green revolution was in many ways a grain revolution— wheat, rice, and maize—but it also was a research revolution that facilitated an extraordinary and sustained general expansion in food production. Bender and Smith note, "Between 1961 and 1994, the number of daily food calories per capita rose from about 1,900 to 2,600 in developing countries, while their populations nearly doubled from 2.2 billion to more than 4.3 billion." Globally in the same period, "average daily per capita food supplies increased more than 20 per cent." The increase in available calories per capita for the developing countries rose 50 per cent from 1948–1952 to 1994–1996. A century-long trend of falling real food prices continued during the period 1950 to 1992 as international food commodity prices dropped 78 per cent in constant 1990 prices.

D. Gale Johnson sums up these achievements in agriculture: "People today have more adequate nutrition than ever before and acquire that nutrition at the lowest cost in all human history." In the 1990s, while population continued to increase, the absolute number of malnourished people in the world declined by forty million, which is a continuation of a trend of the last decades. However, this was considered a "failure" by the UN World Food Programme because it fell short of the reductions that were thought to be possible. To the extent that it is a failure, it is a failure of institutions, not of the technology that is producing the food that could make the programme a success.

So, there are arguments critical of the green revolution. The first argument is that it was a failure. Data such as that just cited above have demolished that argument, although a few still assert it, unaware of the change in the antitechnology party line. Next, we are told that the new varieties may have increased yields, but they led to a worsening of income distribution and the condition of the very poor. As early as the 1970s, studies emerged showing that the poor consumers and farmers benefited proportionately more than other groups as part of the overall improvement in human wellbeing. Some argued that tenant farmers in the Punjab in the north of India were displaced from their land due to high-yielding varieties of wheat. It turns out that they returned to the land as farm laborers and earned more than they did as tenant farmers. Similarly, in southern India, between 1973 and 1994, the average real income of small farmers increased by 90 per cent, and that of the landless— among the poorest in the farm community—by 125 per cent. Further, caloric intakes for small farmers and the landless rose 58 to 81 per cent, and protein intakes rose 103 to 115 per cent. In addition, greater farm income meant more income and employment in the local villages that supplied the farm with goods and services now affordable to those who worked the land.

The green revolution in Indian agriculture is credited with being a major force in poverty reduction in India. From independence to 1970, the per centage of the population of India living below the absolute poverty line remained relatively steady at 55 per cent. In over two decades since the introduction of the green revolution technologies, the per centage of the population of India living in absolute poverty has fallen to 35 per cent. Throughout most of East and Southeast Asia, the early phases of development involved an emphasis on agriculture and food production.

The green revolution technologies in agriculture have played an important role in the dramatic reductions in poverty rates throughout the region, such as in Indonesia from 58 per cent in 1970 to 8 per cent in 1993, and in China from 270 million people in poverty in 1978 to 65 million in 1999. Yao estimates the reduction in rural poverty in China to be from between 596 to 790 million people (75.5 per cent–100 per cent) in 1978 to 57 to 114 million (6.7 per cent–13.2 per cent) in 1996 (447). By either measurement, the magnitude of the reduction in poverty in China is extraordinary and without precedent in human history. In the time period 1969–1971 to 1990–1992, global population was increasing about 45 per cent while the absolute number of people suffering from chronic undernourishment in developing nations decreased from 917 million or 35 per cent of their population to 839 million or 21 per cent. From 1990–1992 to 1995–1997, the absolute number of undernourished people in the developing world continued to decline, falling to 790 million. The prevalent high number should not be minimized, nor should we be complacent, but neither should the progress achieved be ignored. For we can finish the job of eliminating undernourishment by recognizing the forces that have brought us this far, thereby gaining some understanding of what still needs to be done. Unfortunately, many wish to deny the progress that has been made, offering antitechnology solutions, which would reverse these gains.

ALTERNATIVE MEDICINE

Some of the most dangerous ingested products legally sold are found in health food stores. Herbal dietary supplements (including exotic oriental herbs) and amino acids pills have been causing health problems for some of their users. A chemical additive may or may not be inactive, but the herbs and amino acids are taken precisely because they are active and will alter a bodily function. Most of them are untested and unregulated. To the extent that testing is done, there is a critical difference between the approach to herbal medicine and that of pharmaceuticals. Drugs are not permitted to be used until proven to be safe and effective, while the philosophy and practice of alternative medicine has operated on the "presumption that treatments are safe and effective unless proven otherwise".

Alternative medicine is big business. One study estimated that in 1990, "Americans poured $13.7 billion into alternative remedies, and made more visits to alternative healers than to primary-care doctors". An editorial in the *New*

England Journal of Medicine clearly defines the difficulties with alternative medicine: "What most sets alternative medicine apart is that it has not been scientifically tested and its advocates largely deny the need for such testing". Lacking is "the marshaling of rigourous evidence of safety and efficacy" necessary for approval of drugs and for publication in the "best peer-reviewed medical journals". One respected study has found that Americans are spending $27 billion a year on alternative treatments, which is close to their outof-pocket expenditures on modern medical treatment.

Some herbs and treatments were found to be beneficial, but some were not. Investors have found the stocks of the firms manufacturing these dietary supplements and alternative medications to be attractive, indicating a belief that consumers will buy them, whether proved to be safe and effective or harmful. Even for some herbs that have been used in low dosage for centuries, there are no studies on the impact of the mega doses that are often taken. The herbs and alternative medicine advocates have succeeded in blurring the line between food and drugs. Herbal extracts sold as food supplements are being added to packaged, canned, and frozen food products—called functional-foods—with labels that imply health benefits and have sufficient growth potential to have attracted some of the major food producers.

Some of the early advocates of alternative foods and medications, like Rodale Inc. and its publication, *Prevention,* recognize the danger in current patterns of use of herbs and dietary supplements. *Prevention's* national survey on the self-care movement reveals "158 million consumers use dietary supplements for their health and spend $8.5 billion each year." It reported that "widespread use of dietary supplements may cause public health problems". There were risks mixing dietary supplements with prescription drugs and treating oneself without a doctor's supervision. The survey estimated that "11.9 million customers have experienced adverse reactions from using herbal remedies and 6.5 million have had problems of this kind when using specialty supplements". The problems with various herbs are legion, including lack of long-term studies and of any kind of control of dosage intake. Since many who use alternative medical treatment are also taking prescription drugs under the care of medical doctors, there may be substantial health risks, including death, in combining these with herbs and other alternative medications and stopping the use of some herbs after one had been taking them. Serious problems have emerged from ingesting some of the most popular herbs, such as *Echinacea purpurea,* St. John's Wort, or Ginkgo Biloba. Probably the most serious harm can come from the interaction between herbs and prescription medicines. The increasing provision of these herbs by parents to their children is evident in the statistics of the American Association of Poison Control Centers.

One herbal remedy sometimes called ephedra, sometimes called ma huang, sometimes called pseudoephedrine, contains the chemical ephedrine, "a potent drug whether it is synthetically manufactured or produced naturally from a plant". Dr. Robin R. Caldwell says the drug is "chemically almost identical to amphetamine". Not only is ma huang sold in health food stores, it is a listed ingredient in many products sold at truck stops and elsewhere to help drivers stay awake. Pseudoephedrine is in various cold medicines that carry label warnings not to be used by anyone with high blood pressure, heart trouble, or liver problems. Ma huang is also used in diet and pep pills. Because ephedra is classified as a dietary supplement under the 1994 Dietary Supplement Health and Education Act, the FDA can restrict its sale "only if it receives well-documented reports of health problems associated with it." Brody notes, "The agency took four years, and more than 100 reports of life-threatening symptoms and 38 deaths, to act against ma huang, a stimulant that can prove disastrous to people with heart problems."

Warnings have also been posted for Chinese herbal products containing aristolochia acid, which has been shown to cause cancer and kidney failure. Nearly one hundred women with kidney damage from Chinese herbs at a Belgian slimming clinic were studied, and several were found "in dire need of kidney transplants" and many of them were developing cancers in the kidney and bladder, years after taking the drug". Those opposing any regulation of dietary supplements are involved in a contradiction that doesn't seem to bother them. These herbal and other dietary supplements are taken because they have some curative effect from an active ingredient or agent. Stated simply, "if it's strong enough to work, it's strong enough to hurt".

NUTRIENT ADDITIONS TO FOOD PRODUCTS

The following is not an exhaustive review of all food products which are fortified. Food products have been selected for discussion on the basis of their nutritional impact or their applicability to fortification programmes in developing countries.

CEREALS AND CEREAL BASED PRODUCTS

This is one of the most important areas for consideration of food fortification technologies. Foods from this category form the major component of diets around the world. This is especially true in the case of developing countries where dietary diversity is limited. On average, cereals provide 52 per cent of caloric intake globally. For Africans and Asians they represent 60-75 per cent of the caloric intake; for Latin Americans, 50 per cent; and in the United States of America, 26 per cent. In developing countries 95 per cent of the population consume cereals as a dietary staple which also provide about 47 per cent of the per capita protein intake.

Milling of cereal grains prior to their consumption is a common practice. During the milling process a substantial proportion of the nutrients are lost from the refined product. The fortification/enrichment of cereal grains can therefore be rationalised in more than one way. One valid reason is to restore to refined products, nutrients which have been removed during the milling process. Another reason is to improve the nutrient intake levels of target populations which are at risk of micronutrient deficiency.

RICE AND OTHER WHOLE CEREAL GRAINS

Among the cereals, rice presents unique problems in fortification. These are due to the fact that it is most commonly consumed as a whole grain and also in many countries, extensive washing of the grain prior to cooking is the normal practice. The earliest methods of rice enrichment involved the production of parboiled and converted rice. By this means nutrients from the bran layer were transferred to the starchy endosperm. The parboiling process involved the soaking of the rough rice, the application of heat followed by drying and milling.

It was demonstrated that in this way 50-90 per cent of the thiamine was retained. The process for converted rice developed by Huzenlaub was similar to that for parboiling, but also employed pressure differences to facilitate transfer of nutrients. Acid parboiling, described by Kondo was similar to the parboiling except that it was carried out in the presence of acetic acid. After parboiling and converting, the next methods of enrichment involved the actual addition of nutrients to the milled products. Techniques used for this have been classified into two main groups 'powder type' and 'grain type' enrichment.

In powder type enrichment, a powdered pre-blended mixture of vitamins and minerals has been added at a rate of 1, 0.5. or 0.25 oz. per 100 lbs of rice (a w/w ratio of 1:1600, 1:3200 or 1:6400). For white parboiled rice, the normal practice has been to add the premix soon after milling as the heat and moisture at the grain surface at this point facilitates adherence of the powder. A major disadvantage of this method of nutrient addition has been that 20-100 per cent of the nutrients are lost on washing. In the USA, rice enriched in this way must bear a label stating 'to retain vitamins do not rinse before or drain after cooking'.

In the second major type of enrichment, a powdered nutrient mixture has been applied to the milled rice grains followed by coating with a water insoluble substance. A fortified rice premix produced in this way has then been added to milled rice at a rate of 0.5 per cent, to yield an enriched product conforming to the required standard of identity. In the United States of America, this standard requires between 2.0 - 4.0 mg of thiamine, 1.2 - 2.4 mg of riboflavin, 16 - 32 mg of niacin or niacinamide and 13 - 26 mg of iron per 100 lbs. of rice.

It may contain 250 -1000 USP units of vitamin D and 500 -1000 mg of calcium. The requirement regarding riboflavin has been stayed pending final action. The objections regarding the use of this vitamin in the premix have been due to its role in the colour deterioration of the enriched grain. 'Grain type' or 'coated grain' enrichment have also been carried out by spraying the premix solution onto the rice which is contained in a rotary cylinder, followed by hot air drying of the rice, application of a water insoluble sealant, addition of the iron compound and finally a second application of the water insoluble compound. Water insoluble coatings which have been reported include an ethanol or isopropanol solution of zein, palmitic or stearic acid and abeitic acid. Other coating materials have contained ethyl cellulose. They dissolve at the elevated temperatures employed during cooking. Using this 'coated grain' procedure, Cort et al and Rubin et al. were able to successfully enrich rice with niacin, thiamine, pyridoxine, vitamin A, vitamin E, folic acid, iron, and zinc using expanded versions of the procedures already described. The water insoluble vitamins and minerals were added in different layers with intermittent coatings of shellac. Stability to rinsing treatments were reportedly high, with vitamin losses in the range of 0.2-1.1 per cent.

Ricegrowers Co-operative Ltd. (RCL) in Australia revised the HLR rice fortification procedure for the enrichment of rice with thiamine, niacin and iron. The HLR process involved dissolution of vitamins in dilute sulphuric acid, application of vitamin solution to rice, application of a water insoluble substance, application of ferric pyrophosphate and talc followed by reapplication of a water insoluble layer and iron mixture before screening and packaging.

Due to increased cost of raw materials, low factory output, use of hazardous materials, reported problems with browning of enriched grains, RCL modified the procedure to include the application of the nutrient mixture to the rice followed by drying, screening and packing. The nutrient mixture was composed of ferric pyrophosphate suspended in an acid solution of the vitamins.

They reported that the browning found in the HLR premix on storage was due to the formation of ferric sulphates and that this problem was eliminated with the use of an alternative acid. Acid hydrolysis at the surface of the rice grain provided a sufficiently sticky surface to ensure adhesion of the iron compound. Cost of the process was reduced and the factory output was doubled.

There was no mention, however, of the stability of premix to nutrient loss through rinsing. In the Philippines, attempts at fortifying rice by application of vitamin A followed by a water insoluble layer were abandoned since washing losses of between 10 - 20 per cent were recorded. Rice fortification with iron has also been tried in the Philippines using the coating method. The fortificant used in this programme was anhydrous ferrous sulphate. Some discolouration was found after 20 weeks of storage at room temperature and washing losses were 9 per cent.

Ferric orthophosphate (white iron) is a recommended form of iron for use in the fortification of rice. This iron compound is almost water insoluble and has been preferred for mixing with milled rice due to its white colour. When it is oxidised or contains excessive moisture it may become tan, yellow, purple and or black. Hurrel reported that the bioavailability of ferric orthophosphate varied widely from batch to batch and was highly negatively correlated with particle size as was also found with elemental iron. This fortificant compound is more expensive than anhydrous ferrous sulphate. According to Hurrel the cost of ferric orthophosphate was about six times that of anhydrous ferrous sulphate, for the same level of total iron. In Japan, a multinutrient enriched rice has been on the market since 1981. The first step in this procedure was acid parboiling in the presence of thiamine, riboflavin, niacin, pantothenic acid and pyridoxine. The second step involved coating the grain with separate layers of vitamin E, calcium and iron in separate layers and finally a protective coating material. Modified atmosphere packaging, utilising aluminium laminate and carbon dioxide, ensured the stability of vitamin E during storage. The yellow or brown colour of the premix was not a problem in this case as it was used in the fortification of 'brown' rice.

Apart from these well established procedures there have been other innovations regarding alternative methods for rice fortification. Joseph et al. described an enrichment procedure using a premix containing thiamine, riboflavin, niacin and pyridoxine. This procedure involved soaking the milled rice in an acid medium containing the water soluble vitamins followed by the cross linking of starch granules in the enriched grains. The cross linking procedure itself was demonstrated to have caused significant vitamin loss, but the added vitamins were highly cook and wash stable. It is possible that this method could have some utility in the future.

Use of fortified simulated grains has featured prominently in attempts at rice fortification in the developing countries. Murphy et al. described the production of such a fortified rice product for use in the Philippines. The synthetic rice grains were produced by extrusion of rice flour in a pasta machine. The best formulation contained vitamin A stabilised by a mixture of tocopherol, ascorbate and lipids with a low level of unsaturation.

Retinyl palmitate stabilised in an acacia matrix, type 250 SD was the fortificant used. Retention of vitamin A after washing was reported to be 100 per cent. Vitamin retention after cooking, however, ranged from 60 - 94 per cent. The formulations which demonstrated the better storage stability, particularly at high humidity, suffered greater cooking losses. Drawbacks with this technology have been reported to exist with respect to blending with the natural product and in the consistency of the simulated grains after cooking. Field trials in Brazil showed no problems with acceptability of rice enriched in this way.

The enrichment of whole wheat grains with vitamin A has been attempted. The fortificant used was a premix comprised of concentrated vitamin A attached to wheat grains, for mixing at a level of 0.25 per cent with wheat grains. The feasibility of this procedure was not determined. The fortification of whole grain cereals with soluble iron compounds is difficult because they promote oxidation of the lipid component of the grain, thus reducing the shelf life.

FLOURS, CORNMEAL AND BREAD AND PASTA

In the fortification of flour the required nutrient mixture is mixed with an appropriate diluent to produce a premix, which is then accurately metered into the flour. The addition of vitamins B_1 and B_2, niacin, iron and calcium to wheat flour is a common practice in many developed countries. It is technologically feasible to add other vitamins and minerals as well. Cort et al successfully used two premixes to fortify wheat flour; the first comprised vitamin A, pyridoxine, folic acid, tocopherol acetate, thiamine, riboflavin, niacin and iron while the second contained calcium, magnesium and zinc. The vitamin/iron premix demonstrated excellent stability on storage. The flour enriched with both premixes also demonstrated excellent stability on storage at room temperature. Under conditions of accelerated storage at elevated temperature (45 °C), however, there was substantial loss of vitamin A beyond 4 weeks of storage. Parrish et al. also reported good stability of enriched wheat flour stored at room temperature, but about 50 per cent losses in flour stored at 40 °C for 6 months.

The form of vitamin A most commonly used in the fortification of flour was dry stabilised vitamin A palmitate (type 250-sd) powder form. The water soluble vitamins (thiamin, riboflavin, niacin, pyridoxine, folate and calcium pantothenate) are used in pure crystalline form. The mononitrate salt of thiamine is preferred for this use. Iron is normally used in its reduced elemental form.

In the work of Parrish et al. ferrous sulphate was also used as the source of iron and there was no significant difference demonstrated between the flour enriched with ferrous sulphate and that enriched with elemental iron. It has been recommended that ferrous fumarate be used in the fortification of wheat flour within the P.L. 480 programme, due to its greater bioavailability. When mixing dry fortificants with dry foods, careful selection of the physical characteristics of the fortificant compound is important so as to ensure adequate mixing and to minimise segregation on storage.

Rubin et al. investigated the stability of bread made from flour enriched by the 6 vitamins and 4 minerals described by Cort et al. They found that the inclusion of calcium and magnesium adversely affected vitamin A retention during the baking process. There was also an off-flavour detected in bread which included magnesium after 5 days of storage. Crumb colour and grain were also negatively affected by the nutrient multimixes tested by Rubin et al., with the greater affect occurring with the nutrient mix which included magnesium.

Fortification of bread with zinc salts exerted no adverse effect on loaf volume, flavour or any other index of bread quality. Iodisation of bread has been carried out in the Netherlands and Tasmania by the addition of 2-4 ppm KIO_3 to the bread improver which was already in general use. KIO_3 has been used in bread production in the past, not as a fortificant but as an oxidising agent to improve dough quality. The fortification of corn meal with 6+4 vitamin/iron mixture was shown to be technically feasible. Losses of vitamin A on storage at room temperature for up to six months were greater than with wheat flour but still remained below 20 per cent. In many countries pasta or noodles are commonly eaten and these can therefore be important vehicles for fortification. The manufacture of such products involves the production of a dough which is then extruded and dried.

Enrichment can be through the use of enriched flour or alternatively, wet addition of a dispersion of the required vitamins can be carried out at the dough-making stage. Vitamin losses during production depend largely on the drying conditions employed. Dexter et al. enriched durum wheat flour with a vitamin mixture containing riboflavin, thiamine mononitrate and niacin for the production of spaghetti. They reported that high temperature drying treatments resulted in significant losses of riboflavin, whereas the other vitamins were stable to the processing conditions. In all cases cooking losses, estimated between 40 - 50 per cent, exceeded those experienced during processing.

BREAKFAST CEREALS

The fortification of ready to eat breakfast cereals is a wide-spread practice. In the United States the general practice has been to provide 25 per cent of the RDA of thiamine, riboflavin, niacin, pyridoxine, folate, ascorbic acid and vitamin A per 1 ounce. 10 or 25 per cent iron, up to 20 per cent RDA for calcium have also been added; vitamin D has also sometimes been included. Surveys have indicated that fortified breakfast cereals play an important role in ensuring nutritional adequacy of the diets of the U.S. population.

The minerals and the more heat stable vitamins like niacin and riboflavin have been added to the basic formula mix prior to processing. The heat labile vitamins such as vitamins A, C and thiamine are usually sprayed onto the cereals after the high temperature processes such as oven or extruder. According to Johnson et al. two important considerations in the design of the spray system are:

- To ensure uniformity in spray coverage
- To optimise protection of both vitamins A and D.

As the hydrophobic vitamins cannot form true aqueous solutions, they have been sprayed on in the form of an emulsion. The composition of the emulsion has been shown to greatly affect the vitamin stability. The inclusion of 15-20 per cent sucrose into the spraying emulsion stabilised the vitamin A. It acted by forming a protective coating around the vitamin on drying. This fact explained the greater stability of vitamin A included in raisin bran as compared with cornflakes. In the latter case a no-sugar spray was used. Vitamin E can be added to the spray emulsion, it was at least as stable when added under these conditions. Good storage stability of the fortified breakfast cereals was reported. This was partly due to the fact that packaging materials with appropriate barrier properties have to be selected to ensure the maintenance of equilibrium moisture contents below about 5 per cent. Above this level there was textural deterioration of the product which was perceived as 'staleness'. Lund reported almost 60 per cent loss of vitamin C from raisin bran on storage for a 12 month period. This was ascribed to the fact that the moisture levels in this cereal were maintained at a higher level to prevent hardening of the raisins. Improved procedures for the coating of raisins with a moisture barrier material prior to mixing with the material can minimise this problem. The addition of ascorbic acid was found to stabilise vitamin A to a greater extent than did phenolic antioxidants. The addition of tocopherols did not provide any additional stability, possibly because they were already present at optimal levels.

NEW RESEARCH DIRECTIONS IN AGRO-FOOD STUDIES

At the turn of the millennium, even casual acquaintance with the media in advanced capitalist economies reveals the palpable unease and mistrust enfolding the nature-society coproductions more conventionally known as agro-food systems. This unease is more acute in Western Europe, where cases of "extreme food events" and the systemic breakdown of food provision, particularly in livestock production, have occurred with disturbing frequency in recent years.

A litany of these "extreme food events" would include mad cow disease in Britain and its pandemic translation throughout Western Europe in 2000–2001, episodic yet recurrent food-contamination scares, the Belgian dioxin crisis, and 1999 reports of untreated sewage, septic tank residues, and slaughterhouse effluent being used in several animal feed-processing plants in France. Nor is food safety the only register of public disquiet, which would give the entirely misleading impression that this mistrust can be rectified by appropriate regulatory measures and better risk management techniques.

Mistrust of industrial food-provisioning, at least in Western Europe, also reflects ethical opposition to the environmental harms wrought by industrial agriculture and intensive-confinement livestock practices, and fears that the centralizing and homogenizing forces of agro-food globalization are threatening the material and symbolic content of foodways, which are potent bearers of cultural identity. To adapt Jean Brillat-Savarin's aphorism, there is unease about what we eat, how we produce it, and what it means for what we are and might become.

More than ever, food is a signifier for political, social, and cultural struggles over the metabolic reciprocities between nature and society, which are the material and discursive metrics of everyday life. As Daniel Miller and others have realised, personal choices about food can give voice to socioecological commitments whose cumulative expression in biopolitical activism potentially can change the way we live in the world.

Agricultural biotechnologies (ABTs) and genetically engineered organisms contribute to the general disquiet over food provisioning in distinctive ways, exacerbating the instinctive anxieties of the omnivore's paradox, already fully aroused by the events recounted above. ABTs sound the alarm on virtually all registers: food safety, environmental harms, and the further concentration of economic power over the food supply—that is, power over our habitual metabolic interactions with agricultural nature and, hence, over the material and cultural expressions of our corporeal identity.

At this fundamental level, extreme food events and novel, genetically engineered foods create unease because of what they reveal about society's relations with nature and their possible transformation. Starting in Western Europe and some developing countries, but now gathering momentum in the United States and Canada, social mobilization against ABTs and genetically engineered foods is manifest at different scales, from the street protests against the World Trade Organization in Seattle and nongovernmental organizations' efforts to influence the regulation of the Convention on Biological Diversity to the destruction of test sites of GE crops and consumer movements to deny shelf space to genetically engineered food products in supermarkets.

ABTs and the new realities they portend are now being interrogated on a radically more comprehensive scale than at any time over the past two decades. This interrogation already has successfully exposed points of weakness and vulnerability in this technoscientific enterprise: its life-sciences business model has been summarily abandoned, the material and discursive claims of the technology have been called into question, and it is losing ground in national and international regulatory struggles.

In short, the commercial deployment of ABTs and the threatened ubiquity of genetically engineered foods have opened a new front of biopolitical mobilization. Here, if only incipiently, spaces for an ecological politics and forms of social organization are emerging that reject modernist instrumentalist relations with nature.

With this background, I offer some reflections on the main theoretical approaches to ABTs in agro-food studies, as well as discussion of critical engagement with the "new" biopolitics of agriculture and food. For present purposes, the field of agro-food studies is identified with the "critical" rural sociology and the "new" political economy of agriculture, which emerged from the later 1970s. The imprint of classical Marxism and "agrarian question" problematics is still discernible in the agro-food studies literature on ABTs.

This legacy is particularly evident in the conceptual primacy of the labour process and the consequent privileging of production-centreed analytics. Within this conceptual armature, I examine several subthemes that elaborate the "vectors of incorporation" of ABTs by private capitals. Although the labour process-commodification perspective remains preeminent in the agro-food studies literature on ABTs, several contributors recently have begun to explore the advantages of the nondualist, relational ontology of actor-network theory in understanding the new socioecological constellations of human and nonhuman entities associated with agricultural technoscience. These recent developments are reviewed in the second section. A concluding section notes several lacunae in the literature and considers future directions of research on ABTs.

AGRI-BIOTECHNOLOGIES AND AGRO-FOOD STUDIES

Until quite recently, analyses of ABTs in agro-food studies were firmly rooted in the agrarian question problematics and deductivist epistemology of classical Marxism. This importation typically was mediated by neoMarxist debates in development theory and peasant studies, with their focus on agrarian transition and the fate of Third World peasantries as commoditization processes intensified.

In these classical agrarian question problematics, inflected by Third World debates, family-labour forms of production and the immediate labour process constituted the key units of analysis. Theoretical trajectories in the new agro-food studies thus were strongly informed by the recovery of the classical Marxist tradition and embraced its production-centreed analysis of agrarian change. These epistemological foundations and problematics were not seriously interrogated until the later 1980s.

Even so, the labour process, a cornerstone of Marxist political economy, with its embedded ontological and epistemological priorities, has retained an unexplained place in putatively revised, poststructuralist and actor-oriented approaches in agro-food studies.

These comments are a rather circuitous way of recognizing the continuing preeminence of the labour process as a meso-level organizing concept in agro-food studies. The corollary is that social and technical relations of production are privileged analytically. These legacies are easily detected in those "first generation" approaches to ABTs, which attempt to furnish a general framework of analysis.

Although ABTs are represented as a new technoscientific paradigm, the analysis is anchored in the agricultural labour process and its transformation via commodification of seed production and plant breeding. In an italicized passage, Jack Kloppenburg stresses that "the seed, as embodied information, becomes the nexus of control over the determination and shape of the entire crop production process."

The wider structural implications of ABTs in terms of industrial appropriationism and substitutionism, or institutional change and new property forms, for example, are similarly ushered through the privileged gateway of production. This obligatory passage-point into the circuits of capital also is seen in the preoccupation with industrial concentration, reflecting the redistribution of power towards upstream farm-input sectors as agrochemical and pharmaceutical companies take up dominant positions in ABTs and the seed industry. Broadly speaking, this labour process-commodification perspective has provided the general, overarching analytical framework of choice for addressing ABTs in agro-food studies. The "first generation" analyses approached ABTs by problematizing the boundaries between agriculture and industry. This analytic move provided the basis for more systemic and historically informed studies of agroindustrial development by drawing attention to the contingent nature of this division of labour. Nevertheless, the farm labour-process is taken as the privileged locus of the transformative contradictions generated by agrotechnoscientific innovation.

The work of Kloppenburg, for example, is first situated generally on the classical Marxist terrain of agriculture as a "recalcitrant sector" and then takes up the specific theme of the vectors of capitalist penetration of public plant-science and its gradual institutional reconfiguration as "capitalist property".

The commodification of the seed is conceptualized as a process of primitive accumulation, a process whose highlights are the innovation of hybridization—which "functions to uncouple agricultural producers from the autonomous reconstitution of their own means of production"—politically driven shifts in the institutional division of labour between public and private plantbreeding, and changes in intellectual-property regimes to facilitate the private appropriation of plant genetic resources. Kloppenburg's purpose is to explain how the seed, "which is perhaps the element of agricultural means of production most central to the entire farm production process", is commodified and becomes a capitalist force of production. Agricultural biotechnologies are emblematic of this historical trajectory because "what is now occurring in the seed sector is one instance of a much broader technological transformation that is galvanizing changes in the social organization of all production processes in which organic substances or life forms play a significant role".

My colleagues and I apply the commodification approach to on-farm means of production more widely (implements, motive power, nutrient cycles, pest control, seeds, energy) and argue that the biophysical processes of agricultural production and food consumption have constituted natural, though historically contingent, constraints to the industrialization of agricultural use-values.

Unable to reproduce natural production processes fully by direct transformation, industrial capitals have adapted in singular ways to the sectorally differentiated properties of agricultural nature (biological time, photosynthesis, land, climate) and the physiology of human nutrition. These differences, we argue, are analytically significant as a major source of variation in the historical dynamics, specificities, and contemporary configurations of social production in agro-food commodity networks.

The concept of appropriationism is used to designate the historically discontinuous, piecemeal "but persistent undermining of discrete elements of the agricultural production process, their transformation into industrial activities, and their re-incorporation into agriculture as inputs". Elements of natural production processes are progressively internalized by industrial capitals via proprietary science and technology as individual sectors of capitalist accumulation and reproduction.

In brief, in agriculture, where industrial capitals confront a natural production process, agricultural biotechnologies constitute "a generalized advance in the capacity of industrial capitals to manipulate nature". Whether in explicit or implicit terms, the framework adopted by my colleagues and I and by Kloppenburg (1988) posits a contingent, ongoing *tension* between agriculture as a recalcitrant sector and its full assimilation by industrial capital.

Though with differing emphases, these studies broadly explore the cumulative effects and convergence of several interdependent processes of incorporation—cognitive, technoscientific, economic, and regulatory—that progressively extend the commodity form to new spheres of the farm laborprocess. In this context, the technoscientific paradigm constituted by biotechnology is identified as the catalyst of assimilation.

Nature is transmuted into a force of production. This reconfiguration represents the vector of "domestication" of recalcitrant biological processes, hitherto inaccessible to technoscientific manipulation and the reductionist endeavors of industrial capitals. In this framework, biotechnology, actually or potentially, has swept away the *technological* foundations of the recalcitrance or exceptionalism of agriculture.

For example, we formulated the basic question of the agricultural labour process-commodification approach in the following terms: "If biotechnology represents a qualitative breakthrough in that nature can now be reconstituted industrially, does that mean that the food system is open to assimilation within the broader transformations of the industrial system?".

The answer we gave at that time could not be clearer; the dichotomous tension is sundered definitively. "In this perspective, biotechnology ... markfs the end of the pre-history of the food industry and its incorporation within the broader dynamics of the industrial system and post-industrial society".

However, with what now seems fortunate prescience in the light of contemporary biopolitical activism, we suggested that "The frontiers of substitutionism are likely to be defined as much by consumer tastes and loyalty to organic whole foods as technical and engineering constraints".

VECTORS OF INCORPORATION

The labour process-commodification approach clearly has furnished the preferred analytical framework for research on ABTs in agro-food studies. A recent critique and extension of "first generation" analyses attests to the continuing influence of this approach. This same theoretical perspective also frames a number of subsidiary research themes, which emerged in the 1980s and early 1990s as social scientists came to grips with the wider implications of ABTs.

A full account of these complementary studies, whose scope ranges from the institutional intricacies of regulatory change to the ethics of genetic engineering, exceeds the more limited purview of this essay.

However, with some qualifications, these subsidiary studies follow the same analytical threads Kloppenburg (1988) traces in his history of plant improvement: that is, "the commodification of the seed, the changing division of labour between public and private research institutions, and the appropriation of plant genetic resources". A particularly rich vein of scholarship on ABTs is devoted to the constellation of political-economic forces that determine the shifting demarcation line between basic and applied research, and the corresponding realignment of the institutional division of labour between public- and private-sector research.

The changing division of labour in the U. S. agricultural research establishment, and notably the institutional "capture" of the land-grant university system, dominated work on ABTs in the 1980s. These studies grappled with the political-economic and ideological issues raised by the mounting evidence of corporate penetration of American research universities, where "biotechnology was born". As Edward Yoxen puts it, the capitalist incorporation of molecular biology has reached the point where "the industrial exploitation of recombinant DNA research by corporate capital can be serviced directly from the academic research laboratory".

With remarkably few exceptions, the agro-food studies literature has been reluctant to follow Yoxen's lead and venture seriously into the history of science. However, the emergence of the "universityindustrial complex" and the vital contribution of academic scientists to the nascent biotechnology industry are explored fully in Martin Kenney's outstanding and prescient book *Biotechnology: The University-Industrial Complex*.

This study crowned a wave of related papers investigating the emerging social division of labour in agricultural biotechnology and plant-breeding research. Much of this work on institutional change is associated with Fred Buttel and his colleagues, then in the Department of Rural Sociology at Cornell University. The rapidly changing matrix of agricultural research policy, scientific practices, and public research institutions and the international reach of agricultural biotechnologies also distinguish the scholarship of Lawrence Busch, Bill Lacy, and their collaborators in this same period.

A second strong theme of agro-food studies research on ABTs in the 1980s focused on patterns of innovation and industrial concentration, and their impacts on agricultural production and rural social structures. In a pioneering paper, Jack Kloppenburg (1984) assesses the prospective structural consequences of ABTs by extending the historical tendencies already observed in the development and diffusion of hybrid corn. As in this earlier case, "Biotechnology, too, promises to create a vast new space for the accumulation, concentration, and centralization of capital".

This Schumpeterian notion is reinforced by other analytical foci, including the capitalist subordination of public agricultural research and plant breeding, the development of herbicide-resistant crops, the loss of farmer autonomy, acceleration of the technological treadmill, and continued genetic erosion. If the concentration of intellectual-property ownership in corporate hands is added to these foci, a remarkable continuity emerges between the scholarship of the early 1980s and the contemporary political-economic analysis of ABTs.

A further parallel can also be drawn with current activist opposition to the rising corporate control of global agro-food systems, opposition that has captured public attention since the street battles at the 1999 WTO meetings in Seattle. Other contributors approached the socioeconomic impacts of ABTs by examining innovation patterns, typically through the prism of new opportunities for accumulation, the implosion of sectoral barriers to entry, and the industrial reorganization and concentration in the agro-food system as a whole.

These studies focus on the upstream and downstream industrial sectors, notably the chemical-pharmaceutical complex, to draw out the implications of ABT innovation for farming and for social-production relations in farming. For example, John Wilkinson and I suggest that the generic capacity to engineer living organisms "prefigures a new bio-industrial paradigm". "Biotechnologies now threaten to *implode* the long-standing organization of the food system around specialized commodity chains....

There is the capacity simultaneously to relocate agricultural production in factories and industrial production in fields". Although countervailing tendencies are likely to set up tensions between alternative models and trajectories, we argue that ABTs will accentuate the shift towards a more demand-oriented agro-food system.

Biotechnology as the generic catalyst of wider impulses towards bioindustrialization also underpins projected scenarios of change in the social organization of agriculture. These processes include the amplification of the observed trend towards large-scale, intensive, continuous production systems and the introduction of all-purpose agricultural "biomass refineries." Such developments are predicted to lead to greater concentration of land ownership and more widespread contract production and part-time farming.

From the early 1980s, innovation studies with a more action-oriented policy research perspective began to appear in the agro-food studies literature. These studies were galvanized by the field testing and approaching commercial adoption of genetically engineered organisms, which gave concrete expression to concerns about environmental safety, rural structural change, and demands for regulatory processes with greater democratic participation.

Struggles over the licensing of recombinant bovine growth hormone (rBGH) crystallized these issues acutely in the United States. The rBGH controversy also exposed the conflicted politics of ABT research in the U. S. land-grant university system and other public institutions involved in the innovation process. A third and continuing strand of agro-food studies scholarship on ABTs examines the political-economic repercussions and contested politics of changing intellectual-property regimes in plant breeding and genetic resources.

As in the case of rBGH, the academic literature in the 1980s was strongly informed, if not led, by parallel activist contributions, notably from members of RAFI, or the Rural Advancement Foundation International. These interactions have intensified since the 1992 Earth Summit in Rio as part of the ongoing struggles to shape the emerging supranational institutions of global environmental governance, and notably in this context, the Convention on Biological Diversity (CBD) and the Cartagena Protocol on Biosafety, which regulates international commerce in living modified organisms.

Contributors in the 1980s quickly grasped the import of the landmark rulings that extended and consolidated property rights to microorganisms, plant germplasm, and rDNA techniques and other processes involved in plant genetic research. The dramatic institutional shift to allow the cornmodification of life forms—"ownable artifacts, whilst also being a part of nature" — has been analysed from various angles in agrofood studies.

Several accounts trace the institutional origins and evolving structures of intellectual property rights in plant breeding before 1980, when the U. S. Supreme Court's decision in *Diamond v. Chakrabarty* overturned the "product of nature" principle, which had earlier dominated this field of patent law. Other scholars surveyed this same terrain but from specific institutional standpoints to evaluate the implications of these new legal and social conventions for particular groups of actors. This scholarship, for example, investigated university-industry relations and the public and private communities of agricultural scientists and research administrators.

However, in agro-food studies, the issues arising from the patenting of life forms were explored most vigourously at the international level. Academics and activists exposed the institutional dimensions of NorthSouth power asymmetries and the inequalities embedded in the governance structures of international agricultural research, such as the Consultative Group on International Agricultural Research (CGIAR) system, and the associated collection and transfer of germplasm as "common heritage" resources. Jack Kloppenburg deserves much credit for bringing the "North-South seed wars" and questions of political-economic control over international plant genetic resources into agro-food studies, and other scholars soon followed his lead.

More recently, issues of access to and ownership and conservation of international genetic resources have faded from prominence in this literature. This decline, paradoxically, has coincided with the rising salience of these questions on national and international policy agendas as Greenpeace, Friends of the Earth, and other leading environmental groups have reinforced the longstanding campaigns of more specialized, agrienvironmental organizations, such as RAFI and Genetic Resources Action International (GRAIN).

Although incomplete, this survey of the earlier literature on ABTs reveals some strong continuities with contemporary scholarship, in terms of both theoretical perspective and more specific concerns, such as genetic erosion, decline of farmer autonomy, and centralized control of agro-food networks on a global scale. Discursive continuities also persist, despite Fred Buttel's suggestion that the days of extravagant rhetoric characteristic of biotechnology in "its formative years, roughly the mid-1970s to late 1980s," are now past.

In this period, Buttel argues, virtually all the protagonists—leading molecular biologists, venture capitalists, small startup firms, government officials, and environmental and public-interest groups—embraced it as epoch-making and revolutionary. "As late as the end of the 1980s, most academic, policy and activist treatments of biotechnology were essentially agreed on its magic bullet character and transformative potential".

At the turn of the millennium, the implication that this viewpoint is no longer shared seems premature, even if more pragmatic assessments can increasingly be found. The tropes of these formative years are still actively deployed in the rancorous exchanges between supporters and opponents of ABTs. For each group, the discursive starting point remains the promise of the technology: golden rice versus Frankenfoods, precision breeding versus superweeds, and so on. A brave new socionatural world is on offer, which, if realised, would empower the networks of human and nonhuman entities involved in its construction.

Some recent work on technoscience and relational ethics in agro-food studies, particularly extensions of actor-network theory, seek to elucidate such biopolitical choices by adopting an explicitly nondualist framework of analysis. These new approaches represent a reflexive theoretical project to overcome the abstraction of nature in mainstream agro-food studies by interrogating its modernist ontological foundations.

NEW PERSPECTIVES ON AGRO-TECHNOSCIENCE

Although the labour process-commodification perspective retains analytical appeal, it is clearly a product of its times. As Kloppenburg observes, "The model of change that emerges from this analysis is fundamentally dialectical—the forces and relations of production are mutually conditioning". Human praxis is ontologically central, and nature is firmly located on the opposite side of the modernist divide.

At best, nature is endowed with certain qualities of "resistance," which are identified, for example, with "biological barriers" to capital and "secrets of life" that are inaccessible to science. However, in this characterization, nature is figured as a passive entity whose latent properties will be revealed only if it is *acted upon* by industrial technoscience. Following the Enlightenment antinomy between nature and society, nature is other, which industrial society is driven to subjugate and appropriate.

As depicted in *From Farming to Biotechnology*, the agricultural labour process has indeed presented historically contingent biophysical limits to industrial appropriation and substitution. Yet, although nature and industry are drawn as oppositional categories, in analytical terms, industry is the only actor. This unexamined ontological choice precludes the conceptualization of nature as a lively, active, and formative presence and so *underplays* what arguably are the truly revolutionary material dimensions of biotechnology.

Unwittingly and implicitly, with this choice the analysis is captured by the *engineering* metaphor of life propagated by industrial capital, with its omnipotent technocratic discourse of precision, efficiency, and benign evolutionary improvement.

This observation is not intended to deny the incisiveness of the labour process framework as an analytical vantage point to address social *cui bono* issues and rural structural change as nature is reconfigured as a productive force. Clearly, such a denial would fly in the face of historical and contemporary evidence of the salience of production as a terrain of theory and praxis. Nevertheless, the ontological critique of late post-industrial capitalist political economy can fruitfully be extended by bridging the modernist divide between nature and society in order to imagine and construct alternative socioecological worlds.

As already noted, agency in "first generation" approaches to ABTs is uniquely identified with human intentionality and human action on an objectified, but now less recalcitrant, nature. This instrumentalist ontology, with its purified categories of nature and society, is closed to notions of the relational materiality of nature, offers no theorization of the lively entities emerging as social partners in these technoscientific practices, and fails to entertain nonhuman perspectives and shared consequences.

An alternative conceptualization of nature, one that attributes active properties to nonhuman entities, would focus analytical attention on the interrelational negotiation of new socionatural realities, which are now being constituted with the commercial diffusion of genetically engineered crops in the United States.

Such an approach also would make explicit the ethical, environmental, and political choices associated with different human-nonhuman assemblages and "socio-ecological projects". However, as argued elsewhere, mainstream agro-food studies remains transfixed by the modernist ontology. Encouragingly, there are recent signs of a reflexive "turn" to give analytical salience to "the status of nature" and to interrogate the modernist divide.

AGRO-SOCIAL NETWORKS

The analytical richness and potential of actor-network theory (ANT) in agro-food studies have been expressively revealed in these reflexive explorations. However, apart from an illustrative vignette, these analytical resources so far have not been deployed in work on ABTs. This is a major omission in agro-food studies, and efforts to rectify this position deserve high priority on the ABT research agenda.

Lawrence Busch and his colleagues in a recent series of papers on the rapeseed (canola) commodity sector have undertaken the most systematic work on agricultural technoscience from an actor-network perspective. This impressive research elucidates the endogeneity of technical change and the processes by which knowledge-commodity transformations become ecosocially embedded. ANT is used to analyse the rise and institutionalization of rapeseed technoscience by conceptualizing technological innovation as a process of network building. Thus, instead of considering technology as an exogenous factor to the system of commodity production, the network approach does not differentiate between human and nonhuman elements. In such analyses, society does not transform technology, nor does technology cause the transformation of society. Rather, "the very actor-networks ... simultaneously give rise to society and to technology."

Moreover, Arunas Juska and Lawrence Busch find ANT to be particularly helpful in uncovering points both of analytical entry and of political action in cases where, as with ABTs, processes of network formation, extension, and reconfiguration are dynamic and in flux. In this perspective, "it becomes evident that the relationships mediated through the network are contingent in nature: they can be disrupted; they can collapse; they can be organized according to different principles; they can be constantly changed and renegotiated".

In a second paper, Busch and Keiko Tanaka (1996) deploy the concept of symmetry, which ANT extends to nonhumans, to give "an explicit portrayal of non-human actors over time and space" in "the complex networks known ... as commodity chains", such as rapeseed (canola). The product grades and standards constitutive of commodities represent "rites of passage" for both nature and people.

"Thus, by transforming nonhumans and subjecting them to multiple rites of passage, we coproduce nature, society, [and] the capitalist market". The symmetry of these qualification processes also extends to their ethical dimensions, which infuse these modes or "rites" of mutual socialization of humans and nonhumans. In a later analysis of the globalization of rapeseed networks, Busch and Juska contend that the political economy of agriculture perspective and its "categorical apparatus" are "inadequate" to the task at hand.

The political economy approach is underspecified because "in true Baconian style, nature is recast as resources to be transformed", as passive. In contrast, the political economy of actor networks reveals how the strategic positions of human and nonhuman actors and the geometry of power in networks are negotiated, modified, and transformed temporally and spatially.

A more recent paper investigates these questions in finer historical detail by examining the reciprocal and contingent relationships between agricultural research and agricultural production, and the changing strategic positions of nation-states as the rapeseed sector has become more globalized since the mid-1970s.

This brief review of rapeseed technoscience provides some indication of the purview and analytical purchase of ANT. With its symmetrical ontology and method, ANT offers a penetrating conceptual repertoire for the analysis of ABTs. Lively eco-social coproductions and category fusions of the natural and artifactual are particularly in evidence in the everyday practices of agro-food networks in an era of genetically engineered organisms.

Thus Bruno Latour extends an invitation to reject the purified categories of "society" and "nature" and focus instead on "the blind spot where society and matter exchange properties". These transactions are brought centre stage in order to expose the inescapable sociomateriality of the entities mobilized into the heterogeneous associations that hold society together.

The constant interchange of human and nonhuman properties in network formation has created "mixtures between two entirely new types of being, hybrids of nature and culture". These processes of mediation, in turn, constitute the foundation of modern technosciences, which "multiply the non-humans enrolled in the manufacturing of collectives and ... make the community that we form with these beings a more intimate one".

Rather than engage in a lengthy exposition of ANT, its merits and limitations, the point to emphasize is that we have a nondualist, relational, and processual framework in which nonhumans are actively present, performative, and consequential. In this significant respect, although otherwise confessedly modest in its claims, ANT challenges the silences and abstractions of production-centreed analyses.

That is, it furnishes an ontology and conceptual language with which to address the implosions, natureculture hybrids, and newly socialized intermediaries emerging from the heterogeneous engineering practices of agro-food technoscience. ANT's terms of engagement and insistence on "the permeable boundary running between humans and non-humans" resonate strongly with the ethical and relational concerns of biopolitical activism.

An ethical standpoint, a relational moral philosophy, is discernible in ANT's framing of the construction of our world, its insistent reference to crossovers and the exchange of properties between human and nonhuman entities. As Jonathan Murdoch notes, actor-network theorists "force us to look afresh at the categories, divisions and boundaries that frequently divert our attention away from the nonhuman multitudes which make up our world".

This imperative recalls Latour's analysis of Louis Pasteur's research on the anthrax bacillus. That is, the bacilli, once "translated" from the natural competition found in the farmyard to Pasteur's laboratory in Paris, encounter a new environment, an altered state, where they can thrive.

Following Latour, Elizabeth Bird observes that "In the laboratory, nature—in the form of Pasteur's "natural-technical object"—becomes an actor negotiating a *new reality*. In the terms of that context, the microbes become actors in shaping a new environment" (258, original emphasis). Latour continues, "Training microbes and domesticating them is a craft.... Once these skills have accumulated inside laboratories, many cross-overs occur that had no reason to occur anywhere else before".

Similarly, when analyzing Pasteur's discovery of the microorganism responsible for lactic fermentation, Latour draws attention to the change in ontological status that this step involved and observes that "in his laboratory in Lille Pasteur is *designing* an *actor*". The analytical parallels and possibilities of extending this approach to the transgenic "natural-technical objects" of ABTs—such as Roundup Ready soybeans and Bt corn, not to mention the unintended progeny of horizontal gene transfer—are clear.

In a brief vignette on ANT, I suggest that agri-biotechnological innovation can be analysed in terms of the practices deployed to reconfigure or "translate" existing agro-food networks by enrolling genetically engineered organisms (GEOs) and foods. In effect, corporate networks, such as Aventis, DuPont, and Syngenta, are seeking to displace previous socionatural orderings and to realign agro-food networks in ways that support and "naturalize" the diffusion of GEOs into rural environments, crops, and animals.

Thus, "agri-biotechnologies introduce new mediators into the intimate corporeal relations of agro-food networks, promising new corporealities and, quite literally, new bodies". Biopolitical mobilization, notably in Western Europe, has raised material and ethical concerns against this reordering of agro-food networks in marshaling opposition to the environmental release of GEOs and the incorporation of genetically engineered foods into human bodies.

This mobilization directly challenges the industrial, technoscientific problematization of ABTs, variously framed as the answer to world hunger, as an improvement on nature, or in terms of the inevitability of technical change.

In this modernist instrumental perspective, nature is objectified as a field of resources awaiting domination and exploitation by the relentless advance of technoscience. Against this dualist rationality, green biopolitical activism is informed by a relational ethics and by precepts of shared community.

These biopolitics of ABTs reveal the "clash of divergent ontologies" provoked by struggles to realign agro-food networks into new socionatural orderings. By undermining the modernist ideology of nature as externalized and objectified, ANT provides theoretical resources to address nature and its lively materiality directly. This attention to how "socionatures" are constructed broadens critical engagement with capitalist technoscience and political economy and informs our understanding of the heterogeneous associations fostered by this ordering of the socioecological.

The dimensions of this political space are amplified if we interpret ANT as an ethical discourse of how to live in the world. This perception of ethical, as well as material, embeddedness speaks directly to the problematics of biopolitical activism, as noted above.

Even a cursory understanding of ABTs brings recognition of the new socionatural assemblages emerging, under the aegis of capitalist technoscience, to construct new worlds. In turn, this recognition suggests the need for conceptual frameworks that explicitly bring normative judgement and political critique to bear on these new collectives and respond to Donna Haraway's interrogation of technoscience: how, for whom, and at what cost? Acknowledgment of our partnership in these human-nonhuman assemblages would be an initial step towards the development of forms of social organization that encourage democratic choice between alternative orderings and the worlds they bring forth.

For these purposes, we need ontologies that reveal, not abstract, our interactive, relational production of worlds we inhabit with others. The novel socionatural assemblages of capitalist agro-food technoscience, which herald new actors and new realities, underline the significance of theoretical and political choice.

In this re-encounter with agro-food studies research on ABTs, the theoretical and thematic continuities of this literature emerge insistently. Indeed, in theoretical terms, there is a powerful sense of deja vu, of involution even, and alternative perspectives remain firmly on the margins. I emphasized both the primacy of the labour process commodification approach and its ontological limitations for comprehending the new socionatures constituted by ABTs.

This framework constrains efforts to find common ground with the relational ontology and moral economy that inform the biopolitics of environmental movements and Green activists. Such serious limitations, in short, lend urgency to theoretical renewal in agro-food studies and, more generally, to the "greening" projects and explorations of Red-Green rapprochement now under way in various fields of critical social theory.

These projects unmask the political agendas and instrumentalist ethics imbricated in modernist ontological antinomies and their reification. At the risk of repetition, the agro-food studies literature on ABTs has as its analytical focus the ensemble of institutional processes and social relations that have led to the commodification of nature or, as Yoxen (1981) has it, to "capitalizing life".

This labour process conceptualization, with its emphasis on the subsumption of nature and its manipulation as a productive force, sees this transition exclusively from the standpoint of the social. Nature is subsumed by purposive social agency, whose dynamic is to be found in the laws of accumulation and social relations of the capitalist mode of production. There is no place here for the relational materiality of nature, its liveliness, or its "boomerang" qualities.

This framework does not entertain either notions of natural-social coproductions or the consequences of these assemblages for entities with whom we share this world. The labour process perspective, in short, does not lend itself to an assured engagement with the new constituencies of agro-food biopolitics. However, "this is not to dismiss the strengths of this perspective.

Rather, it is ... to observe that this theoretical lens or 'framing' device does not focus directly, for example, on the new socionatural relations, interspecies metabolisms and exotic corporealities unleashed by agricultural biotechnologies.... these new constellations or assemblages of nature-society relations are key catalytic elements of bio-political activism in agro-food networks".

This review also has drawn attention to thematic continuities, although clearly the scale and intensity of social mobilization have grown sharply with the accelerating deployment of ABTs. These thematic links are amply reflected in this volume. Issues of governance, although not known as such in the 1980s, form a recurrent body of concern. Two themes, already identified previously, have retained particular salience.

The first theme concerns the ways in which ABTs heighten the concentration of industrial control in agro-food systems—"a global oligopoly" in William Boyd's estimation. Exploring the "deep structures of monopoly" within a commodification of nature framework, Boyd extends the analysis of ABTs into the era of life sciences multinationals, genomics, and the competitive imperative to capture value by integrating vertically from proprietary intellectual-property platforms to seed marketing and contract farming.

A social constructivist-commodification perspective also frames the contributions of Scott Prudham and Dennis Kelso to this collection. Following Richard White, Prudham adopts the metaphor of nature as an organic machine to describe the transformation of living organisms into technologies and commodities. Within this framework, he is particularly concerned to track the harnessing of public science to private innovation in the development of forestry biotechnology, the concomitant restructuring of university-industry-state relations, and how this trajectory is shaped by the specificity of recalcitrant nature. In warning against technological fetishism, Prudham insists that biotechnologies be seen not as things but as bundles of social relations with historical lineages in order to emphasize that technological change is socially produced.

Although Kelso's analysis of the deployment of biotechnology in commercial salmon aquaculture broadly fits with the labour process-agrarian transition problematic of agro-food studies, his main concern is to reject its technological determinist inflections. This endeavor brings Kelso into closer engagement with issues that have been strangely muted in this literature until recently and especially with the sources, forms, and resources of social resistance to biotechnological innovation.

Kelso's account of transgenie salmon embraces not only mobilization around questions of food safety and environmental risk but also the defensive stance of salmon farmers against the perceived technological threat to the aquaculture industry as presently constituted. In elaborating these questions, Kelso draws attention to the politics of scientific uncertainty and state regulation, highlighting the strategic importance of discursive struggles to form public perceptions of nature and the natural.

This discussion, together with the chapters by Julie Guthman, Frederick Buttel, and Rachel Schurman and William Munro, begins to address important lacunae in the agro-food studies literature. The second thematic grouping is around governance issues and focuses on continued First World-Third World tensions over access to genetic resources and the perceived asymmetries of power articulated by intellectual property rights regimes and, notably, the 1994 WTO TRIPS accord.

These ongoing tensions find expression in the contested politics, shifting alliances, and arcane regulatory processes of new multilateral institutions of global governance—the WTO and CBD—and their disputed mandates in the conjoined policy arenas of international economics, trade, and environment. Environmental governance emerges in another contemporary guise in the chapter by Astrid Scholz.

Her analysis traces the vagaries of the utilitarian rationale for biodiversity conservation as the importance of natural product screening in the R&D strategies of transnational pharmaceutical corporations waxes and wanes, exacerbating the inequities of power embedded in private bioprospecting agreements.

The contribution by Frederick Buttel similarly can be situated within the global governance thematic; however, like Kelso's chapter, it also provides a bridge to some neglected issues. Buttel's wide-ranging account encompasses the institutional architecture of the globalization regime, the power brokering of the GATT-WTO transition and the "spoiling" role of agro-food issues—festering EU-U. S. "food wars" (bananas, beef, GEOs), unilateral trade sanctions, the "environmentalization" of ABTs, the global farm crisis—in revealing potential fault lines and pressure points in this accumulation regime.

Taken in the aggregate, these individually contentious issues have cumulatively resonated with growing force in political, cultural, and institutional domains. Whether or not GEOs prove to be the Achilles' heel of the globalization regime, Buttel is surely right to stress that EU-U. S. disputes, and their potential to galvanize social protest, are less about trade liberalization than about the perceived threat of institutional convergence, and especially the forced harmonization of national regulatory structures and the further erosion of cultural identity.

The bridge in Buttel's account rests on recognition of the crucial nexus formed by the politics of scientific research and policy, environmental risk, food safety, regulation, and consumption. These politics are complexly intertwined and varied in their manifestation, ranging from agro-food movements, environmental coalitions, and many forms of direct action, including street theater and green sabotage, to NGO involvement in regulatory processes and consumer pressure on food retailers and manufacturers.

This terrain remains largely unexplored in agro-food studies, apart from limited incursions into one or other of its arenas, such as the politics of rBGH consumption or the role of ABTs in facilitating transition to demand-driven food systems. This neglect is surprising in view of the considerable prominence of these research themes in other fields of critical social theory. In this context, work on the institutional matrix of science, science policy, and regulation, as undertaken more generally by Yoxen, Kay, Bud, and Wright, could fruitfully be extended to developments in agriculture and food since the 1980s. The chapters by Guthman and by Schurman and Munro set some markers to follow in addressing the research nexus identified above.

Guthman examines the contradictions that GEO labeling and right-to-know legislation in the United States present for movements seeking to build an effective politics of consumption, especially the privatization of risk management for genetically engineered foods and the political disarticulation that labeling implies. In calling for research on "an emerging political economy of risk," Guthman follows Haraway (1997) in emphasizing the centrality of struggles for "the power to define what counts as technical or political".

In this respect, it would be interesting to explore how European consumer groups and environmental organizations have avoided the pitfalls of labeling and convinced national governments and the EU to redraw the boundary between the technical and the political by agreeing to reopen previous regulatory decisions to public scrutiny. Demarcation struggles and boundary changes are the central theme of Schurman and Munro's chapter. The chronology of events in the rising social opposition to ABTs is becoming well-known, but there is a dearth of careful analyses of antibiotech activism, its strategies and modus operandi. Schurman and Munro begin to address this hiatus by examining antibiotech movements in the North, primarily in the United States. In their view, the marked shift in public sentiment against genetically engineered foods is not attributable to inchoate consumer resistance or to spontaneous protest.

Rather, these dramatic changes, and its economic and regulatory repercussions, have been orchestrated by the social-movement organization and mobilization of established civil-society actors. In pressing this argument, the authors provide valuable insights into the wide array of social organizations articulated by this movement, its tactical sophistication, and its networking skills, which facilitate operation across a variety of institutional and spatial scales and in different regulatory spheres and discursive arenas.

Schurman and Munro leave unanswered the thorny questions of representation and the transformative potential of the antibiotechnology movement. However, it is to be hoped that this initial exploration will encourage others to join them in investigating this serious gap in the politics of agricultural technoscience.

In December 2000, completion of the plant genome sequence of thale cress (*Arabidopsis thaliana*) was announced, the first of some 250,000 plant species. Agro-food technoscience, here represented by the public-private consortium, including Monsanto, which comprises the Multinational Coordinated *Arabidopsis thaliana* Genome Research Project, goes marching on.

Now, as in the 1980s, it is vital that critical social scientists join with the antibiotech movement and other progressive forces in the struggle to submit this enterprise and the worlds it would create to democratic debate and public assent.

THE ROLE OF PLANT BIOTECHNOLOGY IN THE WORLD'S FOOD SYSTEMS

At the molecular level, different organisms are quite similar, writes Cornell University Professor A.M. Shelton. It is this similarity that allows the transfer of genes of interest to be moved successfully between organisms, therefore, genetic engineering is a much more powerful tool than traditional breeding in improving crop yields and promoting environmentally friendly production methods. For the past 10,000 years, humans have used the plants nature provided and modified them through selective breeding to have desirable characteristics such as improved taste, enhanced yield and pest resistance.

The result is that the plants we consume today would be largely unrecognizable to our ancient ancestors. Scientists consider the techniques of biotechnology to be an aid in the selective breeding of plants and to have far more potential for providing benefits such as enhanced nutritional properties, more environmentally friendly production methods and improved yields. Already, the techniques of biotechnology have produced tremendous benefits in medicine. Virtually all the insulin used to treat diabetes today is produced through biotechnology and genetic engineering, and many of the medicines used to fight cancers and heart problems are produced through these same methods.

DEVELOPMENT OF PLANT BIOTECHNOLOGY

Corn (maize) originated in Mexico from a grass called teosinte that has a small reproductive structure bearing little resemblance to the ear of corn seen in markets around the world today. Tomatoes and potatoes first appeared in South America - tomatoes as small fruits the size of a grape, and potatoes as knobby tubers with high concentrations of a family of bitter chemicals called glycoalkaloids, which are toxic to humans.

Through selective breeding by our ancestors, the shape, colour and chemical content of these and hundreds of other plants consumed today have been modified to suit consumer preference or to obtain desired characteristics such as high yield, disease and insect resistance, and tolerance to drought and other plant stresses. Not only have these plants changed in appearance and composition, they also have become distributed worldwide through centuries of human migration and trade. For example, cabbage, which originated in Europe, is now grown on every inhabited continent. When today's consumers walk into a market in many parts of the world, they are witnesses to today's global food system where foods produced in one part of the world are daily shipped to local markets.

We now realise our ancient ancestors were modifying the genetic makeup of plants by transferring genetic material from one plant to another. However, it wasn't until Gregor Mendel, an Austrian monk, conducted experiments in the 1800s with garden peas that the basic laws of heredity were first unraveled. Prior to the early 1900s, traditional plant breeding, like that practiced by Mendel, relied on man-made artificial crosses in which pollen from one plant species was transferred to another sexually compatible plant.

The goal was to take a desirable trait from one plant and introduce it into another plant. However, often desirable characteristics either were not present in sexually compatible plants or did not occur in any plant species. This led plant breeders to seek new ways of transferring desirable genes. Beginning in the 1930s, plant breeders developed techniques to allow them to develop plants from two parent plants that could not normally produce viable offspring. An example is the technique called "embryo rescue," in which the new plant embryo is provided with extra care in the laboratory to enable it to survive during its early growth.

In the 1950s, plant breeders also developed methods of creating variation in an organism's genetic structure through what is termed "mutation breeding." Mutations in the genetic makeup of a plant occur continuously and randomly in nature through such events as the sun's radiation and may lead to the occurrence of new desirable traits. Mutation breeding uses similar random processes to cause changes in a plant's genes. Plants then are assessed to determine if the genes were changed and whether the changes provided a beneficial trait such as disease or insect resistance. If the plant was "improved," then it was tested for other changes that may have occurred. Many of the common food crops we use daily have been developed through techniques such as embryo rescue and mutation breeding and virtually all the foods we consume have genes in them.

It is hard to think of an example of a common food crop in the developed world that has not been improved by some form of modern technology, or what can be termed "biotechnology." Simply put, biotechnology is a set of techniques that utilizes living organisms, or parts of organisms, to make or modify products, improve plants or animals, or develop microorganisms for specific purposes. This definition encompasses all human activities conducted on living organisms from the earliest development of plant breeding 10,000 years ago to the present. This is the reason plant breeders consider the term "genetically modified organisms" - or GMOs - to be misnomer since all common food crops of today have been so modified.

THE SCIENCE OF MODERN GENETIC ENGINEERING

Genetic engineering is one form of biotechnology and usually refers to copying a gene from one living organism—plant, animal or microbe—and adding it to another organism. In genetic engineering, a small piece of genetic material (DNA) is inserted into another organism to produce a desired effect. This is in contrast to traditional plant breeding in which all the genes desirable and undesirable contained in the male plant—pollen—are combined with all the genes of the female plant.

The progeny resulting from this cross may contain the gene for the desirable character, but it will also contain many of the undesirable genes from both parents. Genetic engineering has the advantage of being able to transfer only the gene of interest and greatly accelerate plant breeding. But genetic engineering also is more powerful than traditional breeding since it can move genes not only between similar plant species but also from distant relatives, including non-plant species. It is possible to move genes between such seemingly unrelated organisms because all living organisms share the same code for DNA and the synthesis of proteins and other basic life functions. What might seem on the surface to be very different organisms are, in fact, very similar, at least at the molecular level. All living things are more alike than different, and this is one of the reasons that genes can be moved so successfully between such seemingly different organisms as plants and bacteria. Genes are not unique to the organisms from which they came, so there really aren't "tomato genes" or "bacterial genes." It's the collection of all genes in a tomato or a bacterium that makes it a tomato or bacterium, not a single gene. As we learn more about the genetic makeup of all organisms, we see that most plant species differ by only a small percentage of their genes and those even such seemingly different organisms as tomatoes and bacteria have many common genes. These findings suggest that in the long-term evolutionary process even tomatoes and bacteria had some common ancestor. From the discovery 50 years ago of the structure of DNA, scientists soon came to realise they could take segments of DNA that carried information for specific traits—genes—and move them into another organism. In 1972, the collaboration of Hubert Boyer and Stanley Cohen resulted in the first isolation and transfer of a gene from one organism to a single-celled bacterium where it would express the gene and manufacture a protein.

Their discoveries led to the first direct use of biotechnology—the production of synthetic insulin to treat people with diabetes—and the start of what is often called modern biotechnology. Plants were first transformed through genetic engineering in the late 1970s. Mary-Dell Chilton and colleagues used a common soil-dwelling bacterium, Agrobacterium tumefaciens, that attaches itself to plants and transfers some of its DNA into the plant. Chilton and her colleagues added a gene to this bacterium, which in turn transferred the gene into a plant where it became part of the plant's DNA. This bacterium is still commonly used in genetic engineering along with another technique that uses a high-velocity mechanism to inject DNA into plant cells. The result from either technique is the same—the plant cells take up the gene and begin to express it as their own.

THE FUTURE OF BIOTECH FOODS

Since the known benefits of biotech foods outweigh the uncertain risks, the precautionary principle supports the use of such foods. In the 1980s, the precautionary principle migrated from its European roots to the international environmental scene. By 1992, it was ubiquitous: it appeared in the Rio Declaration, the United Nations Framework Convention on Climate Change, and the Convention on Biological Diversity (CBD).

These precedents ultimately spawned the CBD's January 2000 Cartagena Protocol on Biosafety, which uses the principle as a basis for assessing the risk of international transfer, handling, and use of genetically modified organisms. This protocol is considered a major victory for the precautionary principle and its advocates.

A popular formulation of the precautionary principle is the Wingspread Declaration: "When an activity raises threats of harm to human health or the environment, precautionary measures should be taken even if some cause and effect relationships are not established scientifically." This principle has been interpreted by many to require—as precautionary measures—policies to curb technologies that science cannot prove to be absolutely safe.

Specifically, it has been invoked to justify a ban on research, development, and deployment of genetically modified (GM) crops. This application of the precautionary principle acknowledges the public health and environmental risks that a ban might reduce but ignores other risks that a ban might generate or prolong. Therefore, while the operation might be successful, the patient might die.

Moreover, the principle provides no guidance in situations where actions such as a ban on GM crops could simultaneously generate uncertain benefits and uncertain harms. To avoid counterproductive "precautionary" policies, I propose a framework for applying the precautionary principle. This new approach provides a way to evaluate policies that might result in ambiguous outcomes because the harms might offset, in whole or in part, the benefits of GM crops.

Using this framework, the precautionary principle can be used to answer a variety of difficult questions in the debate over GM crops. For example, do the public health and environmental benefits of GM crops exceed their harms? Considering the wide range of consequences of banning such crops, would the precautionary principle, in fact, justify such a ban? Finally, would a ban be consistent with the stated aims and objectives of the CBD and the Biosafety Protocol?

COMPETING UNCERTAINTIES

This framework consists of a set of hierarchical criteria to rank various threats based on their characteristics and likely consequences. The first criterion is human mortality: the threat of death to any human, no matter how humble, outweighs similar threats to members of another species, no matter how magnificent.

Moreover, the other nonmortal threats related to human health—such as blindness, stunting, additional days of sickness, or lowered educational attainment—should take precedence over threats to the environment, although exceptions might be made based on the nature, severity, and extent of the threat. This is the human morbidity criterion. These two criteria can be combined into the public health criterion.

Additional criteria must be considered if an action under consideration results in both potential benefits and harms to public health or the environment. These other criteria include:

- The immediacy criterion. All else being equal, more-immediate threats should be given priority over threats that could occur later. This criterion is justified because people tend to partially discount the value of lives that might be lost in the distant future.

- The uncertainty criterion. Threats of harm that are more certain should take precedence over those that are less certain if their consequences otherwise would be equivalent.
- The expectation value criterion. Other things being equal, precedence should be given to threats with higher expectation values. For example, actions resulting in fewer expected deaths should be preferred over those resulting in more expected deaths.
- The adaptation criterion. To the extent technologies are available to cope with, or adapt to, an impact, then that impact ought to be discounted.
- The irreversibility criterion. Greater priority should be given to outcomes that are irreversible, or likely to be more persistent.

WEIGHING THE BENEFITS

Before making any decisions regarding new technologies, including GM crops, we first need to assess the harms and benefits associated with the technologies with respect to public health and the environment.

Environmental benefits. Diversions of land and water to agriculture are the greatest threats to the world's biological diversity. Worldwide, agriculture accounts for 38 percent of global land area, 66 percent of water withdrawals, and 85 percent of consumptive use of water—water that is no longer available for immediate use elsewhere. Agriculture also affects bio-diversity through water pollution and atmospheric transport—for example, by releasing excess nutrients, pesticides, silt, and greenhouse gases into the environment. Land-based agricultural activities also affect aquatic, coastal, and marine systems. In the future, the world's biggest environmental challenge will be to meet the growing demand of a larger and wealthier population for food, nutrition, fibre, timber, and other natural resource products while conserving habitat and bio-diversity. Biotechnology could be key to reconciling these opposing goals.

Worldwide, there will be a tradeoff between increased agricultural productivity and increased land conversion for the next 50 years. Assuming that the human population will be 8.9 billion in 2050 and that crop production per capita will grow at the same rate between 1997 and 2050 as it did between the early 1960s and late 1990s, and if conventional agricultural productivity increases 1 percent per year between 1997 and 2050, then 325 million hectares (800 million acres) of forests or other habitat will have to be converted to new cropland.

That's a 21.5 percent increase over the 1,510 million hectares of existing cropland. This would inevitably increase pressure on terrestrial biodiversity. Given existing, but unused, opportunities to increase productivity, such increases are plausible with conventional agriculture. However, if biotechnology increases productivity by 1.5 percent per year—corresponding to a cumulative 30 percent increase in productivity by 2050 due to biotechnology alone—then agriculture could return 98 million hectares of current cropland to nature. An increase to 2 percent per year would return 422 million hectares. Numerous GM crops, in various stages between research and commercialization, could make such gains plausible.

GM crops could be bioengineered to tolerate poor climatic and soil conditions, including acidic soils, drought, and high salinity levels, conditions which prevail in many developing countries. For instance, 43 percent of tropical soils are acidic, and one-third of the world's irrigated land has been lost to salinity. GM crops can also make agriculture more environmentally sustainable. Higher productivity leads to less cropland, less soil erosion, less silt, and less carbon in the water and the atmosphere. GM crops could also be bioengineered to fix their own nitrogen, increase nitrogen and phosphorus uptake from the soil, and resist pests. Such crops would lower environmental loading of nutrients, pesticides, and nitrous oxide.

In the United Stares, for example, in 1999, cultivation of corn genetically modified to express the Bacillus thuringiensis (Bt) gene—which repels pests without the need for chemical pesticides-reduced pesticide usage 400 thousand hectare-treatments. Production also increased by 2.3 billion liters, equivalent to the output of 200 thousand hectares. Bioengineering could also enhance feed to reduce phosphorus in animal waste, lowering its run-off into streams, lakes, and other waters. Similarly, in 1999, the use of the herbicide resistant "Roundup Ready" soybean reportedly lowered pesticide usage by 8 million hectare-treatments, while simultaneously increasing yields and facilitating no-till cultivation, a very effective method of containing erosion.

Public health. Sufficient food is the first step to a healthy society. Despite unprecedented progress in the last half century, 825 million people still suffer from hunger and malnourishment, and more than 2 billion people suffer from various micronutrient deficiencies. Worldwide, GM crops could reduce public health problems arising partly or wholly from deficiencies in the quantity or nutritional quality of food, from malnutrition to heart disease.

Cumulatively, in 1995, hunger and malnutrition were responsible for 6.6 million, or 54 percent, of the deaths worldwide in children younger than five years, stunting in 200 million children, and clinical xerophthalmia—a condition of the eye, caused primarily by a lack of vitamin A, that causes thickening and drying of the conjunctiva and cornea and could lead to blindness—in about 2.7 million people. Bioengineering can increase the quantity as well as the nutritional quality of food, as exemplified by "golden rice," which is rich in beta-carotene, a precursor to vitamin A. Crossed with another bioengineered strain, such rice could help reduce vitamin-A and iron-deficiency-related deaths and diseases globally. Bioengineered crops can also help battle the so-called "diseases of affluence." In 1998, these diseases—heart disease, hypertension, and cancer—accounted for 4.8 million deaths in high-income countries, and 14.9 million deaths in the low-and middle-income countries. GM crops that could reduce this toll include low-fat soybeans, higher-protein rice, tomatoes with increased antioxidant content, potatoes with higher starch than conventional potatoes, and vitamin-fortified fruits and vegetables. Finally, pest-resistant GM plants can—by reducing the amount, toxicity, or persistence of pesticides—reduce accidental poisonings and other health problems affecting farm workers.

WEIGHING THE COSTS

In spite of the estimated benefits of deploying bioengineered crops, there has been much hue and cry from concerned scientists and the public about the risks inherent in this new agricultural technology. Some fear that GM crops may encourage pesticide-resistant insects and superweeds, hurt beneficial insects, and harm human health.

Perhaps the most familiar environmental concerns about GM crops centre on pest-resistant and herbicide-tolerant crops. One risk is that target pests will become resistant to toxins produced by pest resistant GM crops, such as corn or cotton. However, recent studies from Arizona, Mississippi, and Australia indicate that bollworm, for instance, did not increase its resistance to Bt toxin produced by a GM Bt cotton. Conventional strategies to deter pest resistance to conventional pesticide, such as maintaining refuges of non-Bt crops, crop rotation, and expanded monitoring, can be adapted for GM crops. Other strategies include developing crops with more than one toxin gene acting on separate molecular targets and inserting the bioengineered gene into the chloroplast—the structure in the cells of green plants that is critical for photosynthesis. These alterations allow more Bt toxin to be expressed, so that more of the pest insects die and fewer live to develop resistance. Another source of risk is that Bt plants could harm nontarget species. Studies in non-field conditions show higher mortality in Monarch butterflies fed with leaves dusted with pollen from Bt corn. Studies have also shown that the larvae of green lacewings that ate corn borers fed with Bt corn were more likely to die. However, whether—and the extent to which—these studies represent real world conditions is debatable.

The U.S. Environmental Protection Agency recently concluded that the weight of evidence indicates no hazard to wildlife from the continued registration of Bt crops, and Bt corn is unlikely to cause widespread harm to Monarchs. [18] More importantly, the inadvertent effects due to airborne dispersal of pollen or leakage into the soil through the roots of Bt crops could be virtually eliminated by bioengineering genes into the chloroplast rather than into nuclear DNA. Bioengineered herbicide or pest tolerant crops might also escape into wild relatives leading to "genetic pollution" and creating "superweeds." Since that would defeat the very justification for such GM crops, farmers and GM seed vendors have substantial economic incentives for preventing the creation of herbicide-tolerant superweeds and, failing that, keeping such weeds in check.

Gene escape is possible if GM crops are planted near wild relatives. However, the most common GM crops—soybeans and corn—have no wild relatives in the United States. Also, conventional breeding has rendered many crops—corn and wheat, for example—"ecologically incompetent," that is, the hybrid forms in common use cant easily reproduce in natural conditions and are unlikely to cross pollinate with other species. A 10-year British study of four herbicide-tolerant or pest-resistant GM crops—oilseed rape, corn, sugar beet, and potato—shows that they are no more invasive or persistent in the wild than their conventional counterparts.

This study also provides reassurance that "superweeds" perhaps can't invade natural ecosystems, and con-firms that such GM plants have no competitive advantage in natural systems. Moreover, had any herbicide-tolerant or pest-resistant weeds begun to spread, available crop management techniques—such as applying an herbicide to which the GM plant had not been engineered to b e resistant—could have been used to control them. Finally, the chances of gene escape from GM to non-GM crops can be further reduced by maintaining a buffer between the two crops.

Of course, gene escape could be limited with greater certainty if the GM plants were engineered to be sterile or prevented from germinating, using, for instance, terminator technology." An alternative approach would be to insert the gene into the chloroplast, which would preclude spread through pollen or fruit, as well as prevent root leakage. Finally, there is a concern that, in the quest to expand yields, herbicide-tolerant and pest-resistant GM plants would work too well and further simplify agricultural ecosystems, decreasing biodiversity. Some fear that the new genes inserted into GM plants could be incorporated into a consumer's genetic makeup. However, plant and animal DNA has always been a part of the daily human diet. In fact, an estimated 4 percent of human diet is composed of DNA, which is naturally present in meats and plants.

Another concern is that GM crops could trigger allergies in unsuspecting consumers. One to 3 percent of adults and 5 to 8 percent of children in the United States suffer from food allergies, and each year food allergies cause 135 fatalities and 2,500 emergency room visits. But GM foods can and have been tested, and most importantly rejected, prior to commercialization for their allergic potential. In fact, the precision of bioengineering could be used to render allergenic crops non-allergenic.

Finally, antibiotic resistant markers used to identify whether a gene has been successfully incorporated into plants could accelerate the trend towards antibiotic-resistant diseases. But this threat is slight compared to that due to overuse of antibiotics in livestock feed and by humans. Moreover, alternative markers have been devised for many crops, with more in the offing.

APPLYING THE PRINCIPLE

Clearly, there are risks associated with using or not using GM crops. The framework presented earlier offers a way to evaluate actions that could result in uncertain costs and uncertain benefits. Ideally, each criterion should be applied separately to the public health and to the environmental consequences of using or not using GM crops. However, since there are variations in the severity, certainty, and magnitude associated with the various competing costs and benefits regarding each of these sets of consequences, I will apply several criteria simultaneously.

The human population could increase 50 percent between 1998 and 2050, and with it undernourishment, malnutrition, and their consequences, assuming global food supply increases by a like amount and all else remains equal. Thus, unless food production outstrips population growth significantly over the next half century, billions of people in the developing world may suffer from undernourishment, hundreds of millions may be stunted, and millions may die from malnutrition.

Based on the sheer magnitude of people at risk of hunger and malnutrition and the degree of certainty attached to their public health consequences, limiting GM crops by limiting the rate at which food production can expand, will almost certainly increase death and disease, particularly among the world's poor. By contrast, the negative health consequences of ingesting GM foods—for example, the effects due to ingesting transgenes—are speculative. Other risks are relatively minor in magnitude, such as a potential increase in antibiotic resistance. Some are speculative and relatively minor, such as increased incidence of allergic reactions. Moreover, as noted, many of these impacts can be contained, if not eliminated.

Thus, based on the uncertainty, expectation value, and adaptation criteria applied either singly or in conjunction, GM crops must be favoured over conventional or alternative agricultural production systems. Hence, the precautionary principle requires that we continue to research, develop, and commercialize—with appropriate safeguards, of course—those GM crops that would increase food production and generally improve nutrition and health, especially in the developing world.

Some have argued that many developed countries are awash in surplus grain, and they do not need larger harvests. But reducing those surpluses would harm public health in developing countries, which currently import about 10 percent of their cereals. Without these surpluses, food supplies would be lower and food prices would be higher, as would undernourishment, malnutrition, and associated health problems. Developing countries deficits are expected to rise because of future population and economic growth and, possibly, global warming. Therefore, developed countries surpluses will continue to be critical for future food security in developing countries, just as they are today.

The above argument against GM crops also assumes that they will provide little or no public health benefits to developed countries. But GM crops are also being engineered to combat diseases of affluence, which claim 4.8 million lives annually in the developed countries. Even a small reduction in these numbers due to GM crops would translate into relative large declines in the death toll. Moreover, the health benefits of "golden rice," for instance, do not have to be confined to developing countries. Thus, even in developed countries, the potential public health benefits of GM crops ought to outweigh in magnitude and certainty the speculative health consequences of ingesting GM foods.

Another argument against developing GM foods is that hunger and malnutrition are rooted in maldistribution and unequal access to food and not in food shortages; therefore, there is no compelling need for biotechnology. This argument tacitly acknowledges that GM crops would boost quantities more rapidly than would conventional agriculture. More importantly, GM crops do not decrease access. In fact, by increasing production, they would lower prices, which would increase access, particularly for the lower rungs of society.

One factor contributing to poor food distribution is crop spoilage. But various GM crops could increase shelf life, reducing spoilage and wastage. Moreover, nutritional quality is just as important as food quantity, and GM foods can enhance the former far easier than conventional agriculture.

Bioengineered crops do not have to be the sole solution for hunger and malnutrition. It is sufficient that they contribute to the solution and are among the most efficient solutions. If the argument that there is sufficient food is truly compelling, then it should be equally valid for arguing against increases in production using conventional technologies. Perhaps, then, developing countries like India and Bangladesh should forego increasing agricultural productivity altogether, and should focus only on improving access and distribution. For obvious reasons, no one makes this argument.

Regardless of the level of demand, GM crops would increase yields per unit of land used. And higher yields mean less land for mankind, and more for the rest of nature. Moreover, reduced land demand for mankind means lower land prices and socioeconomic opportunity costs for sequestering land for conservation and preservation, which would facilitate conservation of species and biodiversity in the wild—a major goal of the Convention on Biological Diversity.

Also, by reducing amounts of cropland, fertilizers, and pesticides, GM crops would reduce soil erosion and releases of nutrients, pesticides, carbon, and nitrous oxide emissions into the environment. Compare these tangible ecological benefits of GM crops to the more speculative environmental benefits of limiting pest-resistant and herbicide-tolerant GM crops even without factoring in the environmental costs of conventional farming. Assuming GM crops actually do harm nontarget pests and weeds more than conventional farming practices, banning GM crops may increase the diversity of the flora and fauna, but this is an uncertain benefit. In addition, the possibility of gene escape to weeds and non-GM crops is also uncertain.

Hence, with respect to the environmental consequences of the use or nonuse of GM crops, one must conclude, based on the uncertainty and expectation value criteria, that the precautionary principle promotes cultivation of GM crops. This would conserve the planet's habitat, biodiversity, and carbon stores and sinks, provided due caution is exercised, particularly with respect to herbicide-tolerant and pest-resistant GM crops. It may be argued that although gene escape to the environment is highly unlikely, under the irreversibility criterion GM crops ought to be banned. However, increased habitat clearance and land conversion resulting from such a ban may be at least as irreversible, particularly if it leads to species extinctions.

Finally, the precautionary principle supports using terminator-type technology because it would minimize the possibility of gene transfer without diminishing any of the public health or environmental benefits of GM crops.

SPEED WITH CAUTION

The precautionary principle has often been invoked to justify a prohibition on GM crops. However, while this justification touts the potential public health and environmental benefits of a ban on GM crops, it ignores any harm caused by delaying or foregoing benefits that GM crops would bring. Worldwide, more than 800 million people suffer from hunger and undernourishment, and more than 2 billion suffer from malnutrition each year. As a result, millions suffer death and disease annually. But in comparison to conventional crops, GM crops, in fact, would increase the quantity and nutritional quality of food supplies. This should reduce mortality and morbidity rates worldwide. And by increasing productivity and reducing chemical inputs, GM crops would increase land available for the rest of nature and reduce global pollution by fertilizers, pesticides, and carbon. Thus, GM crops would also better protect habitat, biological diversity, water quality, and the current climate than would conventional agriculture.

Hence, contrary to conventional environmental wisdom, the precautionary principle—properly applied, with a more comprehensive consideration of the public health and environmental consequences of a ban—argues for a sustained effort to research, develop, and commercialize GM crops, provided reasonable caution is exercised during testing and commercialization of the crops. In this context, a "reasonable" precaution is one whose public health and environmental benefits are not negated by the harm incurred due to reductions or delays in enhancing the quantity or quality of food or the health of the environment. Any such harm to public health would be disproportionately borne by the poorest and most vulnerable segments of society.

The Convention on Biological Diversity is dedicated to conserving biodiversity, preferably through conservation in the wild. But a GM crop ban is more likely to magnify threats to biodiversity, and reduce land and water available for such conservation. Hence, such a ban would be directly counter to the CBD's very reason for existence, and that of any subsidiary agreements such as the Biosafety Protocol. Finally, slowing down commercialization of GM crops would, by reducing the future quantity and quality of food available, arguably abridge the "right to a standard of living adequate for the health and well-being...including food, clothing" and the "inalienable right to be free from hunger and malnutrition" as specified in the Universal Declaration of Human Rights and the Universal Declaration on the Eradication of Hunger and Malnutrition, respectively, thereby violating Article 10 of the Declaration on Human Rights Defenders, which states that "No one shall participate, by act or failure to act where required, in violating human rights and fundamental freedoms." While it might be a mistake to go full steam ahead on GM crops, it might be an even larger mistake to stop them in their tracks. The wisest policy would be to go as fast as possible while keeping a sharp lookout, and staying on the track to improvements in human and environmental well-being.

USING BIOTECHNOLOGY TO ADDRESS POVERTY REDUCTION AND FOOD SECURITY PROBLEMS IN ASIA

A. Linkage Between Biotechnology and Poverty Reduction and Food Security Agricultural biotechnology will contribute to poverty reduction and food security if scientists can develop technologies to increase quality and yields of food crops, and the technologies are adopted by small farmers.

For this to happen, biotechnology R&D will have to meet four conditions:

1. It must address both the problems of small farmers in rain fed areas where most of the poor live, and those of small farmers in irrigated areas, which provide the bulk of food grain supply in Asia.
2. It must focus on crops, livestock, and fish commonly grown by small farmers. Major crops are rice, tropical maize, wheat, sorghum, millet, banana, cassava, groundnut, oilseed, potato, and sweetpotato. Biotechnology R&D should also focus on high value cash crops (*e.g.*, cotton, soybean, and vegetables) that can increase the incomes of small farmers through crop diversification. The prospect for improving these crops is bright due to the large demand for them in urban areas and in international markets. Fish and livestock (cattle, sheep, goats, pigs, and chickens) are also important.

3. The technology to be developed and delivered to small farmers must be simple, low cost, and carry little or no risks to human health and the environment. As in the case of the Green Revolution, the most effective strategy to increase food production is through improved seeds that possess high yield potential, fertilizer responsiveness, resistance to pests and diseases, good agronomic characteristics, and good nutritional quality.
4. Biotechnology development should be accompanied by favourable policy environment; good governance; investments in rural infrastructure; agricultural research and extension; and credit and marketing.

Investment in Agricultural Biotechnology

In much of Asia. yields of major food grains are stagnant or declining in the face of population increases. Pests and diseases cause substantial preharvest and postharvest losses of crops, livestock, and fish. Solutions to many of these problems may lie in the various applications of modern biotechnology. The use of molecular markers to tag specific traits is accelerating the breeding of new varieties of plants and animals. New understanding of plant and animal genes through genomics may offer ways of increasing crop yields. These new developments when used in conjunction with developments in the physical and social sciences, offer more sustainable means for obtaining necessary productivity increases that are less dependent on environmentally damaging inputs of chemical fertilizers and pesticides. Given appropriate policies and necessary human and financial resources, modern biotechnology could make an extremely important contribution to future agricultural growth.

During the next 25 years, Asia will need a Second Green Revolution, often called Biorevolution or Doubly Green Revolution. Conway pointed out that the next technology-driven revolution must be doubly green—it must increase food production at a faster rate than in recent years without significantly damaging the environment. It must also increase incomes and increase access to food by the poor. The differences between the Green Revolution and Biorevolution.

Compared to the Green Revolution of the 1970s, Biorevolution will be characterized by the following features:

- Potentially many crops will be affected as well as livestock and aquaculture.
- Potentially all areas, both irrigated and rainfed, will benefit from biotechnology R&D.
- Technology development and dissemination will substantially involve the private sector with the public sector playing the role of facilitator and regulator.
- Many processes and products will be patentable and protectable.
- Capital costs of research will be high.

Molecular and cell biology expertise will be required in addition to expertise in conventional plant breeding and other agricultural sciences.Modern biotechnology (genetic engineering) is not a silver bullet for achieving food security, but used in conjunction with other techniques it may be a powerful tool in the fight against poverty and food insecurity. Other approaches are available and should be used. Narrowing the yield gap between those obtained from farmers' fields and those from experiment stations using the current technologies is just one example. However, there is concern that some conventional alternatives will not be able to produce the desired results within a limited time. The advantage of modern biotechnology rests on the speed at which desired crop varieties are produced. In some cases, the desirable genetic combination of traits is simply not possible through common breeding methods, and can be done only through genetic engineering.

To increase food production by at least 40 per cent within the next 25 years, Asian countries not only have to move towards the best technological frontier (to push farmers' yields to the optimum level), but keep moving the technological frontier itself. As long as product safety, environmental and ethical concerns, and IP issues are adequately addressed, modern agricultural biotechnology has the potential to significantly increase the quantity and quality of the food supply for developing countries.

APPLYING BIOTECHNOLOGY IN RESOURCE POOR AREAS

The resource-poor, rain fed areas in Asia are home to many poor people, and their population is growing rapidly. Although migration may sometimes be the only viable livelihood strategy, in many such areas sustainable intensification of agriculture may be the best way to achieve poverty reduction and food security. Development policy has often neglected these areas characterized by poor soils, shorter growing seasons, lower and uncertain rainfall, and little infrastructure or access to markets. Yield increases usually lag behind population growth.

Efforts by poor farmers to expand cultivation onto new lands to eke out survival often lead to deforestation and erosion of fragile highland soils. That in turn threatens hydropower, road, and irrigation infrastructure in the lowlands. Policies and investments are required that can achieve food security in ways that protect natural resources, thereby breaking the vicious cycle of poverty, low productivity, and environmental degradation. Creation of nonfarm, rural jobs will also be important.

Sizeable investments in biotechnology in Asia have been made in rain fed areas, although DMCs continue to give high priority to irrigated agriculture. Irrigated areas produce 75 per cent of total cereal production in Asia. Many IARCs such as the Asian Vegetable Research and Development Center, CIMMYT, CIP, ICRISAT, ILRI, and IRRI; and most DMCs have invested heavily in R&D on orphan crops such as rice, tropical maize, sorghum, potato, banana, groundnut, and sweet-potato.

Applicability of Biotechnology to Specific Agricultural Objectives

In Asia, biotechnology has been applied mainly to develop improved varieties adapted to specific environments.

Specific areas where new applications of biotechnology could address poverty reduction and food security are summarized as follows:

- Increasing Productivity and Stability of Crops in Rain fed and Marginal Environments. Producing more food on the same area of cultivated land would reduce pressure to expand cultivated areas to forests and marginal areas. Broadening tolerance of existing HYV cereals for drought, flooding, salinity, heavy metals, and other abiotic and biotic stresses would increase yields in rain fed areas. CIMMYT and IRRI, in cooperation with NARSs in different countries, are focusing their efforts to develop HYVs of rice and maize for rain fed areas.
- Improving Water Use Efficiency in Crops. Future availability of water for agriculture is a major issue. By 2020, there may be a water crisis in Asian agriculture. Crops with higher water use efficiency and tolerance for drought would be an advantage. With the completion of the rice genome, scientists will be able to identify genes responsible for drought tolerance.
- Integrated Pest Management (IPM) Strategies to Reduce Pesticide Use. The insecticides now used in many developing countries are often older, broad spectrum, acutely toxic compounds. Many are now banned in industrial countries except for export. They are a significant health hazard to farmers and farm workers throughout Asia. Over 80 per cent of total pesticide use in Asia is on cotton, rice, and vegetables. In the past, governments encouraged the use of chemical pesticides as part of a yield-increasing package of farm inputs, and some governments still subsidize pesticides. The need is now recognized for more sustainable approaches to IPM to reduce losses to pests without harmful side effects on human health and the environment. New pest- and diseaseresistant crops, produced with the aid of modern biotechnology, may be important components of IPM strategies. Early applications are being seen in the use of novel sources of resistance (Bt genes) to control insect pests on maize and cotton. New pest-resistant cotton varieties are being grown widely in the PRC by 3 million farmers covering 500,000 ha. Similar cotton varieties are undergoing field tests in India, Indonesia, and Thailand. Genetically modified maize resistant to insect pests is being tested in the Philippines.

- Increasing Disease Resistance in Crops. Biotechnology applied to disease resistance in crops would be especially valuable for combating diseases in tropical environments that have not been controlled by conventional means. These include important diseases such as downy mildew of maize; bacterial blight and blast of rice, papaya ringspot virus; banana bunchy top virus; and bacterial wilt of tomato, eggplant, and potato.
- Increasing Nutritional Quality. Enhancing the protein, vitamin, and micronutrient contents of food grains would greatly benefit poor consumers who cannot afford supplementary vitamins and micronutrients. Under the Golden Rice Project, IRRI scientists are undertaking biotechnology research to enhance the vitamin A content of rice by introducing genes from daffodil.
- Increasing Sustainable Production of Livestock. Research to increase the productivity and quality of farm animals, especially of dairy cows, small ruminants (sheep and goats), and water buffalo is promising. Vaccines and disease diagnostics for control of epidemic and endemic diseases of livestock are being developed.
- Increasing Productivity in Fisheries and Aquaculture. Development of vaccines and disease diagnostics for control of epidemic and endemic diseases of shrimp in Thailand is a good example of where modern biotechnology has been used successfully.

BENEFITS FROM AGRICULTURAL BIOTECHNOLOGY

Realised or potential benefits of agricultural biotechnology can be categorized as economic, social, and environmental. A large share of the benefits are concentrated in industrialized countries, where a diversity of applications are widespread in human and animal health care and in many aspects of food production and processing. Some of the potential benefits of biotechnology. As the first transgenic biotechnology product was not commercialized until 1996, little quantitative analysis has been publicly available until recently.

The studies published by ISAAA have helped fill the gap, at least with respect to transgenic crops. ISAAA analysed the global status of transgenic crops, including the area planted and the value of the market for transgenic seed, which was estimated between $2.7 billion and $3.0 billion in 1999. The studies show that the economic benefits were fairly evenly distributed between the companies that developed the technology and the farmers growing transgenic crops. ISAAA studies have also demonstrated positive environmental impacts.

In the cases of herbicide-tolerant soybean and herbicide-tolerant canola, herbicide use was considerably reduced even though weeds were better controlled, and yields increased significantly. In the cases of Bt maize, cotton, and potatoes, targeted insect pests were effectively controlled and yields were increased. Insecticide use for non-targeted insects was also reduced.

Economic Benefits in Developing Countries

Few published analytical studies have attempted to assess the impact of biotechnology in developing countries. Ex-ante studies on transgenic pest- or virus-resistant sweet-potato in Kenya and on transgenic virus-resistant potato in Mexico suggest substantial benefits for both producers and consumers. They point out, however, that smallholders will fully realise benefits only with improved farm-level management and more efficient seed distribution.

The introduction of pathogen-free banana planting material in Kenya illustrates the benefits that accrue to small-scale farmers, mainly woman. Yield increases from the new material have been substantial; in some cases yields almost doubled. This example is significant in that it involved community groups and farmers in the development and field-testing. By the time the new planting material was available commercially, the benefits of the technology had already been demonstrated to farmers, with the result that initially demand exceeded supply. The project won the science and technology prize in a Japan/World Bank-sponsored network of environmentally beneficial development projects. The Kenyan experience is highly relevant to Asia.

Benefits in Asia

A number of biotechnology applications are widespread in Asia. These include biofertilizers and biopesticides, and products from tissue culture and micropropagation (banana, coconut, maize, potato, cassava). They also include diagnostics, vaccines, and embryo transfer for livestock. An ex-post evaluation on the impact of a transgenic crop at the farm level in Asia analysed the impact of Bt cotton in the PRC A sample of 283 cotton farmers from five counties of Hebei and Shandong provinces was studied.

The key findings were:

- The cost to produce 1 kg of cotton was reduced by 20- 30 per cent, depending on the variety and site,
- Net income and returns to labour of all the Bt varieties were superior to the non-Bt varieties.

Smaller farms and those farms that had lower incomes consistently derived larger increases in net income than larger farms and those with higher incomes. The use of Bt cotton substantially reduced farmers' use of pesticides, usually from 12 sprays per season to 3 or 4.

The study also found some preliminary evidence of positive impact on farmers' health. An ex-ante study on the impact of the introduction of biotechnology for the reduction of corn-borer infestation in the Philippines has postulated three kinds of benefits. First, the introduction of transgenic Bt maize would enhance the competitiveness of the Philippines from the point of view of import substitution and export prospects. Second, it would increase farmers' incomes. Third, it would reduce pesticide use.

WEIGHING RISKS AND BENEFITS OF BIOTECHNOLOGY

The evidence thus far available suggests potential economic, social, and environmental benefits from the use of biotechnology. It also suggests benefits for both food and export crops, for commercial and small farmers, and for consumers in the form of nutrition-enhanced products.

The potential benefits will be realised only if a number of conditions are met:

- Public institutions will need to play a key role in directing the research agenda towards social and equity goals.
- Complementarity between the public and private sectors, in both R&D and in technology development and delivery, needs to be maximized so that public investment and research is focused on those areas not of interest to the private sector.
- Governments must be committed to developing technologies for the public good to ensure that the poor, including women, have access to the new technology.

The benefits and risks of biotechnology need to be assessed on a case-by-case basis. The uncertainties and the risks are yet to be fully understood, and the benefits are not yet fully exploited. It seems important not to deny people access to new biotechnology, so long as they are fully informed of the potential risks and benefits in making their choices. We have an ethical imperative not only to keep the technology portfolio open to biotechnology and genetic engineering, but also not to lose time. Every minute lost, every decision delayed, means more deaths from starvation and malnutrition.

4

Biotechnology in Indian Agriculture

BIOTECHNOLOGY INNOVATION SYSTEM IN INDIA

The government plays an important role in building the National Innovation System (NIS) by setting up institutions (public research institutions such as universities, labs etc), developing policies that support innovation and providing regulatory frameworks that can have a considerable influence on sectoral development. In the context of the developing countries, the role of the government becomes more prominent as the majority of the resources for the scientific research and development are being channelised or governed by the government. In case of India, nearly 80 per cent of the gross expenditure for research and development is being incurred by the Government while in case of the industrialized nations such as Japan, USA and UK the Industry spends nearly 75 per cent, 64 per cent and 44 per cent respectively. Though there has been a significant increase in the country's Gross National R&D expenditure in absolute terms from ₹.3974 Crore in 1990-91 to ₹.16199 Crore in 2000-01 to ₹.18000 Crore in 2002-03, the share of the government in national R&D expenditure has remained unchanged.

However, it is interesting to note that internationally all the governments perform the in-house R&D function, including even the highly private-sector-oriented governments such as the US and other OECD countries. A study of the Biotechnology innovation system of the four global leading countries namely USA, UK and Japan and Germany further supports the active involvement of Government R&D in furthering the growth of the Biotechnology sectoral innovation. Biotechnology is being characterised by multidisciplinary fields, high dependence on science and are emerging from industries having large R&D base, high degree of qualified manpower and having close links or interaction with public research institutes as well as universities and thus faces a broad set of different policy interventions by the government underlying the National Innovation System (NIS). The National System of Innovation (NSI) perspective is a convenient framework to understand the process of innovation occurring in an economy and especially within the manufacturing sector. It focuses on the country-specific factors influencing the process of innovation and economic growth. The NSI approach has evolved over the years both analytically and empirically with the contributions from Freeman, Lundvall and Nelson, Edquist (1997, 2002). NSI has became popular as an analytical framework to study innovation at the national, regional and even Sectoral level.

Though there are several definitions of NSI but that best fit in to the present context is that of Metcalfe (1995). NSI is defined by Metcalfe (1995) as a "set of distinct institutions which jointly and individually contribute to the development and diffusion of new technologies and which provides the framework within which governments form and implement policies to influence the innovation process. As such it is a system of interconnected institutions to create, store and transfer the knowledge, skills and artefacts which define new technologies".

NSI framework is comprised of various components such as the government, independent research institutes, firms and higher education system etc and their interrelationships. Innovation or government policy instruments and institutions play a very important role in cementing the relationship among the various components for the success of NSI. An application of this framework shall help in analysing the systemic failure that hamper the generation of innovations. Public policy can be applied to correct such failure. There are several NSI studies on Biotechnology in the context of the developed world but the studies in the developing countries are rather limited. Biotechnology NSI in India has evolved over time and has been instrumental in creating infrastructure and capabilities in promoting biotechnology innovation in the country. There are 241 institutions and 321 firms engaged in various dimension of biotechnology in the country. Indian Biotechnology industry is now among the top 12 countries in the world and like its global counterpart, is dominated by the health care sector. The recently released "Draft National Biotechnology Strategy 2005" by the Indian Government is a step forward in the direction of promotion and development of biotechnology to withstand global challenges and yet, solve the local health care needs of the country. Studies on biotechnology in India by and a recent empirical analysis of the health innovation system of seven developing countries further supports the role of the Governments in providing the sustainable environment for the successful development of health biotechnologies.

According to Acharya *et. al.* government in developing countries is one of the key actors in prioritizing, implementing new technologies and promoting innovation to address their specific health and development needs. The governments of the countries where biotechnology innovation system has been successful in perform or support the following functions/activities such as knowledge support, commercialization of research, collaborative research, financial support, regulation, IPR and resolving ethical issues etc.

Thus it would be interesting to analyse the role of the government in the promotion of the biotechnology in the country with special focus on the health care sector. Hence, the present study traces the role of government or national approaches in terms of policies, institutions and instruments in the promotion of biotechnology innovation system in India with a special focus on the health care biotechnology sector. A systemic framework similar to the EPHOITE (2003) project is being adapted to examine the direction of public policies, institutions and instruments in addressing the issues of knowledge support, commercialization of research, collaborative research, financial support, regulation, IPR and ethical issues in the area of health care biotechnology in Indian context. The significant features of national biotechnology innovation system with focus on health care biotechnology with the help of various S&T indicators. The various government initiatives in terms of various policies, institutions and instruments in the framework of NSI while the lessons learnt along with the policy implications and challenges for the growth of health care biotechnology innovation.

BIOTECHNOLOGICAL APPLICATIONS RELEVANCE TO THE INDIAN FARMER

Increased food production must keep pace with the country's population increase. Not only the quantity of food made available should improve. To keep pace with the present population growth and consumption pattern, India's food requirement has been estimated to cross 225 million tons by 2005 AD. This would mean an annual agricultural growth rate of 6.7 percent, a daunting task considering the rapidly shrinking resource-base and fast declining input-use efficiency in major cropping systems.

Not withstanding the impressive gain in the agricultural production, the vast agricultural potential still remains highly under-utilized. There are serious gaps both in farm yield realization and technology transfer, as the national average yield of most crop varieties of most crop varieties is low.

Not many superior varieties is mot pulses, many oilseeds and vegetable crops have been evolved in recent past. In case of vegetables the varieties which were imported or developed many decades ago are still being grown by the farmers. Their yields are low and their inherent ability to respond to improve production technology is also limited. Farmers have accepted bybrids in my vegetable crops such as cabbage, tomato and capsicum.

RESEARCH AND TECHNOLOGY TRANSFER

The Central and State governments continued to invest heavilyu in research, hence the Indian public sector research system has increased in size in real terms but by mid 90s levelled off. The number of hybrids and varieties developed and released by ICAR Crop Improvement Projects and State Agricultural Universities in major and minor crops are enormous and have been well documented. Private research to develop superior hybrids was initiated by few seed companies in mid 60s. By 1987-88, the number increased to over 12 and in 1997-98 it is estimated at 34. With the governments recognition of provate sector's performance capabilities and with "New Seed Policy" announced in 1988, the number of private companies involved in crop improvement programmes have been steadily increasing (Table below).

NUMBER OF PRIVATE COMPANIES ENGAGED IN CROP IMPROVEMENT PROGRAMME (1987-1998)

For want of financial, infrastructural and trained human resources, most private companies specialize in breeding superior hybrids in few crops characterized by low volume, high value and amenable to hybrid technology. In addition to the ICAR and Agriculture University research system, about 13 private sector seed companies are actively involved in vegetable R&D. Some of them have joint ventures with multinational or companies specializing in vegetable crop research and utilize research facilities available with their overseas partners.

Table: Number of Companies with Crop Improvement Programmes

Crop	*1987-88*	*1993-94*	*1997-98*
Pearl Millet	12	23	25
Sorghum	10	12	15
maize	06	15	20
Sunflower	10	30	20
Cotton	09	21	34
Hibrid Rice	-	4	6

Few have created facilities for field testing imported and public bred varieties. The role of private sector in vegetable research is primarily in the development of F hybrids and few open pollinated varieties.

Table : Hybrid Vegetable and Melon seeds of Private R&D Available in the market (1996-97)

Crop	*No. of Companies*	*No. of hybrid Varieties*
Tomatoes	11	45
Brinjal	10	38
Chilli	8	27
Bhendi	13	25
Cabbage	11	40
Cauliflower	8	21
Ridgegourd	6	8
Watermelon	5	13
Muskmelon	2	7

Number of private/public hybrids in major crops in market during 1995 are compared with the number in 1997.

In pearl millet, sorghum, maize and cotton number of hybrids introduced by private, companies has been steadily increasing. Many hybrids especially in paddy and cotton have regional adaptability. The number of hybrids which are accepted by the farmers and may be grown on 2 percent or more area is much lesser, so also the number evaluated in coordinated project trails or identified for release. Many hybrids recently supplied by the private sector embodies new technology which is either imported or developed by private companies. Private R&D's real investment in research has quadrupled between 1986 and 1997. Subsidiaries or joint ventures with multinational companies account for 30 percent of all private seed industry research in 1997. Both domestic companies and multinationals companies have started research programmes on rice and mustard. Recently, Monsanto has also focused on hybrid wheat.

Table: Number of Private and Public Bred Hybrids in Major Crops in the Market during 1995 and 1997 Number of Hybrids in Market

	1995		1997		
Crop	Private	Public	Private	Public	No of Private Hybrids grown on 2% plus area
Pearl Millet	50	04	60	06	14
Sorghum	22	04	41	05	6
Maize	57	03	67	03	12
Sunflower	47	05	35	06	10
Cotton	73	15	150	15	19
Hybrid Rice	4	4	8	4	-

Few private sector companies have also initiated joint ventures (*e.g.*, Mahyco-Monsanto Biotech, ProAgro-Agro-Evo) to transfer specific genes to their product lines to provide effective protection against pests. Some of them have taken initiative to treat their seed with high-tech chemicals to provide protection against sucking pests (aphids, jassids, thrips) during seedling and grand growth crop stages and diseases (downy mildew in bajra) during flowering.

AGRICULTURAL BIOTECHNOLOGY SECTOR IN INDIA

Agricultural biotechnology sector in India, particularly its plant biotechnology segment, is uniquely poised for a major process of transformation. As the country's agricultural biotechnology sector made strides in the 1990s in r-DNA, transgenics and molecular marker assisted plant breeding process, the Government of India responded with a matching policy support and regulatory framework that was designed to render the path of progress in R&D, sustainable and bio-safe. To a large extent, developments in the policy front have been induced by a vibrant non-governmental sector that intensely intervened on the sensitivities of modern biotechnology. This chapter discusses analysis of the issues and proceeds to further consider the path of progress achieved by agricultural biotechnology companies in India in the area of commercialization of biotechnology products. The special focus of this chapter is on plant biotechnology. The biotechnology sector in India in 2002-03 is estimated to have attracted private investments to the tune of US$ 140 million. Infrastructure and R&D account for nearly 80 per cent of these investments.

During the period from 1986 to 2002 it is estimated that the Government of India invested an amount of US$ 275 million in the biotechnology sector. The Government of India has also extended venture capital support for biotechnology start-ups through the Technology Development Board and the New Millennium Indian Technology Industry Leadership Initiative (NMITLI). Meanwhile the States of Andhra Pradesh, Karnataka, Maharashtra, Gujarat, Kerala and Tamil Nadu have drawn up ambitious sub-national biotechnology policies, which promise fiscal and infrastructure support measures to prospective entrepreneurs. Some State Governments have launched their own venture funds for supporting novel start-up ventures.

To catalyse the process, the Government of India accorded priority to IPR protection. The Indian Patents Act of 1970 was amended to provide for product patents for agro-chemicals, drugs and pharmaceuticals and micro-organisms over a transitory period. A product patent regime would be in place in India by January 1, 2005. The new patent law extends protection term for inventions from 14 to 20 years. To provide for protection of IPRs in respect of new plant varieties, Parliament has enacted a *sui-generis* legislation in the shape of the Plant Varieties Protection and Farmers Rights (PVFR) Act in 2001. These developments have created favourable legal conditions for international partnerships in biotechnology R&D. By prioritizing FDI investments in biotechnology, the Government of India has sent appropriate signals to international investors. Further the fact that these changes have been in consonance with the WTO-TRIPs has lent greater creditability to these changes. Meanwhile, the Government has also brought in regulations over agricultural biotechnology. Apart from the biosafety regulations that form part of the Environment (Protection) Act 1986, the Indian Parliament has legislated the National Biodiversity Act in the year 2002, to provide for regulatory controls over access to biological resources in India.

AGRICULTURAL BIOTECHNOLOGY INDUSTRY

The top ten pesticide and seed firms for 1997 and 1999. Longer-term time series data on the structure of the chemical and seed company are difficult to obtain. In one of the few published papers to provide this data, Ollinger and Fernandez-Cornejo (1998) provide Concentration Ratio data (CR4) for the United States pesticide industry over the period 1972-1989. During this period, the CR4 oscillates, moving from a high of 50 per cent in 1973 to a low of 37 per cent in 1982, and then rising to 48 per cent in 1989. Although detailed data are not available for the 1990s, there is evidence that the structure of the seed and chemical industry remains dynamic.

The consolidation activity of the ten most active biotechnology firms. Mergers and acquisitions are a key component of the new business relationships that have emerged. As Kalaitzandonakes and Hayenga (2000) note, firm entry in the crop biotechnology industry peaked in the early 1980s, with production innovation continuing throughout the 1980s. Oehmke et al. (1999) also discuss the cyclical pattern in mergers in the biotechnology industry. Aggregate market figures mask the much higher concentration that exists in specific markets. For example, in 1998, Monsanto and Pioneer-HiBred (now owned wholly by DuPont) controlled 15 per cent and 39 per cent of the US seed corn market, respectively. These two same companies controlled approximately 24 per cent and 17 per cent, respectively, of the purchased soybean seed market. For the US cotton market, two companies, Delta & Pine Land and Stoneville, had 71 per cent and 16 per cent, respectively, of the seed market.

The determination of the relevant market concentration is not always done on the basis of output markets. The Federal Trade Commission has used innovation competition to assess the impact of mergers. Examining competition in innovation is designed to focus attention on the impact of a merger on innovative activity. Using data on field trials underway each year, Brennan Pray, and Courtmanche (2000) calculate a CR4 ratio at the R&D stage. In 1988, the four largest firms had 87 per cent of the field trials. The CR4 ratio declined to a low of 63 per cent in 1995, then rose steadily over the next few years to reach 79 per cent in 1998. This concentration at the R&D stage is matched by concentration in terms of the number of patents held. The top four firms held 41 per cent of the corn patents (up to 1996), 53 per cent of the soybean patents (up to 1997), 77 per cent of the tomato patents (up to June 1997) and 38 per cent of the Bt patents (up to 1998).

Concentrated markets do not necessarily imply the presence of market power. Baumol, Panzar and Willig (1992) stress that firms will not be able to exercise market power if the market is contestable. The key requirements for market contestability are: (a) Potential entrants must not be at a cost disadvantage to existing firms, and (b) entry and exit must be costless. For entry and exit to be costless, there must be no sunk costs. Sunk costs are expenditures that cannot be recouped once they are incurred—examples include expenditures made to obtain regulatory approval, expenditures on advertising, and expenditures on R&D. If there are no sunk costs, potential entrants can use a hit-and-run strategy in which they enter an industry, undercut the price of the incumbents, reap the profits and exit before the incumbents have time to retaliate. In anticipation of entrants acting in this manner, the incumbents forestall entry by keeping price at average cost. The consequence is that, even in an industry that is highly concentrated, prices can be kept at or near competitive levels. However, if sunk costs are present, firms entering an industry are unable to exit again without losing a portion of their investment.

As a result, hit-and-run strategies are much less profitable and incumbents are able to keep price above average cost. Thus, with sunk costs, markets are not contestable and market power is once again an issue.

VERTICAL STRUCTURE

In addition to becoming larger through growth and horizontal integration, the major agrochemical companies have restructured themselves in other ways. One of the major changes is increased vertical integration—for example, the combining of seed, chemical and biotechnology activities within the same firm. Although some companies initially tried to include pharmaceutical and biotechnology components within the life science model, practically all of them have now divested their agricultural operations. A further element of this internal restructuring is the increased use of strategic alliances.

Many countries have had their own local seed companies that have over the years developed seed for a specific geographical market and operated sales and distribution systems. The major biotechnology companies are increasingly purchasing these seed companies as a source of seed material in which to insert the genes for herbicide or insect resistance. As an example, in 1997, Monsanto acquired a 30 per cent share of the Brazilian corn seed market with the acquisition of Sementes Agroceres.

With its 1998 purchase of Cargill's international seed division, Monsanto now controls over half the Argentine maize seed market. In 1998, Dow AgroSciences acquired Morgan Seeds, Argentina's second largest corn seed company, and Brazil's Dinamilho Carol Productos Agricolas, another key South American corn seed company. Phytogen (majority owned by Dow Agrosciences) acquired a major cottonseed breeding programme in the Chaco province of Argentina. In 1998, Mexico-based Empresas La Moderna (ELM) bought two South Korean vegetable seed companies and Nath Sluis (agricultural biotech company) of India (Action Group on Erosion, Technology and Concentration [ETC], 2000).

FACTORS THAT AFFECT HORIZONTAL INDUSTRY STRUCTURE

Numerous factors are behind the wave of horizontal mergers and acquisitions. Some of these factors are common to all industries and have no specific link to the seed, chemical and agricultural biotechnology industries. Thus, the mergers and acquisitions in the seed and chemical industries are at least in part a result of the need to consolidate costs and rationalize industry capacity, a desire by the management of the firms involved to extend their sphere of influence, and a wish by some firms to pre-empt other firms from taking over valuable assets. MacDonald (2001) provides an overview of concentration levels in a number of US agricultural sectors, many of which have seen higher concentration in recent years.

There are at least two explanations for increasing concentration, however, that pertain specifically to the seed, chemical and agricultural biotechnology industries: (a) sunk costs, and (b) escalation strategies.

SUNK COSTS

Two sources of sunk costs appear to be important in determining industry concentration: (a) intellectual property rights and R&D expenditures, and (b) the regulatory requirements that governments have introduced before the products of the R&D activity can be marketed.

On the empirical side, Ollinger and Fernandez-Cornejo (1998) examine sunk costs and regulation in the US pesticide industry. Using data over the 1972-89 period, they find that research costs and pesticide regulation costs negatively affect the number of companies in the industry, and that smaller firms are affected more strongly by these costs than are larger firms. Research and regulation costs also encourage foreign-based firms to expand into the US market and to force less profitable innovative firms to exit the market. Ollinger and Fernandez-Cornejo also point out that their results on the impact of regulatory costs generally match those found in other industries.

These empirical results can be linked to theoretical arguments concerning sunk costs and the economies of scale and scope that sunk costs tend to create. As Sutton (1991) outlines, the presence of sunk costs means that for firms to be profitable, price needs to be raised above marginal cost, typically by reducing the amount of competition (*i.e.*, the number of firms). Sutton identifies two types of sunk costs: exogenous and endogenous. Exogenous sunk costs are those beyond the control of the firms in the industry. Regulatory costs are a good example of exogenous sunk costs. In contrast, endogenous sunk costs are firm-level strategic variables. Advertising and R&D are examples of endogenous sunk costs. The rest of this section examines the connections between intellectual property rights (IPRs), R&D expenditures, regulatory costs, and economies of scale and scope.

Economies of scale and scope are major factors in propelling industries towards concentration. Economies of scale exist when average costs fall as more output is produced; economies of scope exist when the total cost of producing two outputs together is less than the cost of producing the two outputs separately.

Because economies of scale and scope mean that larger and diversified firms have lower average costs, there is clearly an incentive for firms to get large. Indeed, those that do not get large are vulnerable to being driven out of the market by larger and more cost efficient firms. Of course, there is a limit to how large firms can get. Although development and production costs may fall with an increase in the size of the firm, other costs—particularly those associated with administration—rise. Nevertheless, economies of scale and scope clearly create pressures for consolidation. Economies of scale and scope are created as a result of investment in non-rival goods. Unlike rival goods—such as materials, labour, and energy—which can only be used in one place, by one person and at one time, non-rival goods can be used in more than one place and by more than one person, all at the same time. This feature of non-rival goods—that they can be used over and over again—means that output can be increased without having to increase all inputs. Consequently, economies of scale and scope are created. Much of the expenditure on non-rival goods—examples include R&D and regulation expenditures—can be considered a sunk cost.

Intellectual property, of course, is a good example of a non-rival good. Indeed, ideas generally are considered non-rival goods. It is widely believed that many of the high-technology industries, including the biotechnology and information industries, are subject to increasing returns. Romer (1990) postulates that this is a result of the distinction between physical goods and ideas. Ideas are not scarce, and as such, any industry based primarily on the trade of intellectual property will not face diminishing returns in their primary resource, the idea. The protection of intellectual property, through IPRs, is designed to encourage innovation and to provide incentives for the development and diffusion of new products and technologies.

An example illustrates the connection between intellectual property and economies of scale and scope. Suppose a biotechnology firm has some intellectual property, such as a particular gene that has been isolated. This intellectual property can be used over and over again as the firm expands its activities. For instance, if the company wishes to develop seeds for a new crop, it will not have to invest again in the research that isolated the gene. Although the development of a new seed will require additional lab and greenhouse space, labour, and materials, the expenditure on the technological advancements does not have to be made again.

A similar result occurs if the firm needs to invest substantial amounts of money in obtaining regulatory approval for a seed—although the production of additional units of the seed will require additional costs, the regulatory expenditures do not have to be made again. Once again, large companies typically have an advantage, because they are able to spread the costs of obtaining regulatory approval over more output. Thus, the greater the regulatory requirements in an industry, the more concentrated the industry is expected to be.

Intellectual property may also create economies of scope. Once a specific gene has been isolated—for instance, a gene that confers a resistance to a particular herbicide—this gene can then be put in a number of crops. Once again, the production costs of a number of products together will be less than if the products were produced separately.

To recap, R&D expenditures and regulatory costs are both sunk costs and a source of economies of scale and scope. Because economies of scale or scope mean that larger and more diversified firms have lower average costs, there is clearly an incentive for firms to get large. Indeed, those that do not get large are vulnerable to being driven out of the market by larger and more cost efficient firms. These theoretical results are supported by empirical observations in the US pesticide industry.

ESCALATION STRATEGIES

An escalation strategy is one in which a company spends large amounts on R&D and engages in mergers and acquisitions to achieve a dominant role in the market—that is, the firm tries to leapfrog its competitors to become the dominant firm. Escalation can be a profitable strategy when there is a high degree of substitutability with competitors' products on the demand side, and there are scope economies on the supply side. Both of these factors are present in the agricultural biotechnology industry.

The isolation of a gene that provides particular advantages and which can be inserted into a number of crops means there are scope economies on the supply side. There are also clear supply side scope economies associated with the enabling technologies that are required to use these genes. And on the demand side, herbicide and insect resistant seeds and the accompanying chemicals are clearly a substitute product for traditional seeds and herbicides and pesticides.

Firms engage in escalation strategies in a number of ways. Although internal growth through R&D is one way, the more common method is through consolidation via horizontal mergers and acquisitions. As the theory predicts, the combination of demand substitutability and supply side scope economies present in the seed and chemical industry appears to be linked with escalation strategies.

Monsanto has clearly followed an escalation strategy, with Dow, Novartis, and others following similar strategies of making acquisitions and spending significant amounts on R&D.

BIOTECHNOLOGY FOR SOCIETAL DEVELOPMENT IN INDIA

The application of biotechnology for societal development has always received special attention since it benefits socially and economically disadvantaged sections of the society. Through this programme a number of projects are implemented for economic and technical empowerment of the beneficiaries. Efforts are made through training, demonstration and appropriate R&D inputs for providing tailor made solutions for location specific problems of SC/ST, women and rural population using modern biology and biotechnology. The following pages provide the highlights of the achievements.

PROGRAMMES FOR SC AND ST POPULATION

Biotechnology-based Programme for SC/ST population has benefited around 15,000 people directly and about 45,000 indirectly through 54 projects during the year. The projects have been supported in the areas of cultivation of medicinal and aromatic plants, biofuels, production of biofertilizers and biopesticides, organic farming, aquaculture, mushroom cultivation as well as human health care interventions etc. The universities, public funded institutions, Krishi Vigyan Kendras and voluntary/non-governmental organizations have been involved in the implementation of the projects. The target groups have received the benefit of hands-on training and field demonstrations.

Vermicomposting: MR Morarka GDC Rural Research Foundation Jaipur implemented a project in Jalgaon district of Maharashtra, to benefit 2600 SC/ST target population on economic utilization of waste, agriculture residues and cow dung. Over 1780 MT of waste, agricultural residues and cow dung was utilized for vermicomposting through establishment of 28 community facility centres.

Beneficiaries are utilizing vermicompost on crops like onion, sugarcane, cotton, banana, papaya and wheat. Some farmers switched over from conventional crops to cash crops such as colourful capsicum, gerbera as well as inter cropping of safed musli, turmeric and ginger in banana plantation. Farmers have been trained to use bio-fertilizer, foliar sprays and mulching to use the agro-waste for fertility and management. Marketing linkages have been established with "Organic Shoppe", through which farmers are getting 10-20 per cent premium for their organic produce. About 20 entrepreneurs have taken up vermiculture as commercial activity. extension Material in the local language was distributed to the farmers.

Mushroom cultivation: A project on promotion and production of essential oil, mushroom and vermicompost was implemented by RRL, Jorhat for revenue generation of the ST population of Arunachal Pradesh. Motivation and demonstrationcum-training programmes have been conducted in 13 villages of five districts on cultivation of citronella, production of mushroom and vermicompost. Demonstration units have been established at RRL Branch Naharlagun farm. Mushroom spawn, earthworms and citronella saplings were distributed to the beneficiaries. Several beneficiaries established vermicompost units on commercial basis and are selling vermicompost. The beneficiaries at Ziro, Jairampur, Doimukh, Manmao, Hetlong and Tengman have established oyster mushroom cultivation units. So far, 10 acres of land have been planted with citronella and 7 acres of land with patchouli at Sonajuli village of Papum Pare district.

Medicinal plants: A project on strengthening health care practices and conservation of medicinal and nutritious plants was continued at Herbal Folklore Research Centre, Tirupati for implementation in 100 villages of two Mandals in Andhra Pradesh. Training workshops were organized with the help of Yanadi tribal people for community health volunteers on identification and sustainable harvesting methods of 20 species of locally used medicinal and aromatic species. The reputed healers were encouraged to prepare medicines needed for PHC from the identified plants. About 6000 seedlings were distributed to the beneficiaries for planting in kitchen gardens. Green health kits (samples of herbal medicines) and educational material in local language containing identification, cultivation techniques and preparation of medicines with their therapeutic properties were distributed to the trainees.

A project on creating awareness and promotion of medicinal plants for the production and marketing was implemented by RRL, Jorhat to benefit ST population of Arunachal Pradesh. Awareness and training programmes were organized on the importance, scientific cultivation and marketing of Piper longum, Kaempferia galanga, Andrographis paniculata, Stevia rebaudiana, Rauvolfia serpentina and Mentha piperita. Three nursery-cumdemonstration centers were established at Roing (1000 ft msl), New Bomdila (8160 ft msl) and Ziro (4350 ft msl) and 19000 saplings were distributed to ST farmers/NGOs for introduction and propagation in their fields. At high altitude, villagers were advised to cultivate Aconitum heterophyllym, Coptis teeta and Picrorrhiza kurroa. Beneficiaries have stopped collecting wild Homalonema aromatica, an important but rare aromatic plant of Arunachal Pradesh, and have started its cultivation in their farms. Ten community nurseries have been established with the help of Vivekananda Trust, Roing at ten villages of Lower Dibang Valley District. Twenty farmers were trained on cultivation and post-harvest management of medicinal plants and marketing tie-up has been established. Around 1266 local ST people participated in the awareness programmes.

A project on organic cultivation of medicinal and aromatic plants for income generation and sustainable development in Uttarkashi District of Uttaranchal was undertaken through Forests Conservation Programme of World Wide Fund for Nature, New Delhi. Demonstration-cum-Resource Material Production Centre established for training on vermicomposting and cultivation of species.

Construction of poly house, vermi-composting units and planting of seeds of Rauvolfia serpentine completed. Awareness and motivation camps were organized for the farmers. Extension material brought out in the local language for cultivation of Rauvolfia serpentina, Coleus barbatus and Pelargonium graveolens were distributed to the farmers.

A project on developing skills and economically sound vermicompost units utilizing distillation waste of Mentha arvensis (menthol mint) was implemented for the benefit of SC/ST communities of Lucknow and Barabanki Districts of Uttar Pradesh. Around 200 farmers were trained on menthol mint nursery raising, transplanting, distillation and practices followed in vermicomposting. Demonstrations were carried out in farmers' field using high yielding variety

'Kushal'. Farmers trained on organic farming started vermicompost production using distillation and other agro-waste. The yield data obtained from the demonstrations showed the possibility of reduction of chemical fertilizers by 75 per cent without affecting the yield.

BIOFUELS

A project was continued on cultivation of Jatropha in arid lands of Namakkal District in Tamil Nadu for production of biofuel and other products for SC/ST community by Centre for Research in Social Sciences, Coimbatore. A high tech nursery was established and plantlets of Jatropha were raised. Training programmes were organized for 223 members on cultivation of Jatropha in 31 acres of land. About 30,000 plants were supplied for cultivation. Awareness camps, training and demonstrations were conducted on Jatropha cultivation, oil extraction and soap making. SHG groups were trained in nursery techniques. Women self help groups were trained in various activities of jatropha cultivation and soap making. Extension material bought out in the local language was distributed to the beneficiaries.

Biocontrol agents: A project was implemented on development of eco-preneurship among the coastal community of Pondicherry by MS Swaminathan Research Foundation, Chennai. Sixteen SHGs involving women and youths were formed in Ariyankuppam, Nonankuppam, Kakayanthope and Veerampattinam villages. Demonstrations and trainings were conducted to educate trainees on basic aspects of IPM, importance of biocontrol agents, production and marketing of Trichogramma, importance of human nutrition, role of edible mushrooms and production and marketing of oyster mushroom. Micro-enterprise was developed for mushroom and Trichogramma production. Technologies were disseminated on biopesticide (Trichogramma), oyster mushroom production and vermicomposting to recycle the spent waste of oyster mushroom. Extension material brought out in the local language was distributed to the villagers.

AQUACULTURE

A project was undertaken on enhancing livelihood of SC/ST people in Orissa and Tamil Nadu by MS Swaminathan Research Foundation, Chennai for introducing various aquaculture enterprises to 250 participants in backyard ornamental fish culture and carp seed production. Fingerlings produced were distributed to the entrepreneurs for stocking. Various aspects on ornamental fish breeding, carp nursery and gold fish culture were disseminated to the self help groups, which helped them in earning a revenue of ₹. 200/month/household and an income of about ₹. 3200/from carp nursery and ₹. 8500/from vegetables and flowers as bund crops. Community ponds were effectively utilized by stocking with rohu, catla, mirgali, Chinese and Indian carps and marketable size fish was harvested. Technologies transferred to the entrepreneurs on community pond aquaculture, carp nursery and ornamental fish culture and extension material in local language were distributed to the beneficiaries.

A project on utilization of integrated marine living resources by SC youth from Tuticorin region of Tamil Nadu was undertaken by College of Fisheries, Tuticorin. Lobster fattening and crab culture techniques were standardized and demonstrated to the beneficiaries, who have grown lobsters in cages for about 4 months and produced 250-350gms size for marketing. The beneficiaries were helped in availing credit facilities and bank loan. In the project around 120 people of target population have been trained. A manual on crab and lobster fattening technology was brought out for the distribution to the beneficiaries.

A project on freshwater prawn culture was implemented in village Naubasta Kalan in Allahabad District by University of Allahabad. Demonstrations were conducted on prawn hatchery technology and standardized methods for biocontrol of prawn disease for integrated giant freshwater prawn (Macrobrachium rosenbergii) culture. Trainings were conducted on scampi culture, breeding and seed production and microbial inputs were utilized for feed silage preparation using indigenous technology. Apiculture activity was introduced and disseminated to the target population and extension material was distributed to the beneficiaries.

GENETIC COUNSELING

Tribal population of northeast region was screened for anaemia and nutritional deficiencies through a programme implemented at Vivekananda Institute of Medical Sciences, Kolkata. The tribes of Arunachal Pradesh, Assam and Tripura displayed high incidence of anaemia. Awareness programme was organized for school children on inherited anaemia in pre-marital and pre-natal population. More than 1500 families from 80 villages were educated on anaemia, thalassaemia and hemoglobinopathies and counseling was provided to them. Leaflets were distributed for awareness of anaemia with special reference to Thalassaemia in local languages.

A project on molecular epideomiology of alcoholism was implemented jointly by Centre for Cellular and Molecular Biology, Hyderabad and Tribal Research Centre, Udagamandalam, Nilgiri District to study Kota tribe population in Nilgiri hills. General health awareness camps were arranged in seven villages for periodical medical check up. Blood samples were collected from 620 individuals and clinical data on health profile revealed that the disease morbidity is more for hypertension, heart related problems and diabetes. Genealogies indicated that the alcohol drinking habit is running in the families for the past two to three generations. Study undertaken on the health awareness camps and medical camps were found beneficial in counselling the tribal population.

FLORICULTURE

Over 180 women from Shillong, Itanagar and Chandigarh were trained in cultivation of commercially important and hybrid species of orchids. A total of

118 women of Shillong have raised 6 species *viz.* Dendrobium fimbriatum var. oculatum, D. longicornu, D.lituiflorum, C. pendulum, C. aloifolium and C. giganteum in their backyard. These plants have started blooming. The beneficiaries have started selling their harvest in local market. Through S&T Council, Bhopal 200 women were trained in cultivation of gladiolus, marigold, mogra, rose and tuberose.

FOOD PROCESSING

At College of Horticulture, Vellanikara, Thrissur, 88 women were trained in various aspects of entrepreneurship in preparation of cocoa products including molding, packaging and marketing. Twenty-two women have already set up their production unit and are selling their products in local market.

In another project at Fisheries College, Tuticorin, 500 women were trained in cook-chill technology for seafood and various parameters for quality assurance *viz.* estimation of moisture content, bacteria load, presence of bacterial species, the nutritive value, shelf-life etc. The trained women were grouped in 48 SHGs for establishing market for seafood products.

At the Central Food Technological Research Institute, Mysore data on biochemical and microbial analyses for nutritionally important factors like protein, amino acids, isoflavones, genistein, diadzein and methods for improvement of shelf life were collected on for 'Tofu' preparation. Also a byproduct of Tofu preparation, Okara, was found to be rich in protein and amino acids has been used as a food supplement. In another project at CFTRI, 17 women were trained in preparation of various food products of high nutritive value, methods of removal of anti-nutritional factors like phytic acid and trypsin inhibitor found in soya bean, better texture and improved shelf life. The food items included fermented products of soya bean and fruit pulp. Also 34 women were trained in spirulina production and its application as health food. Through CFTRI, technologies for rural bakery are being disseminated to rural women in three villages.

POULTRY AND LIVESTOCK FARMING

The 'Action Research on Health and Socio Economic Development', Bolangir, Orissa, identified 100 women and trained them in poultry rearing. The beneficiaries were provided with parent birds of Carigold (dual purpose) and Vanaraja varieties. The beneficiaries were able to earn ₹. 1000/per month through the sale of eggs and birds. The beneficiaries were also trained in feed formulation using local ingredients, management practices and entrepreneurship skill. The trained women formed SHGs and are approaching a bank for setting up their poultry farms. The market is established through state owned apex cooperative known as OPOLFED.

Through three projects supported at Tamil Nadu Veterinary and Agricultural Sciences University and its Extension Centres at Namakkal,

Chennai and Tiruchirapally, over 1200 women were trained in poultry farming including feed formulation, farm management / health care and preparation of various food items such as quail egg pickle and ready-to-eat food products etc.

Aquaculture: Under a project at the North Eastern Hill University, Shillong, techniques were standardized for induced breeding for common carp and 4,00,000 fish seeds were reared up-to fries for distribution to the beneficiaries. Selected 92 women were trained in fish farming techniques including water quality testing, pond maintenance, induced breeding, harvesting and feed formulation using slaughter house waste. They were provided with fish fries for rearing in their own pond.

Over 470 women from Bhopal, Shimla, Solan and Bhubaneshwar were trained in polyculture fish farming. Interested women were also provided advance training in composite carp culture, integrated fish farming with duck/ poultry, pond management and value added products including fish processing. These women are selling their products in local market and are earning ₹. 3000/per harvest/family.

Considering the opportunities in ornamental fish, a project was supported at the University of North Bengal. The identified women, 57 from Darjeeling and 54 from Jalpaigudi district were trained in rearing and breeding of guppy (Poecilia reticulate) in cemented cisterns and angel (Pterophyllum scalaer) in ponds. The women are being trained in management skill and marketing.

ORGANIC FARMING / VERMICOMPOST

At the Sri Sathya Sai Institute of Higher Learning, Anantapur Campus, Andhra Pradesh, neem varieties with higher level of azadirachtin were identified in 5 villages. Thirty trees were selected and 40 women from two villages were trained in fruit collection, drying and storage. Specialised training on maintenance of neem processing gadgets, processing of neem fruits for production of cake, oil or powder, maintenance of aza-rich plantlets, hardening, harvesting and application of neem based biopesticide was provided to five resource persons

At the University of Agricultural Sciences, GKVK Campus, Bangalore, five villages of Bangalore and Kolar were surveyed for the availability of organic waste. The waste materials were mostly sugarcane trash, coconut coir pith and vegetable wastes. Demonstration programmes for 20 to 25 farmwomen from each identified village were organized. Six interested women were given onfarm training. The technique of Trichoderma enriched vermicomposting was perfected and demonstrated at University campus.

At MSSRF, Chennai, phosphate-solubilising local strains were collected and screened. Simultaneously several awareness programmes on application of biofertilizer were organized for the villagers of Rediarchatram block. About 300 interested farmers were trained in large-scale production of biofertilizer.

Under the project support at Dangoria Charitable Trust Hyderabad, apart from training women in processing various items of food, 100 women from 13 villages were trained in production of vermicompost, neem-based biopesticide and their application in organic farming, About 100 vermicompost units were set up using the ring method. The beneficiaries produced over 130 quintals of compost. The beneficiaries were encouraged to raise nurseries of drumstick, papaya and creeper spinach at their backyard using vermicompost. The beneficiaries were able to sell excess vermicompost and planting material.

HORTICULTURE

Through the Tamil Nadu Agricultural University, Coimbatore, 54 progressive farm women were identified for training on cultivation of hybrid rice. The women were provided with quality seed their cultivation over an area of approximately four acre. The yield for the first harvest (December to May) was recorded at 2000 kg/ha. The beneficiaries were able to earn an additional income of ₹.20,000/per acre/year through three crops in an year. At Supi farms, Uttaranchal non-traditional vegetable crops were grown on farmers' land in a small way. An herbal garden was established. Incidence of pests to various crops was evaluated and farmers were trained on production of vermicompost and application of neem based biopesticide in horticultural crops.

MUSHROOM CULTIVATION

In an attempt to empower rural women through transfer of mushroom cultivation technology 93, women were trained at Himachal Pradesh Krishi Pradesh Vishwavidyalaya, Palampur and 750 at Tropical Botanic Garden and Research Institute, Kerala in mushroom cultivation. In addition, 15 women were trained in spawn production at TBGRI. The beneficiaries were given 15 days training in cultivation of Oyster and Agaricus species using wheat straw. The first harvest of about 1000 kg was sold @ ₹. 16/kg. The beneficiaries were also trained in various recipies of mushroom. Market linkage has been established for their products.

Under the project support at Dangoria Charitable Trust Hyderabad, apart from training women in processing various items of food, 100 women from 13 villages were trained in production of vermicompost, neem-based biopesticide and their application in organic farming, About 100 vermicompost units were set up using the ring method. The beneficiaries produced over 130 quintals of compost. The beneficiaries were encouraged to raise nurseries of drumstick, papaya and creeper spinach at their backyard using vermicompost. The beneficiaries were able to sell excess vermicompost and planting material.

SERICULTURE

A total of 550 women were trained for various activities including the organic cultivation of mulberry and rearing of cocoon and value added products from waste at the Central Sericultural Research and Training Institute, Mysore and at Vidyasagar University, Midnapore, West Bengal. The beneficiaries are getting an additional income through the sale of value added products and surplus vermicompost.

HEALTH CARE

Kidwai Memorial Institute of Oncology, Bangalore, Karnataka organized over 50 awareness camps on Carcinoma cervix to educate 2071 women for its prevention. Blood samples collected from 1830 women based on their medical history, 1812 were found negative and 18 positive cases were referred for treatment. A project on mental retardation including awareness generation among rual women, screening for early detection and genetic counseling was supported to Manovikas Bio-Medical Research. Over 7,000 families from three Gram Panchayats were surveyed. Ninety-one of these showing signs of mental retardation were shortlisted. The families were informed about the possible risk factors in future. Analysis of blood samples from these for genetic carrier showed only 4 cases of low level of fragile X chromosomes.

A project is supported at Saraswati Dental College, Lucknow for dental fluorosis. Water samples from 12 villages around Lucknow and Kanpur were analysed for fluoride content. Health camps were organized in 4 villages with a team of dental surgeon. A total of 2860 people were examined for dental fluorosis. The severely affected persons were advised for regular dental examination at SGPGI, Lucknow. They were also provided with calcium and cholicalciferol supplement. The blood samples from the severely affected boys and girls within the age group of 15 years and pregnant women were examined for their blood calcium level. Under another project supported at Mahavir Hospital And Research Centre, Hyderabad, Andhra Pradesh, blood lead levels in children and adults including men and women were measured to study environmental genotoxicity in women and children occupationally exposed to lead, a heavy metal.

AGRICULTURE IN INDIA

India's record of progress in agriculture over the past four decades has been quite impressive. The agriculture sector has been successful in keeping pace with rising demand for food. The contribution of increased land area under agricultural production has declined over time and increases in production in the past two decades have been almost entirely due to increased productivity.

Contribution of agricultural growth to overall progress has been widespread. Increased productivity has helped to feed the poor, enhanced farm income and provided opportunities for both direct and indirect employment. The success of India's agriculture is attributed to a series of steps that led to availability of farm technologies which brought about dramatic increases in productivity in 70s and 80s often described as the Green Revolution era. The major sources of agricultural growth during this period were the spread of modern crop varieties, intensification of input use and investments leading to expansion in the irrigated area. In areas where 'Green Revolution' technologies had major impact, growth has now slowed.

New technologies are needed to push out yield frontiers, utilize inputs more efficiently and diversify to more sustainable and higher value cropping patterns. At the same time there is urgency to better exploit potential of rainfed and other less endowed areas if we are to meet targets of agricultural growth and poverty alleviation. Given the wide range of agroecological setting and producers, Indian agriculture is faced with a great diversity of needs, opportunities and prospects. Future growth needs to be more rapid, more widely distributed and better targeted. These challenges have profound implications for the way farmers' problems are conceived, researched and transferred to the farmers. On the one hand agricultural research will increasingly be required to address location specific problems facing the communities on the other the systems will have to position themselves in an increasingly competitive environment to generate and adopt cutting edge technologies to bear upon the solutions facing a vast majority of resource poor farmers.

In the past agriculture has played and will continue to play a dominant role in the growth of Indian economy in the foreseeable future. It represents the largest sector producing around 28 percent of the GDP, is the largest employer providing more than 60 percent of the jobs and is the prime arbiter of living standards for seventy percent of India's population living in the rural areas.

These factors together with a strong determination to achieve self-sufficiency in food grains production have ensured a high priority for agriculture sector in the successive development plans of the country.

An important facet of progress in agriculture is its success in eradication of its critical dependence on imported foodgrains. In the 1950's nearly 5 percent of the total foodgrains available in the country were imported. This dependence worsened during the 1960's when two severe drought years led to a sharp increase in import of foodgrains. During 1966 India had to import more than 10 million tonnes of foodgrains as against a domestic production of 72 million tonnes. In the following year again, nearly twelve million tonnes had to be imported. On the average well over seven percent of the total availability of foodgrains during the 1960s had to be imported.

Indian agriculture has progressed a long way from an era of frequent droughts and vulnerability to food shortages to becoming a significant exporter of agricultural commodities. This has been possible due to persistent efforts at harnessing the potential of land and water resources for agricultural purposes. Indian agriculture, which grew at the rate of about 1 percent per annum during the fifty years before independence, has grown at the rate of about 3 percent per annum in the post independence era.

AGRICULTURE – SUB-SECTORS

Indian agriculture broadly consists of four sub-sectors. Agriculture proper including all food-crops oilseeds, fibre, plantation crops, fruits and vegetables is the largest accounting for nearly 70 percent of the agriculture sector as a whole. The rapid growth in this sub-sector through exploitation of wastelands and fallows, spread of irrigation and adoption of production enhancing technologies was critical in transforming India from a country vulnerable to food shortages to one of exportable surplus. Although this sub-sector has made impressive progress its share in the sector as a whole has declined from 78 percent in 1960-61 to less than 70 percent by early 90s.

Table. Agriculture sub-sectors – value of output (%)

Sub-sector	1960-61	1970-71	1980-81	1993-94
Agriculture	78.0	79.3	75.3	68.5
Livestock	16.6	15.1	17.2	24.5
Forestry	4.4	4.4	5.9	3.8
Fisheries	1.0	1.2	1.6	3.2
	100.0	100.0	100.0	100.0
Total value at Current price (Rs. Billion)	89	218	615	2740

Correspondingly the share of livestock sector has increased considerably. The livestock industry has grown from ₹. 15 billion in early 1960s to ₹. 100 billion by 1980-81 and ₹. 672 billion by 1993-94. In nominal terms the sector grew at almost 15 percent per annum during 1980s. Milk production, which was almost stagnant for two decades ending 1970, grew by over 5 percent per annum in the 80s. Similarly, production of eggs increased at the rate of about 6.5 percent during the same period. As a result the share of livestock increased from about 17 percent till early 80s to 25 percent by 1993-94.

Though it plays relatively a minor role within the sector as a whole, fishing sub-sector activities have been on the rise. The sub-sector has grown from only ₹. 3 billion in 1970-71 to nearly ₹. 90 billion in 1993-94. The growth was particularly rapid in 70s and 80s. Value added increased at over 5 percent per annum during this period. In real terms forestry and logging activities have been on the decline since mid seventies. As of 1993-94, the size of the industry in terms of value of output was 103 billion.

Table. Foodgrains Production

Year	*Area (million ha)*	*Production (million tonnes)*	*Yield kg/ha*
1950-51	97.32	50.82	522
1960-61	115.58	82.02	709
1970-71	124.32	108.42	872
1980-81	126.67	129.59	1023
1990-91	127.84	176.39	1379
1998-99	124.0	195.25	1574

Over the past three decades, the country has successfully transformed itself from a food deficit economy to one which is essentially self sufficient in availability of foodgrains and other essential commodities, albeit only at the prevailing level of effective demand. Annual aggregate foodgrains production, which averaged about 82 million tonnes in 1960-61 increased to 123.7 and 172.5 million tonnes for the trienniums ending 1980-81 and 1990-91 respectively. Current (1998-99) production level is 195 million tonnes and the country has been able to accumulate substantial, (35 million tonnes) stocks of foodgrains to cope up with any sudden difficulties arising from drought or a similar situation in any part of the country.

Increased outputs, have been achieved chiefly by adopting, since mid sixties, a strategy aimed at increasing foodgrains production by concentrating public sector efforts and resources in regions with a high potential for quick and substantial productivity gains through increased cropping intensity and average yields. These were the areas favoured by agroclimatic resource conditions and where irrigation facilities already existed or could be developed relatively rapidly. The main elements of this strategy were:

(i) Expansion of irrigation coverage,

(ii) Increased provision and utilization of key inputs – mainly high yielding varieties (HYVs) of crops, mainly of wheat and rice and chemical fertilizers and plant protection chemicals,

(iii) Expansion and improvement of institutional support services such as research and extension and

(iv) Price policies favourable to producers of major foodgrains.

The success of this strategy was made possible by development and availability of replicable production technology packages, so called 'Green Revolution' technologies. Irrigation facilitated double cropping and widespread adoption of HYVs. The HYVs performed particularly well under irrigated conditions, were highly responsive to fertilizers and their short duration permitted increases in cropping intensities.

Irrigation development was the cornerstone of the strategy. Undivided India was amongst the largest irrigated areas in the world. With partition nearly one-third of the irrigated area went to Pakistan. At the time of independence the net irrigated area was 20.9 million ha (gross irrigated area 22.6 million ha). Recognizing large-scale development of irrigation facilities as critical to rapid agricultural growth, the country has spent about ₹. 45,000 crores on irrigation development in the first four decades after independence. During the period 1950-51 to 1965-66 development of irrigation through government canals grew from 7.2 million ha to 9.8 million ha – a growth rate of 2.1 percent per annum. During 1970s this pace dropped slightly to 1.9 percent. In 1980s the rate of increase dropped significantly to 1.1 percent per annum. The growth of tube-well irrigation, however, increased rapidly from 4.5 million ha in 1970-71 to 9.5

million ha in 1980-81 and then to 14.3 million ha by 1990-91. The net irrigated area increased from 31 million ha in 1970-71 to 53.5 million ha in 1995-96 which corresponds to 22 percent of the net sown area in 1970-71 and 37.63 percent in 1995-96 *(Tables 3, 4)*. With improvements in irrigation efficiency the gross irrigated areas has increased to 71.51 million ha. The percentage of gross cropped area service by irrigation increased from 18.3 percent in 1960-61 to 23.0 percent in 1970-71 and to over 38 percent at present.

Table. India – irrigated area

Year	*Net*	*Gross*	*Net as % of net sown*	*Gross as % of gross cropped*
	––– *million ha* –––		––––– % –––––	
1970-71	31.10	38.19	22.17	23.04
1980-81	38.72	49.78	27.66	28.83
1990-91	48.02	63.20	33.61	34.03
1995-96	53.51	71.51	37.63	38.33

Table. Sources of irrigation

Year	*Net Irrigated Area (million ha)*	*Sources*				
		Canals	*Tanks*	*tube-wells*	*Other wells*	*Other Sources*
		(% net irrigated area)				
1960-61	24.80	10.4	4.6	0.2	7.2	2.4
1970-71	31.10	41.28	13.22	14.34	23.88	7.29
1980-81	38.72	39.49	8.22	24.62	21.08	6.59
1990-91	48.02	36.34	6.13	29.69	21.73	6.11
1995-96	53.01	32.04	5.81	33.52	22.16	6.47

Fertilizers have constituted yet another key input in addition to expanded irrigation and spread of HYVs in achieving goals of high production and productivity. India currently occupies third position in the world, after China and USA, in terms of fertilizer production and consumption. Consumption of fertilizers has increased from 1.54 million tonnes in 1967-68, representing the pre green revolution era, to 17.31 million tonnes currently (1997-98).

The average per hectare use of fertilizers currently around 85 kg per hectares is the lowest among several Asian countries.

However, rice and wheat account for a major fraction, around 65 percent of the total fertilizer consumed in the country, with very little fertilizers going to the rainfed areas.

According to some current projections, fertilizer's use will need to increase to 30-35 million tonnes to meet the foodgrains need of 2020. The demand for nutrients will stretch by almost another 15 million tonnes if requirements for horticulture, vegetables and plantation and commercial crops are included. At present domestic production of N and P fertilizers (13.42 million tonnes) falls short of consumption by over 20 percent. In addition the entire requirement of K fertilizer is imported.

Most agricultural development programmes initiated in 1960s were concentrated in regions of high potential. Thus five states, Punjab, Haryana, Uttar Pradesh, Andhra Pradesh and Tamil Nadu account for 50 percent of the country's net irrigated and 53 percent of the gross irrigated area. The combination of expanding irrigation coverage and widespread adoption of short duration HYVs led to significant increases in cropping intensities. Acreage cropped more than once per year increased from 13 million ha in 1950-51 to about 44 million ha at present.

Average cropping intensity for the country as a whole rose from 115 percent in 1960-61 to 131 in 1993-94. By 1993-94 cropping intensity has risen to 187 percent in Punjab, 167 percent in Haryana and 142 percent in Uttar Pradesh.

Table. Cropping intensity

Years	*Net sown area*	*Gross cropped area*	*Area cropped more than once*	*Cropping intensity*
	million ha			*(%)*
1950-51	118.8	131.9	13.1	111.1
1960-61	133.2	152.8	19.6	114.3
1970-71	140.3	165.8	25.5	118.2
1980-81	140.0	172.6	32.6	123.3
1990-91	143.0	185.7	42.7	129.9
1993-94	142.0	186.4	44.4	131.2

An important consequence of the strategies adopted since sixties has been to boost production of, chiefly, two crops rice and wheat. Their share in total foodgrains production went up from 57 percent in 1970-71 to more than 75 percent in 1990-91. Production of foodgrains other than rice and wheat did not increase significantly and in the eastern region even the yield of rice did not increase.

Agricultural production and income rose substantially in the north-western states of Punjab, Haryana, Western Uttar Pradesh, parts of Rajesthan, Tamil Nadu and Andhra Pradesh. By contrast productivity and output growth have been modest in eastern and central India and in deccan plateau. Progress was particularly slow in rainfed areas, which account for over 60 percent of the cropped area and where a great majority of rural poor are concentrated. An important impact of the strategies pursued in the 'Green Revolution' period has been intensification of regional disparities and imbalances in agricultural development and food availability and hence levels of food security. For the country as a whole while per capita availability of cereals has increased substantially, that of pulses has decreased significantly.

In summary an annual increase in foodgrains production of 3.22 percent during fifties was mainly because of expansion in area. Sixties recorded a low annual growth rate of 1.72 percent necessitating large-scale imports of foodgrains.

Table. Rice and Wheat Production Production (million tonnes)

Year	*Rice*	*Wheat*	*Foodgrains*	*R+W as % of Foodgrains*
	million tonnes			
1970-71	40.80	20.86	108.4	57.4
1980-81	49.91	34.55	129.6	65.4
1990-91	72.78	53.03	176.4	75.0
1996-97	81.74	69.35	199.4	75.0

Annual growth of 2.08 percent was recorded during seventies. This decade was the turning point in India's foodgrains economy leading to self-sufficiency through significant productivity increase first in wheat and later in rice in the eighties. An annual growth of 3.5 percent in foodgrains in eighties was the hallmark of green revolution that enabled India to become self-sufficient and even a marginal exporter. The pace of growth slowed in nineties barely making or even slower than the population growth rate. This is a matter of concern.

Table. Annual compound growth rate of foodgrains production

Crop	**1950-51 to 1959-60**	**1960-61 to 1969-70**	**1970-71 to 1979-80**	**1980-81 to 1989-90**	**1990-91 to 1997-98**
Rice	3.28	- 8.05	1.91	4.29	1.53
Wheat	4.51	5.90	4.69	4.24	3.67
Coarse cereals	2.75	1.48	0.74	0.74	- 0.49
Total cereals	3.00	2.51	2.37	3.63	1.84
Pulses	2.72	1.35	- 0.54	2.78	0.76
Total foodgrains	3.22	1.72	2.08	3.54	1.66

Table. Per capita net availability of cereals and pulses per day (g)

Year	*Cereals*	*Pulses*	*Total*
1951	334	61	395
1961	400	69	469
1971	418	51	469
1981	417	38	455
1991	468	42	510
1998	451	33	484

HORTICULTURE

The diversity of physiographic, climate and soil characteristics enables India to grow a large variety of horticultural crops – fruits, vegetables, flowers, spices, aromatic and medicinal plants, plantation crops etc. India is the largest producer of fruits in the world and second largest producer of vegetables. The area under fruits estimated at 1.45 million ha in 1970-71 grew to 2.8 million ha in 1991-92 and then more rapidly to 5 to 6 million ha by 1994-95. This sector is likely to grow rapidly in the future both on account of internal demands and export opportunities.

ANIMAL HUSBANDRY AND FISHERIES

Animal husbandry and dairying sub-sector plays an important role in overall economy and in social development. The contribution of the sub-sector is estimated to be about 25 percent of the total value of output of agricultural sector.

Table. Production of milk, eggs and fish in India

Year	*Milk production / availability*		*Fish*	*Eggs*
	million tonnes/annum	*g./day*	*(000 tonnes)*	*(million)*
1950-51	17.0	124	752	1832
1960-61	20.0	124	1160	2881
1970-71	22.0	112	1756	6172
1980-81	31.6	128	2442	10060
1990-91	53.9	176	3836	21101
1998-99	75.0	210	5388	28400

The sector also plays a significant role in supplementing family incomes and generating employment in the rural sector particularly among the land-less, small and marginal farmers and women besides providing nutritious food. Production of suitable cross breeds and their wider adoptions has contributed to increasing country's milk production, which has now reached 75 million tonnes annually. Similarly, genetic improvement and better management practices have markedly pushed up production of poultry and eggs.

Table. Number of holdings (million)

Category of Farmers	*1970-71*	*1980-81*	*1990-91*
Marginal	35.7 (50.6)*	50.1 (56.4)	62.1 (59.0)
Small	13.4 (19.0)	16.1 (18.1)	20.0 (19.0)
Semi-medium	10.7 (15.2)	12.5 (14.0)	13.9 (13.2)
Medium	7.9 (11.2)	8.1 (9.1)	7.6 (7.3)
Large	2.8 (3.9)	2.2 (2.4)	1.7 (1.6)
All groups	70.5 (100)	89.0 (100)	105.3 (100)

*Figures in parenthesis represent the percentage of holdings.

Through the overall contribution of fisheries sub-sector is small, development of prolific and fast growing forms of several common fish species coupled with breakthrough in breeding under captivity, fish seed production and multi-layer fish culture has resulted in registering a very high annual growth of 11 percent in aquaculture production during the past decade.

Table. Area operated - % of total area

Category of Farmers	*1970-71*	*1980-81*	*1990-91*
Marginal	9.0	12.1	14.9
Small	11.9	14.2	17.3
Semi-medium	18.5	21.1	23.1
Medium	29.7	29.6	27.2
Large	30.9	23.0	17.4
All groups (million ha)	**162.1 (100)**	**163.8 (100)**	**165.6 (100)**

5

Use of Genetically Modified Crops in Developing Countries

GENETICALLY MODIFIED CROPS

In October of 2000, the detection of the genetically engineered StarLink corn in Tacos and several other food products caught the attention of American consumers. During the last week of February, it was reported that most of the corn seeds ready for planting this year are contaminated with StarLink corn. Farmers and federal regulators, consequently, face a dilemma about what corn to plant. Americans are also increasingly aware of protests around the world objecting to genetically modified foods, identifying them as frankenfood or as environmentally unfriendly. These recent news items have re-ignited public debate over the technology of genetically altering crops. This and the subsequent two articles in future issues of this newsletter will attempt to address some common issues related to Genetically Modified Foods and the technology of genetic engineering. Part I will delve into the background of genetic engineering and the future of genetically altered foods; Part II will address issues related to the benefits and drawbacks of this technology; and Part III will delineate, in simple terms, the basics of how genetically modified organisms are created.

Selective plant breeding is not a new concept. Casual selection of observed desirable traits by our ancestors essentially tamed wild plants and made them suitable for agriculture. In the past, if pests devastated a field of crops and a few plants stayed alive and healthy, the seeds from these healthy plants were used to generate the next crop. Thus the beneficial factors that made the plants resistant were transferred to the next generation, making the new generation of crops slightly more resistant to the same pests. Such selections have been used for over 10,000 years, since the beginning of agriculture and have resulted in significant advances for humanity with increased yields, disease resistance and, overall, greater productivity. A good example is corn; the original crop was Teosinte with very small seeds and very few seeds in each cob. Over the centuries people selected for various traits thus improving the size of the corn kernel, the number of kernels in each cob, the stronger attachment of the kernels to the cob so that it could be harvested and so on, resulting in the corn that we are familiar with today.

It has since become evident that all traits, beneficial or otherwise, are conferred to living organisms by genes. Genes are small segments of the DNA (deoxyribonucleic acid) that code for specific proteins, which in turn regulate the various traits in all living organisms. There are thousands of genes in plants that regulate the activities governing the growth, stature, colour, and all other aspects of plant growth and survival. Selection through traditional breeding involved the transfer of numerous genes from one generation to the next, including the genes for the beneficial trait as well as the undesirable traits. It then took years of self-breeding and selecting to get a plant with both the normal characteristics as well as the beneficial trait desired.

The advent of genetic engineering greatly enhanced this process of transferring a beneficial trait into plants by directly transferring the gene/s responsible for the beneficial attribute. So in one generation, or one planting season, a plant can be created that is the same in all respects except the addition of the beneficial trait. Since genes in all living organisms code for similar proteins and properties, it is possible to transfer a gene from, say, one good corn variety to another corn variety or from fish to strawberry plants.

We can now define a genetically modified organism: It is any organism that has been modified by altering one or more genes by recombinant techniques. A recombinant technique is the method used to transfer a gene of interest from one organism to another. A genetically modified food therefore is any food that is produced from plants or animals that have been genetically altered using this method. The first genetically altered plant created was a tobacco plant with resistance to antibiotics in 1983. It was almost 10 years later when the first commercial genetically altered crop, a delayed ripening tomato, "Flavr-savr", was commercially released. This was not a commercial success, however, for reasons related to production and marketing strategies. This was soon followed by the release of several crops including Roundup Ready soy and corn. Corn and soy are the two most commonly used food crops that have been genetically altered. They have been primarily altered so that the plants can resist pests, diseases or chemicals used to destroy weeds in the field.

Such alterations that improve the health of the plant and potentially benefit farmers are commonly termed as input traits; that is farmers use fewer inputs to grow their crops, be it pesticide, herbicide or chemicals to prevent diseases. There are other alterations that are possible that alter the property of the oil or starch in the seeds. These are termed output traits; that is the seeds produced by the plant have altered properties either by way of improved yields, nutritional content or higher levels or quality of starch, proteins or oils.

Genetically altered foods are very prevalent, at least in the United States and the Western world. More than 60 per cent of the foods we purchase from the supermarket today have ingredients derived from genetically modified crops. Most of these are either from corn or soybeans, which are the base for numerous ingredients manufactured for the food industry, including starch, oils, proteins and other ingredients.

Despite this prevalence, a recent USDA consumer focus group survey revealed that most consumers were unaware of the use of biotechnology in foods. Furthermore, the benefits of biotechnology were viewed as skewed towards producers and manufacturers, with little benefit to the consumer. There was also skepticism related to the long-term health effects and impact on environment. It is therefore essential that we disseminate the information about the technology of genetic engineering so that we can have an informed debate on the merits and shortcomings of this technology.

DISEASE RESISTANT CROPS

CUCURBITS AND SQUASH

Squash belongs to the Cucurbitaceae, a family of crops that is an important source of food throughout the world. While not as important as the cereals or oilseed crops, such as soybean, nor other vegetable crops such as potatoes and tomatoes, cucurbits are nevertheless grown on significant acreages worldwide.

In 1998, cantaloupes and other melons were grown on 1 044 672 hectares, producing 17 764 188 metric tonnes of food; cucumbers and gherkins were grown on 1 567 389 hectares, producing 26 673 943 metric tonnes; while pumpkins, squash, and gourds were grown on 1 136 083 hectares, producing 14 169 983 metric tonnes. Cucurbit crops, including squash, are used for food in all regions of the world, but are also used for oil and other non-food uses, such as musical instruments or household implements.

IMPACT OF VIRUS DISEASE ON SQUASH

Squash, like the other major cucurbit crops, is susceptible to infection by several viruses. The most damaging are cucumber mosaic virus (CMV), papaya ringspot virus watermelon strain (PRSV-W), watermelon mosaic virus 2 (WMV2) and zucchini yellow mosaic virus (ZYMV). While the extent to which viruses cause economic damage in cucurbits, and squash in particular, is not well documented, the problem is extensive enough that diligent effort to breed squash resistant to CMV, PRSV-W, WMV2 and ZYMV has been a goal of breeders for many decades.

DISEASE BREEDING IN SQUASH

Sources of genes for resistance or tolerance to viruses have been reported in various relatives of squash and in *Cucurbita pepo* itself. However, introducing useful levels of resistance to more than one virus into commercially acceptable cultivars has been difficult to achieve. While releases of zucchini-type and English marrow-type varieties carrying resistance to a single virus (CMV) has been achieved, breeding for multiple virus resistance has only recently been successful. A summer squash cultivar, Whitaker, has recently been developed by researchers at Cornell University.

This cultivar is resistant to multiple pathogens, including viruses and fungi. The difficulty of this accomplishment is emphasized by the fact that Whitaker was the result of work conducted by a team of researchers for more than a decade. Because the sources of resistance were related species not normally sexually compatible with *C. pepo*, these researchers had to employ a commonly used breeding tool, embryo rescue, to overcome the barriers to crossing between species.

SQUASH AS A SPECIES

Cucurbita pepo is a crop native to North America. It consists of several subspecies and varieties that are completely interfertile, and are therefore presently considered part of a single species. *C. pepo* ssp. *pepo* includes cultivated pumpkins, marrows and a few ornamental gourds; *C. pepo* ssp. *ovifera* var. *ovifera* includes cultivated crookneck, scallop and acorn squashes, and most ornamental gourd cultivars; *C. pepo* ssp. *ovifera* var. *texana* includes free-living populations found in Texas; *C. pepo* ssp. *ovifera* var. *ozarkana* includes free-living populations found in the Mississippi Valley and the Ozark Plateau; *C. pepo* ssp. *fraterna* includes non-cultivated populations found in northeastern Mexico. Populations of *C. pepo* exist today under non-cultivated conditions in the United States.

These free-living populations range from northeastern Mexico and Texas, to Alabama, and through the Mississippi Valley to Illinois. Some free-living populations, especially in Arkansas, Louisiana and Mississippi, occur as weeds in soybean and cotton fields, and are difficult to control. Gene exchange between free-living and cultivated populations has been documented.

Furthermore, some free-living populations may have originated as escapes from cultivation, which might have undergone subsequent introgression with other nearby cultivated, or other free-living populations.

GENETIC ENGINEERING OF SQUASH

Because of the need to obtain useful virus resistance in a commercially acceptable cultivar, an alternative to breeding was attempted: the direct introduction of genes conferring resistance into already known commercial genotypes. The genetic engineering approach was to introduce the coat protein genes of the viruses in question, a strategy first successful in tobacco, and subsequently shown to be effective in cucurbit species. Constructs containing viral coat protein genes were introduced into proprietary inbred lines of *C.*

pepo ssp. *ovifera* var. *ovifera* via *Agrobacterium*-mediated transformation. The engineering of the constructs used in the transformations, as well as the transformation procedure, are described in Tricoli *et al.*.

Based on greenhouse testing and small-scale field tests, two lead lines, ZW20 and CZW3, were chosen as parents of commercial hybrids. The characteristics of these lines are described in detail by Tricoli *et al.*

ZW20

Line ZW20 was transformed with disarmed *Agrobacterium* harboring a binary Ti-plasmid. This plasmid contained expression cassettes consisting of the coat protein genes of zucchini yellow mosaic virus (ZYMV) and watermelon mosaic virus 2 (WMV2) under the control of the cauliflower mosaic virus 35S promoter. The T-DNA also contained, as a selectable marker, a neomycin phosphotransferase II (NPTII) gene from *E. coli* transposon Tn5, under the control of the cauliflower mosaic virus 35S promoter. The transformation event contained multiple insertion events, some of which were structurally complex. Among these insertion events were two loci containing genes encoding both the ZYMV and WMV2 coat proteins, but not the NPTII gene.

One locus provided a moderate level of resistance to the two target viruses, while the other provided a high level of resistance. Advanced breeding lines were made homozygous for the high-resistance locus, but were segregating for the moderate-resistance locus. During this process, the NPTII marker was also eliminated by selection of lines that not only were homozygous for the high-resistance locus, but had also lost the selectable marker through segregation.

CZW3

Line CZW3 was transformed with the plasmid pPRBN-CMV73/ZYMV72/WMBN22, containing expression cassettes for the coat protein genes for ZYMV, WMV2 and cucumber mosaic virus (CMV), as well as the NPTII selectable marker. Line CZW3 was a simpler transformant, consisting of only one insertion event, containing all three coat protein genes, as well as the selectable marker. Advanced breeding lines were homozygous for the transformed locus.

Field Results

Field trials at a number of locations showed that the lines ZW20 and CZW3 were effective in controlling infections by ZYMV, WMV2, and - in the case of CZW3 - CMV as well. Tricoli *et al.* mechanically inoculated two different ZW20 sublines, including one homozygous for the high-resistance locus. This subline showed only 20 per cent infection after inoculation with WMV2, and no infection at all after ZYMV infection, compared with 100 per cent infection of non-transgenic controls. The same researchers studied the resistance of CZW3 after mechanical inoculation with CMV, ZYMV and WMV2 individually, and in a mixed infection. Segregating inbred lines were tested, as well as a hybrid in which the segregating inbred was used as a parent. After individual and mixed inoculation of transgenic segregants, only one plant out of a total of 146 showed mild symptoms, while all of the 103 non-transgenic segregants as well as all of the 259 non-transgenic control plants became severely symptomatic.

Further tests were carried out in the field, comparing virus resistance in these lines after mechanical and aphid inoculation. Other tests studied the economic benefit of the resistance provided by lines ZW20 and CZW3 either as an inbred line or as the parent to a hybrid. These tests confirmed that virus resistance conferred by the coat protein genes was able to dramatically increase crop yield and economic return under the pressure of naturally vectored disease in the field.

USDA EXEMPTION

The determination of efficacy against viral disease under field conditions led to the preparation of a petition for a determination of non-regulated status from the United States Department of Agriculture, Animal and Plant Health Inspection Service (USDA-APHIS). USDA-APHIS is the primary agency responsible for regulating work, in the greenhouse and field of transgenic plants. It also has the authority to grant non-regulated status to transgenic plants that in its opinion no longer present a plant pest risk. The main responsibility for conducting an assessment of safety rests with USDA-APHIS.

The assessment relies upon a review of the information package supplied to the agency by the petitioner, but the agency can and does seek out information on its own, and relies on the expertise of the reviewers on its staff. All relevant information is used by the agency to conduct an environmental assessment, and - if the assessment allows - grant non-regulated status. One guiding principle followed by the USDA-APHIS in assessing risk is to compare the characteristics of the transgenic plant with the characteristics of plants that are produced by traditional plant breeding. Plants produced by traditional breeding are not regulated, because history provides sufficient assurance of safety.

Thus, if the transgenic plant were determined to pose a risk different from that posed by plants already existing in nature or plants that would be produced by traditional breeding, then non-regulated status would be withheld. If, on the other hand, the risks presented by the transgenic plant are not different from those that are presented by traditionally bred varieties, then non-regulated status can be granted. Since USDA-APHIS relies heavily (but not exclusively) on the information contained in the petition for non-regulated status, information in such a petition should include the following:

- A description of the biology of the plant prior to the transformation. The biological description should include detailed taxonomic information.

- Relevant experimental data and publications relating to the work conducted on the transgenic plant.
- A description of the differences in genotype between the transgenic plant and the corresponding non-transgenic plant. Information must be provided on the organisms from which the transgenes or other control elements were obtained, the method of transformation and the vectors used during the transformation process, as well as the inserted genetic material and its product(s). USDA-APHIS, the Canadian Food Inspection Agency, and Health Canada, have standardized the types of molecular and expression data submitted in a petition for non-regulated status and the criteria by which they should be judged.
- A description of the phenotype of the transgenic plant, including any known and potential differences from the corresponding non-transgenic plant, in order to substantiate the claim that the transgenic plant is unlikely to pose a greater plant pest risk than the non-transgenic plant from which it was derived. This information should include the following: plant pest risk characteristics, disease and pest susceptibilities, expression of the gene product, new enzymes, or changes to plant metabolism, weediness of the transgenic plant, impact on the weediness of any other plant which is sexually compatible with the transgenic plant, agricultural or cultivation practices, effects of the transgenic plant on non-target organisms, indirect plant pest effects on other agricultural products, transfer of genetic information to organisms with which it cannot interbreed, and any other information which might be relevant. Any information indicating that a transgenic plant may pose a greater plant pest risk than the corresponding non-transgenic plant must be included.
- Field test reports for all trials of the transgenic line in question that were conducted under USDA-APHIS permit or notification. The field test reports should include any observations of deleterious effects on the environment caused by the transgenic plant.

Based on information received by USDA-APHIS from various sources including the petition, the Agency granted non-regulated status to ZW20 and CZW3, and published their support for this decision. In keeping with their goal of considering all information that might be available to them, USDA-APHIS made use not only of data supplied by the petitioners, but also relied on the literature, external expert opinion, and comparison with traits that were already familiar to agriculture. This approach acknowledged that scientifically valid risk assessment could be conducted by relying on information obtained from a wide variety of sources.

WEEDINESS

Of particular concern in determining the plant pest risk of transgenic squash was the potential for the crop itself to become a weed, or for new weeds to arise as a consequence of gene flow from the crop to sexually compatible wild populations. This question was especially important because the transgenic squash would be released in a region that was known to be the centre of origin for *C. pepo*. After evaluating all the information available, USDA-APHIS concluded that ZW20 and CZW3 were unlikely to cause a weed problem or a threat to genetic diversity of free-living squash if released from regulation.

With regard to causing a weed problem, the agency concluded that the transgenic lines were unlikely to become weeds themselves, and that the virus-resistance traits were unlikely to cause free-living squash to become weeds. The conclusion that the transgenic lines themselves were unlikely to become weeds was based on three reasons. First, the inbred lines used in the transformations belonged to the yellow crookneck squash type (*C. pepo* ssp. *ovifera* var. *ovifera*), which did not exhibit weedy characteristics and was not considered to be a weed in standard texts and weed lists. Second, introduction of CMV (for CZW3) as well as ZYMV and WMV2 resistance into the crop was not different from other disease-or pest-resistance traits already present in the *C. pepo* gene pool or introduced into it by traditional breeding.

These traits had already been introduced into squash, and even into some commercial cultivars, yet these cultivars had not become weeds. Finally, the two transgenic lines did not exhibit any other characteristics that were more 'weedy' than those exhibited by traditionally bred squash cultivars. Aside from considering the potential problem caused by the transgenic lines themselves, USDA-APHIS also evaluated the possibility that traits transferred from ZW20 and CZW3 to free-living relatives would cause those free-living plants to become more 'weedy'.

USDA-APHIS concluded that this problem was unlikely to happen. The agency assumed that the transgenic lines and their cultivated progeny would be able to pollinate free-living squash plants and transmit the virus resistance traits to them. This undoubtedly had occurred in the past with other disease- or pest-resistance traits that had been introduced into *C. pepo*. However, despite this opportunity, no weedy hybrid progeny had emerged. Since there was no evidence to suggest that the transgenic traits would behave differently from the traditionally bred traits, the historical lack of weed problems provided a sufficient level of assurance that crosses between transgenic and free-living squash were unlikely to create a weed problem as well.

Furthermore, the cultivated squash genotypes that were used for transformation possessed characteristics that placed those plants at a selective disadvantage in a non-cultivated habitat. The transmission of these traits to free-living relatives along with the transgene would place recipient plants at a disadvantage, further reducing their weedy potential. Finally, surveys of free-living squash populations taken by the applicant failed to detect infections of CMV, WMV2 or ZYMV. Therefore, there was no evidence that viruses limited the populations of free-living squash.

Consequently, it was not likely that virus resistance would release these populations from control, nor was it likely that virus resistance conferred a significant selective advantage in free-living habitats. The lack of selective pressure exerted by viruses was also cited by USDA-APHIS as the reason for concluding that genetic diversity in free-living populations would not be reduced. Without selection pressure, there was no reason to expect selection of a subset of genotypes into which the virus resistance would have been crossed.

TRANSCAPSIDATION/RECOMBINATION

The commercial introduction of viral sequences via transgenic plants - and in particular viral genes encoding viral coat proteins - present the potential for interactions between the transgenes or their products, and the viruses to which the transgenic squash might remain susceptible. These interactions include transencapsidation of heterologous viral RNA by coat proteins expressed in the plants, phenotypic mixing between viral coat protein subunits expressed by the transgenic plant or by the transgenic plant and infecting viruses, and recombination between transgenic viral components and other viruses infecting the transgenic plants. DeZoeten pointed out that these phenomena have the potential to produce new viruses. USDA-APHIS concluded that the risk of transencapsidation and phenotypic mixing in transgenic virus-resistant plants would be no greater than their likelihood of occurrence in non-transgenic plants already in agriculture. Data submitted by the petitioners, and other information gathered by the agency, showed that the amounts of viral coat protein produced in transgenic plants resistant to CMV, ZYMV and WMV2 were less than the amounts found in susceptible, virus-infected plants.

With respect to this set of data, USDA-APHIS noted that ZYMV and WMV2 coat protein levels increase when ZW20 plants were inoculated with ZYMV and WMV2 singly or combined. The agency speculated that this increase might be due to limited replication of the viruses in inoculated ZW20 plants, or because the coat proteins produced by the transgenes were stabilized by limited replication of the inoculated viruses.

EPA EXEMPTION FROM TOLERANCE

As is the case with USDA-APHIS, the US Environmental Protection Agency (EPA) also has the primary responsibility of conducting a safety assessment, based on information supplied by the developer of a transgenic crop and other information that the agency might obtain, including that of external and internal experts. When the first transgenic squash line (ZW20) was ready for commercialization, the EPA did not have a final rule covering transgenic plants. However, consultation with the EPA took place during the time the agency was formulating its proposed rules.

These rules extended the agency's jurisdiction to all plants expressing disease-or pest-resistance traits, since they could be viewed as producing a substance that was pesticidal - that is, they mitigated the effect of a pest. While the agency claimed jurisdiction over these plants, it recognized that certain broad categories of plants could be exempted from regulation. For example, plants produced through traditional breeding, or plants that repelled pests or disease through the modification of some structural feature could be exempted since they usually presented no new pesticidal exposure to humans consuming those plants. Furthermore, the EPA proposed to exempt coat protein produced in transgenic plants from regulation under the Federal Insecticide, Fungicide, and Rodenticide Act (FIFRA) and the Federal Food, Drug, and Cosmetic Act (FFDCA), again because they presented no new exposures to humans consuming them. Even though the rules were only proposed, consultation with the EPA led to the decision to submit a petition for exemption from tolerance (the legal limit for a pesticide chemical residue in or on a food), as required by the FFDCA prior to the commercialization of any pesticide.

In order to obtain this exemption, it was important to show that the establishment of a tolerance level was not required to protect the public health. This goal was achieved for line ZW20 by measuring the amount of viral coat protein in samples of various melon and squash fruit on sale at major grocery stores in one representative locality in the United States. Results showed that levels of viral coat protein in transgenic fruit were within the range found for most of the fruit in the sample. For some samples, the fruit on the market had many-fold higher levels of coat protein than in the transgenic squash.

These data were used to support the claim that viral coat proteins in general, and ZYMV and WMV2 coat proteins in particular, were not substances that fitted the two main criteria which the EPA planned to use to determine whether a pesticidal substance would require review. First, in order to trigger EPA review the pesticidal substance would be one that was not derived from a known food source. Second, the substance would present the following changes in pesticidal exposure to humans: it would be found in significantly higher levels in the transgenic plant than that normally found in food, it was being transferred to a different crop than that in which it was already found, it was not already found in edible tissue, and it was not already found in unprocessed food.

Other arguments advanced in the petition in support of the exemption from the requirement of a tolerance were based on the premise that a demonstrated history of safe use was sufficient to assess safety. Examples from the literature showed that humans were already exposed to viral coat proteins in general, not only through consumption of naturally infected produce, but also through ingestion of food obtained from crops that had been subjected to cross protection (the intentional inoculation of a crop with a mild strain of a virus in order to protect the crop from subsequent infection by a more virulent strain of the virus).

The petition also argued that supported the claim that expression of ZYMV and WMV2 in ZW20 would not present any changed exposure of humans to allergens. Because ZYMV and WMV2 coat proteins are commonly present in cucurbit fruit, anyone who might be allergic to viral coat protein in squash (a phenomenon for which no documented cases were known) would have the same reaction to the transgenic fruit as to non-transgenic fruit they had already consumed. The petition also addressed exposures due to dermal contact and inhalation. The petition argued that a history of safety had been established through the lack of reports of toxicity or allergenicity from researchers who had undoubtedly been exposed to these viruses by both the dermal and inhalation routes.

Finally, the petition pointed out that viral coat proteins were ubiquitous in the environment, and therefore a natural component of it. Infected cucurbit fruit that are not marketable are routinely plowed back into the soil, thus adding large amounts of viral coat protein to the environment. Furthermore, studies in the literature have reported the recovery of plant viruses from ground water run-off in Europe and North America. The EPA considered the data and the arguments presented in the petition and granted an exemption from tolerance for the ZYMV and WMV2 coat proteins produced within ZW20 plants.

Subsequently, the EPA granted a more categorical exemption from tolerance for ZYMV and WMV2 coat proteins in all food commodities. The decisions and supporting reasoning are summarized in the following paragraphs, which are based on the published decision. A detailed review of this decision is especially instructive because it provides insight into the issues that by law must now be considered by the EPA. A detailed review also provides another example of the reliance of US regulatory agencies not only on data submitted by petitioners, but also on the literature and other sources of information. In granting the broader exemption from tolerance, the EPA considered several issues that by law they were obligated to address, especially as a result of the passage of the Food Quality Protection Act of 1996 (FPQA), which amended the FFDCA.

The amended FFDCA allowed EPA to establish an exemption from the requirement for a tolerance only if EPA determines that the tolerance is 'safe'. By law, 'safe' meant that there was 'a reasonable certainty that no harm will result from aggregate exposure to the pesticide chemical residue, including all anticipated dietary exposures and all other exposures for which there is reliable information'. This includes exposure through drinking water and in residential settings, but does not include exposure due to its use during application or other occupational activities. In deciding on exemptions from tolerance the EPA is also required to pay special attention to exposure of infants and children. The EPA examined two areas of concern in determining the risks from aggregate exposure to pesticide residues.

First, the agency examined the toxicity of the 'pesticides' (ZYMV and WMV2 coat proteins). Second, it examined exposure to the pesticide through food, drinking water, and through other exposures that occur as a result of pesticide use in residential settings. In considering toxicity, petitioners are required by law to submit data regarding acute toxicity, genotoxicity, reproductive and developmental toxicity, as well as subchronic and chronic toxicity. However the petitioner requested, and EPA granted, waivers from these data requirements. The waivers were based on the long history of mammalian consumption of the entire plant virus particle in foods, without causing any deleterious human health effects.

The EPA once again acknowledged that virus-infected plants currently are - and have always been - a part of both the human and domestic animal food supply, and that there is no evidence that plant viruses are toxic to humans and other vertebrates. Furthermore, the EPA pointed out that plant viruses are unable to replicate in mammals or other vertebrates, thereby eliminating the possibility of human infection. As additional assurance of safety, only the coat protein would be expressed in the plant, and this component alone would be incapable of forming infectious particles. The EPA also had to consider the toxicity of the introduced DNA (coding and regulatory regions) directing the synthesis of the coat proteins. The EPA acknowledged that DNA is common to all forms of plant and animal life and that it was consumed as a safe component of food.

The specific nucleic acids that were used to transform the squash plants were characterized by the petitioners, and a review of the nucleic acid sequence led EPA to conclude that no mammalian toxicity was anticipated from the genes encoding the coat proteins of ZYMV and WMV2. In considering exposures, the EPA had to consider available information concerning exposures from the coat protein in food and all other non-occupational exposures, including drinking water from groundwater or surface water, and exposure through its use in gardens, lawns or buildings. With respect to dietary exposure in food, the EPA concluded, as they did before, that the use of viral coat protein in squash would not result in any new dietary exposure to plant viruses, because:

- Entire infectious particles of ZYMV and WMV2 are found in the fruit, leaves and stems of most plants,
- Viruses are ubiquitous in the agricultural environment at levels higher than will be present in transgenic plants,
- Virus-infected food plants have historically been a part of the human (including that of infants and children) and domestic animal food supply, with no known adverse effects to human or animal health.

Moreover, the EPA concluded that there was no evidence to indicate that harmful effects occurred as a result of exposure to viral coat proteins through other means, such as non-food oral, dermal and inhalation routes. They arrived at this conclusion by relying not only on data submitted by the petitioners, but on the expert opinion of its Scientific Advisory Panel, which concluded that the levels of virus in the agricultural environment are much higher than those present in transgenic plants, and that the existing presence of viral coat proteins in the food supply provided a scientific basis for exempting transgenic plants that express viral coat proteins from the requirement of a tolerance. With respect to drinking water exposure, the EPA concluded that potential exposures through this route would be negligible.

The agency reasoned that viral coat proteins produced in plants are an integral part of the living tissue of the plant, and are therefore subject to rapid degradation and decay. Consequently, viral coat proteins would not accumulate in the environment or in the food chain. Therefore, because of their presumed transience, these proteins would present only negligible exposure in drinking water obtained from surface or groundwater sources. The EPA used similar reasoning to conclude that exposure of engineered coat proteins as a consequence of their use in such areas as around homes, parks, recreation areas, athletic fields and golf courses, would be negligible. The EPA concluded that the likelihood of exposure to plant viral coat proteins, including ZYMV and WMV2 coat proteins, through any dermal or respiratory routes would be unlikely.

Although physical contact with the plant or raw agricultural food obtained from the plant may present some opportunity for dermal exposure, the potential amounts involved in this exposure were judged to be negligible compared with exposure through the diet. Furthermore, the EPA reasoned that the skin would provide a barrier that would not be crossed by viral coat proteins. Similarly, exposure through inhalation was judged to be negligible compared with potential exposure through the diet. The most likely opportunity for respiratory exposure through inhalation was through pollen. However, the EPA concluded that it was again unlikely that the amount of exposure to the coat proteins in the pollen would be significant. Viral coat proteins, when present in pollen, would be part of the pollen grain and would not cross the barrier presented by the mucous membrane of the respiratory tract.

Therefore, this route would not add any additional exposure to that presented by the dietary route. A new requirement imposed by the Food Quality Protection Act involved the assessment of available information, indicating that there might be special concerns regarding exposure of infants and children to viral coat proteins. The EPA concluded that viral coat proteins were present in all foods, including those foods consumed by infants and children. It had no evidence to indicate that the amount of exposure was different between adults and children, nor that the effect of viral coat proteins differed between the two groups.

Most importantly, the EPA had no evidence suggesting that exposure of adults or infants and children was in any way harmful. Finally, the EPA had to consider whether ZYMV and WMV2 coat proteins, or the DNA needed to produce them in plants, would have an effect on the immune and endocrine systems. Because the agency had no information indicating that such an effect could be expected, it did not require that the petitioner submit information regarding that question. The decision to grant an exemption from tolerance for ZYMV and WMV2 coat proteins in ZW20 and subsequently in all raw commodities also cleared the way for commercialization of CZW3. EPA then granted an exemption from tolerance for the coat protein of CMV in all raw commodities, based on the same safety assessment it conducted for the other two coat proteins. Line CZW was included in this exemption and was therefore cleared for commercialization.

FDA FOOD SAFETY ANALYSIS

In contrast to USDA-APHIS and EPA, the United States Food and Drug Administration (FDA) makes the safety assessment of new plant varieties, and the foods made from them, the responsibility of the developer or manufacturer. In 1992, the FDA published guidelines to clarify the obligations of any producer of new plant varieties, whether produced through traditional plant breeding, or through techniques of recombinant DNA. The principle governing the guidelines is that assessment of the safety of a product (in this case food produced from a new plant variety) should focus on the product's characteristics rather than the method by which it was made.

The authority of FDA to regulate food is a post-market authority. Thus, a food producer can market a food containing a new component that it has assessed to be generally recognized as safe (GRAS). If this component is not GRAS, then a food additive petition must be filed and granted before food containing that new component can be marketed. The 1992 guidelines were meant to help producers in determining whether a food safety concern existed. If a concern existed, a consultation with FDA would take place to determine whether a new component could be considered GRAS, or whether a food additive petition was required.

In practice, consultation with FDA regarding new transgenic plants that will become food usually occurred regardless of safety concerns. The consultation assures producers that FDA is aware of the developer's safety assessment process, and serves to identify those parts of the safety analysis that might not be acceptable to the Agency. This process enables the producer to proceed towards commercialization with confidence that no unanticipated safety concerns will be encountered at late stages of product development.

Recently, the FDA has published its intention to make this consultation process mandatory.

The assessment applies equally well to CZW3. In general, the safety assessment specified by the guidelines examines the host plant (the plant which has been modified by introduction of foreign DNA), the organisms from which the foreign DNA has been obtained, and the new compounds produced as a result of the introduction of the foreign DNA.

First, the guidelines direct the developer to determine whether there is a history of safe consumption of the species that was transformed. In the case of squash, the species has had a history of safe consumption for at least 3000 years, and is consumed in many forms throughout the world. The particular genotypes transformed to produce ZW20 and CZW3 have been in use commercially for several years.

The guidelines also ask the developer to consider whether characteristics of the host species, related species, or the lines used in transformation warrant analytical or toxicological tests. If so, they also ask whether the results of the tests provide evidence that the toxicant levels in the new plant variety do not present a safety concern. Squash indeed has characteristics warranting analytical testing, since the plant is known to produce cucurbitacins, which are extremely bitter toxic alkaloids. Squash breeders are aware that cucurbitacins must be eliminated from commercial varieties. Therefore, standard breeding practices take into consideration the potential presence of this compound, especially if genes from wild or free-living relatives are crossed into cultivated squash.

The FDA guidelines indicate that the types of testing currently used to screen for known toxicants have been effective in managing the risk of such compounds, and are therefore appropriate to apply to the case of transgenic crops. For the cucurbitacins, tasting fruits for bitterness has been an effective test method for the detection of this compound. The documented sensitivity of the human sense of taste for this compound is as low as 1-10 parts per billion, depending upon the particular form of cucurbitacin. At 10 parts per billion, this detection level is 34 000 times less than the reported oral LD50 in mice. Therefore, tasting fruit is an appropriate analytical test for the presence of cucurbitacin, and the presence of these compounds can be detected long before toxic levels are reached.

The non-bitter taste of ZW20 and CZW3 indicates the absence of cucurbitacins in these lines, as they were in the original lines before transformation. This result CMV, ZYMV and WMV2 coat proteins are almost identical to the coat proteins (presumably edible) found in cucurbit fruit. Squash containing viruses is not known to be allergenic, nor is it known to be toxic. Intake of viral coat proteins from the fruit of transgenic plants will be comparable to the intake of these proteins in non-transgenic infected fruit. Finally, the amounts of viral coat protein expressed in transgenic ZW20 and CZW3 are too low to present a possibility of becoming a major dietary constituent. These data were presented to the FDA in order to inform the Agency of the justification for the judgement that no food additive petition was warranted for ZW20 and CZW3 (they were therefore GRAS). Since no objections to these analyses were raised by the Agency, commercialization proceeded.

The assessment of safety of transgenic crops continues to be a topic of scientific and public concern. A review of the issues that surround the debate over the safety of these crops reveals that the issues addressed seven years ago during the commercialization of virus-resistant squash - which was the first commercialized virus-resistant crop - are still unresolved.

The appropriate perspective on safety is one that was expressed well before transgenic crops were at issue:

A thing is safe if its risks are judged to be acceptable. By its preciseness and connotative power, this definition contrasts sharply with simplistic dictionary definitions that have 'safe' meaning something like 'free from risk.' Nothing can be absolutely free of risk ... Because nothing can be absolutely free of risk, nothing can be said to be absolutely safe. There are degrees of risk, and consequently there are degrees of safety.

This is the perspective that is taken by the agencies responsible for regulating transgenic crops in the US. They all judge the level of risk of these crops by comparing them to the existing level of risk, such as risks presented by crops developed by traditional breeding or risks presented by natural phenomena. Furthermore, even though the risk presented by these crops might be greater than those presented by current unregulated activities, the acceptability of that risk also determine whether these crops are safe enough to commercialize and use. It is this latter issue that adds the dimension of public perception and acceptance to the debate.

The deployment of transgenic crops will most likely continue and increase with time. The potential benefits to the consumer, the farmer, as well as to the companies developing these crops will serve as a potent force for the exploitation of this technology. As this continues, scientific enquiry into the potential risks should continue. These investigations will help to further clarify areas of real concern. By focusing attention on these areas, which will probably be very few, safe use of this technology will be further improved.

CHALLENGES TO CROP GENETIC DIVERSITY

Farmers have traditionally depended upon their own skills and resources to develop the crops that they need. The process began with the earliest domestication of wild species and has continued through the careful selection of plant materials adapted to varied (and often changing) growing conditions and preferences. The result has been a complex and continually evolving collection of local crop varieties, often referred to as landraces, that reflects interactions with wild species, adaptations to changing farming conditions, and responses to the economic and cultural factors that shape farmers priorities.

The richness and range of the diversity of these landraces is now under threat because of the changing nature of agricultural production. One important factor is the widespread adoption of modern varieties (MVs) that are the products of formal plant breeding. These MVs often provide yield increases and other advantages that result in their being sown over large areas. Technological change, such as the use of fertilizer and irrigation, also lowers the demand for landraces adapted to marginal growing conditions.

In addition, farm production is increasingly market- oriented, making farmers less inclined to select for crop characteristics that once were important for local customs and culture. If the adoption and use of MVs and other agricultural technology brings significant benefits to farmers, why should we be concerned with the loss of landraces and the preservation of crop genetic diversity? There are several important reasons.

First, many advances of modern plant breeding have been possible because of the wide range of genetic source material provided by landraces. The very success of modern plant breeding now threatens the source of genetic diversity on which further progress depends, as farmers find it less rewarding to maintain the diverse mixture of landraces developed by their ancestors. Second, the widespread use of MVs raises questions about the stability of crop production and the threat of disease or pest attack. The threat to stability does not necessarily derive from the MVs themselves. MVs are more uniform than landraces, which may increase their susceptibility to pests or pathogens, but many MVs owe their acceptance to the fact that they are more resistant than the varieties they replace. The principal threat to yield stability from MV use is the increasing uniformity, and continuous cropping, that their use engenders. Large areas planted to a single variety are potentially a cause of concern, no matter what the source of the variety.

In addition, the preservation and utilisation of crop genetic diversity is of particular importance to the more marginal, diverse agricultural environments where modern plant breeding has had much less success. Farmers in these areas tend to be poorly served by public research and extension systems. These areas are often centres of diversity for many crop species, but increasing poverty is forcing many of these farmers to place more dependence on non-farm sources of income, with consequent reduction in their capacity to grow and maintain the range of local varieties they have been accustomed to manage.

PLANT GENETIC DIVERSITY IN THE PERUVIAN ANDES

Farmers in the Andes of southern Peru depend on both potatoes and maize as basic staples. A study in the area of Paucartambo illustrates the complexity of genetic diversity maintenance for these crops and provides examples of the criteria that farmers use to select their varieties. Potatoes are the most important crop in the region. Local potato varieties are classified first by use category, divided among those appropriate for boiling, soup-making or freeze drying.

Further sub-classification is done on the basis of morphological differences, and nomenclature reflects a rich cultural symbolism. Most farmers manage between 3 and 6 separate potato fields in which a mixture of varieties of a single use category are planted together. Women are in charge of selecting planting material; for any one field, they specifically select only a few of the most important varieties and then make a representative selection of the remaining material to provide the desired diversity. Observations in fields of boiling potatoes revealed a mean of 21.1 varieties, with a standard deviation of only 5.1, indicating a remarkable consistency in maintaining field diversity. Maize is also classified by the farmers of Paucartambo into use categories, distinguishing among boiling and parching. In contrast to potatoes, sub-classification of maize varieties is based more on specific choices for use, yield, and time of planting. Also in contrast to potatoes, maize varieties of different use types may be mixed in the same field. The average number of varieties per field is much lower than for potatoes, and the women who select the seed for planting make much more precise choices

Finally, the maintenance of a wide and evolving range of local crop landraces is threatened by the advent of intellectual property protection for crop varieties, accelerated by the formation of the World Trade Organization. The increasing application of plant breeders rights has several implications for plant genetic diversity. For a new variety to be legally protected, it must be subject to very precise description, including the requirement that it be distinct, uniform and stable (DUS). This is a disincentive to the promotion of intrinsically diverse landraces or of varietal mixtures. An additional debate concerns farmers privilege, the ability of farmers to save seed of a variety, to exchange it with neighbours, and to adapt it to their own growing conditions. These practices could be challenged by seed companies when crop varieties are sold under strict legal protection. It is even possible to envisage situations where varieties that originated in farmers fields may be legally protected and then denied to the farmers responsible for having developed them. Hence the advent of plant variety protection lends added urgency to the search for solutions to the conservation of plant genetic diversity.

INNOVATIONS AND INITIATIVES

There are several important strategies that address the conservation of plant genetic diversity.

Genebanks

The conventional solution to the conservation of plant germplasm has been through genebanks. These are based on collections of genetic material, from centres of crop origin and elsewhere, that are stored in controlled conditions and periodically regenerated. Genebanks have been important sources of material for plant breeding programmes and other research activities, and will continue to be a basic element in conservation programmes.

Genebanks are not without their problems, however. Although some of the larger genebanks contain impressive numbers of accessions, the costs of characterising and cataloguing genebank material are considerable. Only a small proportion of genebank material is actually used by plant breeders, in part because of inadequate characterisation of accessions. In addition, farmers usually do not have easy access to the materials they have donated. The control of genebanks is also an issue of debate; there are concerns that a country that hosts an important international collection can deny another country access for political reasons.

These issues of control are beginning to hamper efforts at germplasm collection, as countries are increasingly concerned with the possibility that their local landraces may be collected, utilised commercially, and even legally protected, without any compensation. More resources are needed to support national plant genetic conservation efforts, and more attention needs to be devoted to making these collections relevant to farmers interests and capacities. In some cases, national genebanks are being redirected to encourage more farmer participation in their development and utilization.

Crucible Group

The Crucible Group is a collection of 28 individuals including representatives of grassroots organisations, agricultural researchers, intellectual property specialists, trade negotiators and agricultural policy analysts from South and North which met to discuss the issue of the intellectual property protection of plant genetic resources. Not surprisingly, the group s discussions produced varied points of view, but also developed a consensus on a number of important issues. Some of the group s conclusions include:
The Crucible Group stresses the primacy of specific national conservation strategies for plant genetic resources that invite the participation of local communities as well as private companies. Holders of ex situ germplasm collections should develop equitable partnerships with indigenous and rural societies and make their collections available to them.

Under the pressure of possible exclusion from an encompassing global trade agreement, many countries feel pressed to adopt some form of intellectual property protection for plant varieties. The Crucible Group concurs that compulsion is inappropriate and that countries, have every right to protect their environment and the well-being of their peoples if they feel that trade rules threaten their security.

In Situ Cconservation

An option that is attracting increasing attention is the possibility of carrying out genetic resource maintenance in the field. This is often referred to as in situ conservation, in contrast to the ex situ alternative of genebanks. Arguments in support of in situ conservation include the importance of recognising the roles of both environmental factors and farmer intervention in landrace development.

In areas of high crop genetic diversity, landraces often evolve through crossing with wild or weedy relatives, and farmers play a crucial role in selecting and adapting new material. It is this process of landrace development that in situ conservation attempts to enhance and preserve. It is also argued that in situ conservation programmes would provide more opportunities for community conservation, where farmers would be better able to recognise and utilise the plant genetic resources that are available in the community.

But there are several decisions that must be made in designing in situ conservation projects. First, because landraces are not static entities, decisions have to be taken with regard to the nature of human intervention in the selection process. Whose criteria are to be used in the selection and adaptation of new materials? Are local farmers criteria the only ones to be applied in deciding what is conserved, or should scientists interests also play a role in determining the direction of conservation? In addition, farmers generally seek germplasm from other sources to complement their own landraces. To what extent should this be allowed or encouraged in an in situ conservation project?

To what extent is the objective to build a fence around an area of genetic diversity in order to protect traditional crop development processes from outside influences, and to what extent is the objective closer to that of community development? In the latter instance, resources and information are provided to farming communities to empower them to make more informed decisions about the management of local varieties and the utilisation of MVs.

In situ conservation projects will undoubtedly approach these decisions in a variety of ways. At this stage it is important to encourage a range of in situ programmes and to learn from experience. The International Plant Genetic Resources Institute (IPGRI) in Rome is currently developing a large multi-country collaborative project to test and develop techniques for in situ germplasm conservation.

Farmer-participatory Plant Breeding

There are also a number of efforts under way to encourage a wider scope for farmer participation in formal plant breeding. Possibilities include greater farmer representation in priority setting for crop breeding programmes, more explicit attention to the crops and varietal characteristics of importance to these farmers, the transfer of significant aspects of plant breeding research to farmers fields, and the enlistment and training of farmers to take a more active part in the variety testing and selection process.

There is a growing literature on methods to encourage farmer participation. The innovations include rapid rural appraisal techniques to understand farmer varietal preferences, the organisation of various types of adaptive on-farm research to test varieties under farmers conditions, the wider use of landraces in formal breeding programmes, and the establishment of mechanisms for contact between farmers and experiment station personnel. Some plant breeders see the possibility of an integrated system that incorporates the strengths of both formal and informal plant breeding techniques.

Farmer Participation in Gene Banks

The Plant Genetic Resources Centre/Ethiopia (PGRC/E) is involved in several activities that involve farmer participation in conservation activities. The Unitarian Service Committee of Canada helps support a PGRC/E project in which materials collected from nearby areas are given to farmers to plant and to carry out simple mass selection to improve their characteristics. Farmers select plant types on the basis of characteristics important to them, including pest or disease resistance or maturity. Farmers receive assistance from plant breeders in this activity, and PGRC/E scientists establish standard descriptor lists for the materials.

Farmers evaluate their selections in experimental plots and compare performance and yields with the original seed stock. The plots are also used for on-site maintenance of landraces. In another project, the PGRC/E is collaborating with Debre Zeit Research Centre to maintain indigenous material of tetraploid wheat. Wheat germplasm collected by PGRC/E is tested and multiplied by breeders and then provided to farmers for further multiplication and selection. This allows farmers to experiment with landrace lines while the indigenous populations are maintained in the genebank.

KNOWLEDGE GAPS

Although there is undeniable evidence of the erosion of crop genetic diversity, and several innovative responses have been developed, there are important gaps in knowledge that limit our capacity to decide among the various alternatives. Some of these gaps involve our technical understanding of the nature of biodiversity, while others are concerned with our understanding of its socio-economic correlates.

Appropriate measures for diversity still need to be developed in order to better characterise the current situation and to evaluate changes in the future. Farmers crop variety classification systems are one place to start, but we do not know how these correspond to actual genetic differences. With few exceptions, there is little research that correlates the variation in folk taxonomy with genetic differences. The capacity to evaluate genetic material in the laboratory is growing rapidly, but these are still expensive techniques, and more robust markers and measures are required to follow the progress of genetic resource conservation.

When better measures of genetic diversity are devised they will contribute to a clearer understanding of what exactly needs to be conserved. Currently, there are only very general ideas about what portion of a plant population needs to be maintained in order to conserve particular genetic traits. This information is crucial to the efficient design of in situ conservation projects, for instance, in order to know if efforts might be limited to a fewfarmers fields, or should instead sample more widely from an agroecosystem.

More study is also required to understand the causes of plant genetic erosion. In some cases, the introduction of MVs has led to the replacement of large numbers of local varieties. On the other hand, several studies of varietal adoption in centres of crop diversity have shown that MVs often complement, rather than replace, local varieties. The availability of inputs and crop management techniques that reduce the need to tailor varieties to difficult growing conditions has also contributed to the demise of many local varieties.

The cultural value placed on crop diversity and local selection techniques is also diminishing in many areas, and the skills that have contributed to landrace evolution are consequently disappearing. As farmer variety selection skill is a threatened resource, there is also a need to understand how to conserve and enhance it. Although farmers deserve credit for landrace development, there is little knowledge of what the genetic consequences of their selection techniques are, or what specific effects these have on enhancing diversity. Farmers select materials for practical reasons that may not always be compatible with the maintenance of genetic diversity. Very little interdisciplinary study has been devoted to understanding the biological outcome of the application of indigenous technical knowledge and skills in variety selection. There is also little understanding of whether attempts to improve local selection capacity should focus on individual farmers or on communities. Indeed we know very little about how varieties, and information about them, are exchanged within communities.

POLICY IMPLICATIONS

The challenges of crop genetic resource conservation highlight once again the dilemma of balancing between development and conservation. The development contribution of modern plant breeding weakens diversity by replacing many local varieties, and thus threatens the possibilities of further progress. Areas of high genetic diversity, on the other hand, are often distinguished by their poverty and the declining viability of their agriculture, and are desperately in need of innovation and change.

The dilemma is evident in choices of conservation strategies as well as in the design of development programmes. The identification of an optimum mix of development and conservation initiatives is one of the most difficult tasks to face policy makers in the next decade, and the necessity to develop location-specific strategies adds to the complexity of the challenge. Much more effort is required to develop adequate analytical tools to enable policy makers to explicitly address the trade-offs and consequences of particular decisions.

The Integration of Farmer and Formal Plant Breeding

It should be possible to strengthen farmer breeding efforts by linking them to elements of the formal system. If the tools and methods of formal plant breeding were made available to traditional systems, an integrated plant breeding model could be the result.

The choice of appropriate plant genetic resource conservation strategies also requires clear policy decisions about the appropriate mix of public, commercial and voluntary contributions to variety development and seed distribution. Commercial seed enterprises are now replacing many public or parastatal seed operations, and are also making an increasing contribution to plant breeding and variety development. In many instances, seed companies will be able to respond more effectively to farmers needs than the public sector.

But the commercial sector will not be likely to address many of the minor crops or special growing conditions that are important to resource-poor farmers, nor will they be likely to play a prominent role in conservation activities. Adequate legal protection for commercial varieties should be balanced against the assurance of farmers privilege to save and adapt their own seed. Too much current seed policy advice focuses on the negative experience of public sector seed production and ignores the new set of responsibilities that public seed policy and public research will have to assume in order to assure more equitable agricultural development.

As is clear from the discussion on in situ conservation and new approaches to plant breeding, plant genetic resource conservation will increasingly involve farmer participation. This will require innovative interaction among public sector and voluntary or community organisations. Policy makers need to devote more attention to fostering long-term, productive collaboration between public agricultural research and organisations active at the community level.

A particularly challenging feature of conservation programmes is their necessarily long-term nature. Both national policy makers and external donors who wish to support conservation programmes must assure that funding is available for an extended period to include the necessary research, training, and implementation. Piecemeal efforts at genetic resource conservation that do not feature a long-term commitment will accomplish little. This puts a premium on careful planning and coordination among participating agencies, and implies a significant re-orientation to the way that donor funding for agricultural development is currently managed. Finally, good policy always depends on good information, and this is particularly true for genetic resource conservation. Despite rapid progress, serious gaps in our knowledge are likely to constrain the informed management of plant genetic diversity for at least the next decade. The issues demand interdisciplinary collaboration among social and biological scientists.

They also require increasing collaboration between researchers and the members of grassroots development initiatives. Policy must direct national research and academic institutions to give priority to collaborative research on conservation issues, and should take responsibility for establishing appropriate fora where different perspectives can be presented, debated, and synthesised.

GENETICALLY ENGINEERED CROPS AND PESTICIDE USE

The major genetically engineered (GE) crop varieties commercialized since 1996 in the United States have been designed to help control a damaging class of insects and simplify herbicide-based weed management systems. Over the first nine years of commercial use, 670 million acres of crops expressing GE traits have been planted, or about 23 per cent of the total 2,970 million acres of crops harvested across the country during this period. Crops engineered to tolerate applications of herbicides, or so-called "herbicide-tolerant" crops (HT), account for the largest share of GE acres.

About 487 million acres have been planted since 1996, or 73 per cent of total GE crop acres. Herbicide-tolerant soybeans are the most widely planted GE crop technology and account for more than half the total acres planted to GE varieties since 1996. The vast majority of HT crops are engineered to tolerate glyphosate (trade name "Roundup," or referred to as "Roundup Ready"), the herbicide introduced to the market in 1972, by Monsanto.
Corn and cotton have been genetically engineered to express the bacterial toxin *Bacillus thuringiensis*, or Bt. This transgenic trait allows plants to manufacture within their cells a crystalline protein that is toxic to most Lepidopteran insects (moths and butterflies). Some 183 million acres of Bt transgenic corn and cotton have been planted since 1996, representing 27 per cent of total GE crop acreage.

The top five countries growing transgenic crops in 2005, according to The International Service for the Acquisition of Agro-Biotech Applications (ISAAA) were the United States, Argentina, Brazil, Canada, and China. Fourteen countries were ranked in the first tier as major adopters of the technology.

A checklist to aid prospective U.S. growers of transgenic crops in interpreting conditions imposed by the agribusiness licensee, including company technology agreements, is published online by RAFI-USA. No uniformly accepted definition of biotechnology exists, according to the National Center for Agricultural Law Research and Information (NCALRI). The center provides several definitions and commentary.

Under the broadest definition, the use of biological sciences to develop products—conventional plant and animal breeding techniques, conducted since the dawn of civilization—fall under biotechnology. In the popular press, biotechnology generally refers to newly developed scientific methods used to create products by altering the genetic makeup of organisms and producing unique individuals or traits that are not easily obtained through conventional breeding techniques. These products are often referred to as transgenic, bioengineered, or genetically modified because they contain foreign genetic material. Agriculture is one of the first industries radically affected by this new technology on both a fundamental production level and a legal level.

Table. Global Status of Commercialized Transgenic Crops 2005.

Rank	Area	Area (mil. Ha/A.)	Crop
1	USA	49.8/723.0	Soybean, maize (corn), cotton,canola, squash, papaya
2	Argentina	17.1/42.2	Soybean, maize, cotton
3	Brazil	9.4/23.2	Soybean
4	Canada	5.8/14.3	Canola, maize, soybean
5	China	3.3/8.2	Cotton
6	Paraguay	1.8/4.4	Soybean
7	India	1.3/3.2	Cotton
8	So. Africa	0.5/1.2	Maize, soybean, cotton
9	Uruguay	0.3/.7	Soybean, maize, cotton
10	Australia	0.3/.7	Cotton
11	Mexico	0.1/.2	Cotton, soybean
12	Romania	0.1/.2	Soybean
13	Philippines	0.1/.2	Maize
14	Spain	0.1/.2	Maize
15	Colombia	<0.1/.2	Cotton
16	Iran	<0.1/.2	Rice
17	Honduras	<0.1/.2	Maize
18	Portugal	<0.1/.2	Maize

1 ha = 2.47 a. (results rounded to.0)

The focus of this publication is on crop varieties created through transgenic modification, or genetic modification (GM). The products of transgenetic engineering are often called genetically modified organisms, or GMOs. All these terms refer to methods by which biologists splice genes from one or more species into the DNA of crop plants in an attempt to transfer chosen genetic traits. The method is known as recombinant DNA technology.

Genes are segments of DNA that contain information that in part determines the end function of a living organism. Genetic engineers manipulate DNA, typically by taking genes from one species—an animal, plant, bacterium, or virus—and inserting them into another species, such as an agricultural crop plant. An intermediate organism or virus can be used to "infect" the host DNA with the desired genetic material. Microparticle bombardment technology is also widely used to deliver exogenous nucleic acids (DNA from another species) into plant cells. The desired genetic material is precipitated onto micron-sized metal particles and placed within one of a variety of devices designed to accelerate these "microcarriers" to velocities required to penetrate the plant cell wall. In this manner, transgenes can be delivered into the cell's genome. New DNA can also be inserted into a host cell using electroporation, in which a jolt of electricity is applied to cells to create openings in the plasma membrane that surrounds a cell. A (typically antibiotic-resistant) marker gene is included in the package to verify degree of effectiveness in introducing the foreign DNA. Gene stacking is becoming more common, adding a whole array of traits at once into the host organism. The whole process can be illustrated by its application in the engineering of transgenic rice, using electroporation.

Steps in electroporation and other methods of gene transfer:

- The DNA sequence for the gene that will be altered is identified and obtained from a donor organism (bacterium). This can be done by referring to known information pertaining to the sequence of the gene which is to be selected, followed by the removal of the gene from the donor organism.
- The desired gene is removed from the donor organism through the use of site-specific enzymes known as restriction enzymes.
- The desired gene is then subject to polymerase chain reaction (PCR), a method to amplify DNA and produce a workable amount of the gene.
- Once acquired, there are several ways to transfer the donor gene into the cells of the target organism. In rice, a somewhat advanced process is utilized. This process is electroporation, wherein special wall-denaturing enzymes remove the plant cell wall. The cells become protoplasts, which are plant cells stripped of the cell wall but still encapsulated in the cellular membrane. In the next step of electroporation, a very high voltage electric charge is sent through the protoplast-containing solution. This charge causes the membrane to temporarily deteriorate, forming small pores. Through these temporary pores, the donor gene's DNA is injected. The DNA is injected in the form of transfer plasmids that migrate to the chromosome and become incorporated in the plant's DNA. Shortly after the charge and injection, the cell membrane reforms. The cell wall also reforms in a reverse process.
- The newly altered cells are then placed in a culture to reproduce the unique cell types that compose the organism.
- The resulting cells are then transferred to a regular growth environment where the newly incorporated gene will be expressed.

With the advent of genetic engineering of plants around 1983, it appeared that transgenic manipulation might benefit and even revolutionize agriculture. The transfer of desirable genetic traits across species barriers offered potential promises to solve problems in the management of agricultural crops, provide new possibilities to improve human and animal health, and provide a new revenue stream for farmers through contract production of pharmaceutical and industrial crops.

Potential environmental benefits included reduced toxic pesticide use, improved weed control resulting in less tillage and soil erosion, and water conservation. Furthermore, the new technology promised increased yields. Transgenic crops were also patentable. Technology agreements or engineering would insure that seed could not be saved over for planting the next year. The developer's intellectual property rights were thereby protected, which offered the potential to increase profits and theoretically garner a monopoly over the transgenic seed supply.

UNINTENDED EFFECTS

Current methods of gene transfer are not precise. While scientists can control with relative exactness the "trait gene" (or its synthesized analog) to be inserted into a host plant genome, they cannot entirely control its location, nor the number of copies that get inserted. Location of genetic material is important because it controls the expression of biological traits, just as genes themselves do. Also, inserted DNA frequently contains multiple stacked genes for different traits (eight in the New Leaf potato), increasing chances of undesirable interactions. A common and unpredictable occurrence is "silencing" of either the inserted genetic material or adjacent native genes. Present scientific knowledge is still a long way from being able to precisely control the traits the host plant will express and to guarantee genetic stability in subsequent generations.This potential for instability can lead to unpredictable and undesirable effects, examples of which include plant infertility, production of toxins and allergens, and reductions in yield and plant fitness. The transgenic seed industry consistently counters that since genes from no known allergens are incorporated, adequate care has been taken to guard against this contingency.

Transgenetic engineers who rely on the simple model of gene expression—the position that one gene equals one effect—habour an outdated interpretation of genetic theory, and one that could have serious implications. Pleiotrophy is the understanding that one gene may control multiple traits in an organism. Pleiotrophy multiplies the uncertainty surrounding transgenic crops.

A gene identified as controlling a desirable trait may in fact control multiple traits in a variety of ways. Pleitrophy is common, and the interactions of genes with each other and with the environment add complexity. To accurately predict the effects of new genetic combinations is nearly impossible. The introduction of a novel life form into an ecology can trigger effects perhaps too great to be understood during our time. While it is true most mutations don't survive, those that do can profoundly affect human and other life forms. For instance, transgenic soy strains appear to exhibit unintended effects. Field observations reported to the University of Georgia and University of Missouri noted physiological problems affecting yields. Research published by University of Arkansas scientists in 2000 noted that glysophate disrupts the nitrogen fixation process in Roundup Ready soy. The current marker and promoter genes of choice also may create new hazards. The antibiotic-resistant marker genes carry the potential to increase the variety of bacteria resistant to antibiotics. The viral promoter genes could combine with other infecting viruses, or be scrambled by the plant, to create new viral proteins.

The cauliflower mosaic virus (CaMV) is a very powerful promoter and is commonly used. The CaMV can potentially cause the inserted DNA package to be expressed out of proportion with the rest of the genetic code. When inserted with a particle gun, the CaMV promoter can jump out of the DNA package and land somewhere else in the host genome, causing disruption. The bacterial and viral vector genes could recombine to form active pathogens—either new ones, or old ones with renewed virulence, or with broader host specificity.

The offering specific advice for an extensive education campaign in support of biotechnology, at the same time called for research studies to be carried out on several key safety issues raised by the public. An 18-member Biotechnology Implementation Task Force convened and issued an update in July 2001. However, nothing more has been heard from this committee.

Fig. A Retooled Gene in Endless Summer Tomatoes Controls Ripening to give Better Flavour and Shelf-life.

A main regulating agency for transgenic technology in the U.S. is the Animal and Plant Health Inspection Service (APHIS), a division within the Department of Agriculture. Rather than conduct safety studies, APHIS appears to accept risk-management tools such as "performance-based regulatory standards" and "science-based risk assessment policies and procedures." This approach allows for an acceptable level of possible collateral damage, as long as it is far enough down the road. Each piece of the inserted gene package described above carries with it the potential to disrupt non-target portions of the host plant's DNA, to create instability in the new genetic construct, or to result in unpredictable combinations that can create new substances, viruses, or bacteria. What this adds up to is the possibility, again, of unintended effects—particularly in subsequent generations of the engineered plant. To date, no known replicated studies have been conducted that confirm or disprove potential long-term effects on human health. No known mechanism was proposed or included to identify undesirable side effects of the engineering process.

A December 2005 USDA assessment of APHIS protocols for monitoring GE trial crops criticized oversight lapses. APHIS countered that it was relying on an accepted risk/benefit assessment process, while USDA took the position that oversight should be strengthened, on the assumption there is significant risk—until the new technology has been proven safe beyond doubt.

COMMERCIAL TRANSGENIC CROPS AND THEIR TRAITS

While increased yields and improved nutritional value are among the promised benefits of transgenic crops, most now planted worldwide are designed either 1) to survive exposure to certain herbicides (called herbicide-tolerant, or HT), or 2) to kill certain insect pests (called pesticidal or insecticidal). The transgenic tomato was designed for long shelf life. It is unclear whether the increased beta-carotene in transgenic "Golden Rice" (derived from the daffodil) is in a usable form for human nutrition, especially in the absence of dietary fats and proteins.

Transgenic herbicide-tolerant crops have been altered to withstand being sprayed with broad-spectrum herbicides, with the idea that one application will take care of most types of weeds without killing the crop. Insecticidal crops contain genes of the soil bacterium *Bacillus thuringiensis* (Bt). These Bt genes cause the plants to produce a chemical toxic to the European corn borer, the cotton bollworm, and other caterpillars. (Caterpillars are the larvae of insects in the Lepidoptera order, which includes moths and butterflies.)

As of 2005, about 87 per cent of world transgenic acreage was in the U.S. Herbicide-tolerant crops accounted for about three quarters of the acreage planted, worldwide, to genetically engineered crops in 2005. Pesticidal crops, or a combination of pesticidal and herbicide-tolerant crops, accounted for most of the remaining acreage. Acreage devoted to crops with stacked genes intended to express a variety of traits is increasing.

With an overwhelming amount of U.S. commodity programme crop acreage devoted to transgenic versions, seed for conventional varieties is becoming scarce for those who choose not to plant transgenic crops. Traditional seed scarcity can affect farmers who wish to return to non-transgenic corn, soya, or cotton. Cotton seed is controlled by two large suppliers working with a large public research institution. Development of the non-transgenic organic/specialty cotton sector, which accounts for the 37 per cent non-transgenic cotton acreage in Texas, has been hampered by concerns about cross-pollination and boll-weevil control. Soybeans and corn (often planted in rotation in the Upper Midwest) cover the most transgenic acres. There may be some new evidence that field workers working with Bt cotton are developing allergic reactions.

One other large-acreage North American transgenic crop is canola (a low erucic acid form of European rapeseed). Canola is a major oilseed crop in Canada, but only a minor crop in the U.S. However, until recently, it was thought that acreage of both canola and rapeseed would increase in the near future in the Pacific Northwest. On May 9, 2006, a proposed large production facility at Gray's Habour, Washington, announced that it would produce biodiesel from Asian palm oil, thus bypassing the "seed crushing hassles" of canola/rapeseed.

Proposals to plant substantial acreages of canola and rapeseed (*Brassica napus, B. rapa*)— much of it transgenic varieties—in Oregon's Willamette Valley to produce raw material for biodiesel production caused considerable concern among small-acreage vegetable seed producers. A preliminary 2006 Oregon State University Extension study predicted a high potential for gene flow between *B. napus* canola and other *B. napu*s crops (rutabaga and Siberian kale). Likewise, *B. rapa* rapeseed holds the potential for gene flow with its closely related vegetable crops (Chinese cabbage, pai-tsai, mizuna, Chinese mustard, broccoli raab, and turnip). Potential for crossbreeding between the two oilseed crop types was rated high, as well. Potential of crossbreeding with wild (*Raphanus raphanistrum*) and cultivated (*R. sativum*) forms of radish was considered low. More study was called for regarding outcrossing of canola with *B. oleracea* vegetables (cabbage, cauliflower, Brussels sprouts, kohlrabi, collards, and kale).

Table. Percentages of U.S. 2005 Crop Acreage Planted to I0nsecticidal, Herbicidal, and Stacked-gene Varieties.

	Insect resistant	Herbicide resistant	Stacked-gene
Soy	0	87	0
Corn	26	17	9
Cotton	18	27	34

Oregon Extension concluded that "genetically modified canola [and rapeseed] present the greatest risk to vegetable crucifer seed crops.... The presence of the gene would make the seed crop unsuitable for markets that have strict tolerance on GMO contamination"—*i.e.*, organic, identity preserved (IP), and European exports. Furthermore, "transgenes are relatively easy to detect at very low levels, so it is likely that their presence could be detected even if only a few interspecific hybrids were found in a vegetable seed lot."

While acknowledging the risks to the producers of the nation's garden seed crops located in the Willamette Valley, researchers suggested that the vegetable seed producers could pack up and move.

Most transgenic cotton is herbicide tolerant, though some varieties have the Bt trait; transgenic canola is herbicide-tolerant. The first transgenic wheat, initially planned for commercial introduction in 2003, is Roundup-tolerant. On

May 10, 2004, Monsanto announced that it was discontinuing all research and field trial activities on Roundup-Ready wheat. After seven years of development, the release said, efforts to win over farmers and the international wheat market had failed. A 2005 study published by the Western Resource Council showed that introduction of genetically modified wheat would lower income for wheat growers and the wheat industry. The report projects costs per bushel and per acre for farmers adopting Roundup-Ready wheat and for nonadopters under best-case and worst-case scenarios. Either way, farmers were projected to lose money from introduction and use of the Roundup-Ready wheat. Other traits engineered into commercial transgenic varieties include disease resistance, high pH tolerance, and several nutritional, taste, texture, and shelf-life characteristics—primarily through gene stacking.

In the absence of transgenic labeling, the average U.S. consumer may not realise that ingredients derived from transgenic corn, soya, and oilseed are in 70 per cent of the foods found in U.S. retail food outlets. Most prevalent is high-fructose corn syrup, which is replacing other sweeteners in a wide variety of mass-produced food products. The Biotech Industry Organization agrees that transgenic oils and ingredients derived from corn and soya are pervasive in conventional processed foods. Now that transgenic horticultural crops are in the marketplace, no one will know for sure—in the absence of labeling—whether fresh produce or processed shelf products contain engineered crops.

Five years ago introduction of transgenic fresh produce appeared imminent. Winter squash and a limited amount of sweet corn are now being retailed. However, after the Flavr-Savr tomato was withdrawn and Starlink® feed corn caused a recall of taco shells, the subsequent paths of crops such as tomatoes, potatoes, sunflowers, peanuts, and sweet peppers diverged. Field trials were conducted from 1993–2001 on transgenic peanuts, all in the U.S. Field trials were conducted from 1993–2002 on sunflowers—in Australia, three European countries, and the U.S.

Sweet bell peppers have been joined by rice, alfalfa, cabbage, carrots, cauliflower, sweet corn, cucumber, lettuce, mustard—and most recently, eggplant—on the list under development for commercial release. Transgenic fruits for which field trials are currently underway (some in the U.S.) are apples, cherries, cranberries, grapefruit, kiwi, pears, persimmons, pineapple, plum, and strawberries. Transgenic papaya, raised in Hawaii, has been commercialized for several years, and plum has recently been deregulated by APHIS. (Plum is, of course, the source of prunes.)

Transgenic Potato in the U.S.

More than 700 field trials of transgenic Bt potatoes were conducted in the U.S. from 1989–2002 by a single company. In 1996 Bt potatoes were made available to commercial growers, but after 2000 the Bt potato programme was abandoned due to lack of consumer acceptance. Large fast food chains, snack food manufacturers, and potato processing conglomerates eliminated transgenic potatoes from their products. There are no other types of transgenic potatoes currently approved for sale in the U.S.

One variety of transgenic flax was approved in the U.S. in 1999, but transgenic flax is reportedly not being grown because of consumer resistance and market rejection. Flax seed oil and flax seed are popular nutraceutical products. Transgenic rice trials in Missouri were halted by public protests. So far Iran is the only known country producing transgenic rice for human consumption. Despite indications in 2002 that lack of public acceptance of transgenic food would cause transgenic firms to change course, it has turned out that transnational corporations have changed tactics—conducting trials overseas, keeping U.S. trials strictly secret (perhaps even from regulatory oversight by APHIS). The companies also lobby industry groups, such as the wheat boards, and seek to develop indirect markets such as processing aids and minor ingredients. Transgenic processing aids—enzymes and ingredients used to improve the colour, flavour, texture, and aroma of manufactured foods—and preservatives, stabilizers, vitamin additives, and a vast number of minor ingredients are currently being derived from transgenic corn or soy.

The industry currently takes the position that the public has been consuming highly processed, transgenic foods for several years and that this large-scale experiment with the American food supply has been a success. Corn, oilseeds, cotton, and wheat are the North American crops with the most acreage and profit potential. For a more complete list of current and future commercial transgenic crops and their traits, see the APHIS list. Many disturbing unanswered questions remain about transgenic crops and their potential benefits, costs, and risks. In fact, according to an independent survey of research data on transgenic crops, conducted by the Winrock Foundation's Henry A. Wallace Center for Agricultural and Environmental Policy, "The varieties and uses of genetically altered crops have grown much more rapidly than our ability to understand them." This study reveals that only four per cent of total federal agricultural biotech funding is dedicated to environmental assessment. It should also be noted that there is even less research dedicated to human and animal health impacts of the technology.

Unresolved issues in transgenic agriculture:

- Food safety
- Farm management
- Crop yield, costs, and profitability
- Marketing and trade
- Organic industry impacts
- Influence on public research
- Industry concentration and farmers' right to save seed
- Regulation of transgenic crops and apportionment of liability

6

Biosafety Concern and Regulation of GM Crops

BIOTECH CROPS AND FOODS ASSESSED FOR SAFETY

Commercially available foods and crops made using biotechnology have been subjected to more testing and regulation than any other agricultural products, and have all been found safe. This note is an introduction to the assessment processes that have been used in determining the food and environmental safety of these products. Although biotechnology broadens the scope of genetic changes that can be introduced into plants used for food, it does not inherently result in foods that are less safe than those produced by other techniques. This means that previously established principles for assessing food safety still apply for products of biotechnology. Moreover, these products can be judged on their individual safety, allergenicity, toxicity and nutrition rather than their method of production.

The safety assessment of biotech foods is based on the principle that they can be compared with traditional foods that have an established history of safe use. This comparison can be based on an examination of the same risks that have been established for traditional foods, keeping in mind that many conventional foods may present specific risks depending on conditions of processing, or to individuals within a population. The intent of this comparative approach is to establish whether the new food is "as safe as" its conventional counterpart.

International discussion and expert consultations have resulted in a consensus on the specific safety issues that should be considered when evaluating these new foods.

They include the following:

- *The Parent Plant*: Knowledge of the biology of the plant and its history of safe use as a food are important for understanding the natural range and variation of key nutritional components, such as vitamins and minerals; naturally occurring toxicants; anti-nutrients; and potential allergens.

- *The Gene Source*: Information about the natural history of the source organism of a new gene is important in assessing whether it could be a potential source of allergens or toxins.
- *Nutrition*: All plant breeding methods have the potential to change the nutritional value of plants or to lead to unexpected changes in levels of natural toxicants or anti-nutrients. Food safety assessments take into account the potential for any change in nutritional composition, especially in key components that have a significant impact on the diet. Laboratory *analyse*s are used to compare the profile of the new plant with its conventional counterpart for constituents such as protein, fat, fibre, micronutrients, amino acids, fatty acids, vitamins, toxins and anti-nutrients.
- *Allergens*: The potential of accidentally introducing a new allergen (a protein that causes an allergic reaction) into a food is an important safety concern. Fortunately, food allergens have common characteristics, such as remaining stable during digestion and food processing, and they are usually abundant in foods. Internationally accepted approaches for evaluating allergenicity have been established and none of the new proteins present in biotech foods share the properties of allergens.
- *Toxins*: The possibility that new toxins may have been introduced into a food is also tested by comparing any newly produced proteins with known protein toxins using laboratory analyses and animal studies.Like allergens, protein toxins have very well characterized properties, which are not shared by any of the new proteins in biotech foods.

In assessing these safety issues, the processed version of the food and its impact on specific population subgroups, such as infants or the elderly, is also evaluated.

ENVIRONMENTAL SAFETY

Assessing the environmental safety of a biotech plant requires an understanding of the biology of the plant itself and the practices used in its cultivation. This knowledge is important in identifying and evaluating potential environmental risks and also in designing any appropriate risk management measures.

Most countries use similar environmental risk assessment approaches, which include:

- Evaluating the role of the introduced gene in the plant and any changes in the plant's characteristics
- The possible unintended secondary effects on non-target organisms
- The possibility that the modified plant could persist longer in the environment or invade new habitats

- The likelihood and consequences of the potential spread of newly introduced traits to related plants
- Potential impacts on biodiversity

These environmental safety concerns are not unique to plants produced using biotechnology and are also important when evaluating new varieties produced through conventional plant breeding. The objective of environmental safety assessment is to identify and evaluate any additional risks associated with the release and cultivation of these new plants in comparison with a conventional crop variety that has a history of safe use.

In addition to considering the potential risks associated with the introduction of new biotech crop varieties, consideration must be given to the risks associated with not using biotechnology to achieve desired goals. For example, the biodiversity of tropical rainforests or other ecologically sensitive areas can only be maintained if these natural ecosystems are not destroyed because of the expansion of the agricultural land base. Biotech crops can alleviate pressure to expand agricultural areas by increasing yields with improved pest resistance and increased tolerance to drought or saline soil conditions.

DEVELOPING A BIOSAFETY SYSTEM

All countries with functioning biosafety regulatory systems have developed these systems gradually, usually beginning with voluntary guidelines and standards developed cooperatively by stakeholders in academia, industry, and government. Over time these were incorporated in laws, either under existing legislation covering food and agricultural products or new legislation dealing specifically with biotechnology. Even in countries with long-established systems, biosafety policy and its implementation continue to evolve and it is not unusual to have a mix of voluntary and mandatory measures. For countries seeking to develop a national biosafety system, it must be emphasized that there is no model for a single best approach. The issues to be considered can be broadly divided into six elements, which are briefly discussed below.

NATIONAL INVENTORY AND EVALUATION

An inventory and evaluation of national priorities, agricultural policies, existing regulatory regimes, and national scientific and technical capacities, is an ideal prerequisite to the development and implementation of biosafety-related policies and regulations. This national appraisal provides a means to identify and characterize available resources and regulatory infrastructures, assess their adequacy for supporting a biosafety system, and identify gaps where capacities need to be strengthened.

NATIONAL POLICIES AND STRATEGIES

A national biosafety policy or strategy provides a set of principles to guide the development and implementation of a biosafety system and should describe the goals and objectives of the regulatory framework. Direction on many of the fundamental issues and public policy choices that must be considered during the development of regulations can be provided by such a strategy. Examples of these issues include the extent to which social, ethical, and economic factors should be considered, the social acceptability of biotechnology and its products, and linkages with other national policies on food, agriculture, and economic development.

SCIENTIFIC KNOWLEDGE, SKILLS AND CAPACITY BASE

The human resource environment that both enables and limits biosafety implementation is shaped by the scope and quality of: competency in the biological sciences; expertise in information acquisition, communications, and management; and, experience in critical thinking, analysis, and decision-making. These capacities have an overriding influence on the development and implementation of a biosafety system. Addressing capacity needs is the top priority for many developing countries. Building a strong base of scientific knowledge in support of the regulatory system, and development of core competencies in biotechnology product evaluation, are fundamental to any national biosafety system. These activities allow an improved scientific basis for assessments of potential risks and/or benefits, and they strengthen the scientific capabilities for risk management, inspection, and monitoring.

DEVELOPMENT OF REGULATIONS

Decisions on an appropriate regulatory framework should be informed by the national inventory and evaluation, and through extensive consultation with stakeholders, including the public. This is particularly true if a country chooses to incorporate non-safety issues into its decision-making process.

IMPLEMENTATION OF REGULATIONS

The central issues around the implementation of biosafety regulations involve the establishment of appropriate mechanisms for risk assessment, risk management, and risk communication within existing financial, technical, and human resource constraints. Decisions made during the implementation phase directly affect the costs associated with assessing and managing risks and ensuring compliance with regulations.

CROSS CUTTING ISSUES

Cross cutting issues are those that are common to each of the five preceding elements and they are often the most challenging factors to address and resolve. They are, however, the issues that will ultimately dictate the scope of a national policy on biosafety, and the conversion of policy into practice. Cross cutting issues affect the implementation of the system designed to assess biosafety, and perhaps more importantly, those non-technical factors that are crucial to public acceptance and confidence in the decisions that are made by government on behalf of the people. The twin issues of public information and participation have to do with the degree of transparency in a regulatory system, and the degree to which the public has input either into the formulation of regulatory policy or into specific regulatory decisions. Transparency refers to the extent to which governments provide information on why and how certain products are regulated, how risk assessments are performed and decisions made, and as well, the conclusions and decisions that have been reached. Transparency can also involve the perceived independence and objectivity of the regulatory decision-makers. Human, financial and infrastructure resources largely determine the scientific and administrative capacity of any country; they obviously influence any biosafety related policy or programme. Funds must be available to develop and implement a national biosafety system; to support the infrastructure required, such as buildings, labs, equipment, and computers; to facilitate communication and public participation; to train scientific and regulatory personnel; and to foster the research required to assure that risk assessments are sound.

The development of an effective national biosafety system is important to encourage the growth of domestic biotechnologies; to ensure safe access to new products and technologies developed elsewhere; and to build public confidence that products in the marketplace are safe. The absence of a suitable framework affects the ability of the public and private sectors to invest in biotechnology and to make the products of biotechnology available so that the benefits they afford can be realised.

BT CROPS - MODE OF INSECT CONTROL

The bacterium *Bacillus thuringiensis* (Bt) is distinguished from other bacilli by the production of large crystals during sporulation composed of one or more insecticidal proteins. These proteins have served as the principal active ingredient of numerous commercial bacterial insecticides used worldwide for more than 40 years to protect crops against caterpillar pests, and more recently to control coleopterous insects and mosquito and blackfly vectors of human and animal diseases. The first genes encoding insecticidal Bt proteins were cloned early in the 1980s.

This led quickly to their use to construct recombinant bacterial insecticides containing novel combinations of these proteins and insect-resistant Bt crops. The first Bt crops became commercially available in the United States in the mid-1990s, and have been widely adopted by farmers despite higher seed costs in comparison with conventional crops. During the 1999 growing season, farmers in the United States planted over 20 million hectares of insect-resistant Bt transgenic crops, including over 10 million hectares of Bt maize (corn), 1.5 million hectares of Bt cotton, and about 30 000 hectares of Bt potatoes.

This acreage is expected to grow to about 15 million hectares of corn and 3 million hectares of cotton within five years, representing, respectively, about a third of the corn and half of the cotton acreage in the United States. Bt crops are also being grown in China and Argentina, and their potential deployment is being assessed in several other countries. In addition, more than 30 other Bt crops, including rice, sorghum, most major vegetable crops as well as many tree crops used for fruit, nut and fibre production are under development.

For farmers, Bt crops offer advantages over conventional crops in that the insecticidal proteins are produced directly by the plants, and continuously during most of the growing season, thereby reducing the material and application costs of using synthetic chemical insecticides. Growers have reported improved profit margins over conventional varieties averaging $60 to $100 per hectare during the first few years of Bt cotton plantings in the southeastern United States.

Consumers and the environment benefit from reductions in the use of synthetic chemical insecticides. For example, it is estimated that the use of Bt cotton in the United States reduced chemical insecticide applications in cotton for lepidopterous pests by 2.7 million pounds in 1999. This represented a reduction of 15 million insecticide applications (22 per cent) in comparison to the number used in 1995 prior to the use of Bt cotton. These reductions preserve populations of insect natural enemies such as spiders and predatory and parasitic insects, and eliminate exposure of non-target vertebrates, including humans, to these chemicals.

The agronomic and environmental advantages of Bt crops are widely recognized in the agricultural community. In fact, the United States Environmental Protection Agency has declared that Bt crops are an "environmental asset". However, concerns about the long-term impact of their use have been raised by environmentalists and scientists in regard to the development of resistance in insect populations, and the safety of insecticidal proteins produced by Bt crops to humans and other non-target organisms.

Important questions have been raised about the safety of Bt crops, but most of the specific test protocols and resultant data relevant to Bt proteins produced by these crops are scarce and are only now being published in refereed scientific journals. This is because these data were developed primarily to fulfill governmental registration processes required in most countries to sell products, processes which do not require the data to be published in journals. Nevertheless, an extensive database exists in the literature on Bt strains, proteins and products and this information is reviewed here along with data available on Bt crop safety. To address the issue of safety, it is important to understand the role that Bt's insecticidal proteins play in its general biology and the mechanisms by which these proteins exert their toxic effects on insects. Bt and its insecticidal proteins have a high degree of specificity that accounts for their safety to most organisms, especially in comparison with synthetic chemical insecticides. An understanding of Bt's mode of action, specificity and toxicology, only rarely treated in discussions of safety, is important in assessing any risks. Thus, to place the safety of Bt insecticidal proteins and Bt crops to non-target vertebrates and invertebrates in perspective, I will first summarize our knowledge of the most critical aspects of Bt relevant to understanding this issue, and then deal with the data on the safety of Bt crops.

THE GENERAL BIOLOGY OF BACILLUS THURINGIENSIS

Bacillus thuringiensis (Bt) is a common Gram-positive, spore-forming bacterium that can be readily isolated on simple media such as nutrient agar from a variety of habitats including soil, water, plants, grain dust, dead insects and insect feces. Its life cycle is simple. When nutrients are sufficient for growth, the spore germinates, producing a vegetative cell that grows and reproduces by binary fission. The bacterium continues to multiply until one or more nutrients, such as sugars, certain amino acids and oxygen, become insufficient for continued vegetative growth. Under these conditions, the bacterium sporulates to produce a spore and parasporal body, the latter composed primarily of insecticidal proteins. These are commonly referred to in the literature as protein toxins or endotoxins.

Although Bt can be isolated from many environmental sources, and is often referred to as a 'soil bacterium', several features indicate that its principal ecological niche is insects. The original isolations of Bt, for example, were made from diseased caterpillars, and these remain a good source of isolates. More importantly, Bt produces a range of protein toxins and toxin synergists that are very effective at killing certain species of insects, especially larvae of lepidopterous insects, providing a rich substrate for Bt's reproduction.

Representative Endotoxin-Containing Parasporal Bodies

The principal toxins are protein -endotoxins, the '' designating a particular class of toxins, and endotoxin referring to their localization within the bacterial cell after production. Many isolates of Bt also produce the â-exotoxin (a competitive inhibitor of messenger RNA polymerase), insecticidal proteins produced during vegetative growth (Vips), the synergist zwittermicin A, and enzymes such as phospholipases that enhance the activity of the -endotoxins. In addition, the spore itself can synergize the activity of Bt proteins in some insects, especially those not very sensitive to -endotoxins.

In many insect species, especially grain-feeding lepidopterans, Bt reproduces to very high levels after insect death, with millions of spores being produced per insect. Thus, Bt's ecology and reproductive biology suggest that its toxins and toxin synergists evolved to debilitate or kill directly a range of insect species, thereby providing a substrate for reproduction of this bacterium.

The Diversity of Bacillus Thuringiensis Subspecies

While commonly referred to in the singular as 'Bt', *B. thuringiensis* is actually a large group of subspecies, all characterized by the production of one or a few parasporal bodies per cell during sporulation. Each parasporal body contains one or more protein endotoxins, typically as crystalline inclusions, and each of these is typically toxic to a limited range of insects. The endotoxins occur in the parasporal body as protoxins, which after ingestion, dissolve and are converted to active toxins through cleavage by proteolytic enzymes in the insect stomach (midgut). The activated toxins bind to receptors on the midgut microvillar membrane in sensitive insects, lysing the cells and destroying much of the midgut epithelium, causing insect death.

In essence, Bt endotoxins are stomach poisons selective for insects and certain other invertebrates. At present, there are more than 70 subspecies of Bt, distinguished from one another by immunological differences in flagellar (H antigen) serotype. Each subspecific name corresponds with a specific H antigen number. For example, *B. thuringiensis* subspecies *kurstaki* is H3a3b, whereas *B. thuringiensis* subspecies *morrisoni* is H8a8b. In the literature, the term 'variety' is also used in place of subspecies, as is occasionally the term 'strain'. Because the H antigen serotype-subspecific name often does not correlate with insecticidal properties, acronyms and numbers are often used to designate specific isolates, especially those with important insecticidal properties.

For example, HD1 (isolate number 1 from Howard Dulmage), is used to designate a specific isolate of *B. thuringiensis* subsp. *kurstaki* (H3a3b) that produces four major endotoxin proteins and has a broad spectrum of activity against lepidopteran pests. Another isolate of *B. thuringiensis* subsp. *kurstaki* (H3a3b) is HD73. This produces only a single endotoxin protein (Cry1Ac) and as a result has a much narrower spectrum of activity against insects than HD1. Historically, HD1 was the first Bt isolate developed commercially for the control of lepidopterous pests, and remains the most widely used in commercial products today. This isolate has also been the source of much of our knowledge of Bt genetics and molecular biology as well as endotoxin genes used in transgenic bacteria and plants.

Despite the large number of Bt subspecies that have been described, the taxonomic validity of these as well as maintaining *B. thuringiensis* as a species separate from *B. cereus* has been in question for many years. At the species level, the primary problem is that the only phenotypic character that clearly differentiates *B. cereus* from *B. thuringiensis* is the parasporal body synthesized by the latter species during sporulation.

In most Bt subspecies, the information for endotoxin production is encoded on large transmissible plasmids. When these plasmids are lost naturally or cured from Bt subspecies by growing cells at 42°C, no endotoxins are produced. Cured Bt strains that hosted these plasmids cannot be reliably distinguished from *B. cereus*. At the subspecies level, phenotypic biochemical differences among many are minor, indicating that many subspecies may not be valid using accepted standards of differentiating bacterial subspecies. In other cases, the differences in biochemical properties and insecticidal activity are so significant that some subspecies could be viewed as distinct species.

From the standpoint of Bt's use as an insecticide, taxonomic studies, especially flagellar serotyping, have aided isolate classification but have proven unreliable as accurate predictors of insecticidal activity. Perhaps the best example of this is found in *B. thuringiensis* subsp. *morrisoni* (H8a8b), which includes isolates active against lepidopterous (isolate HD12), coleopterous or dipterous (isolate PG14) insects. Another example of this is found in the occurrence of the 128-kb plasmid that encodes mosquitocidal Cry and Cyt proteins among numerous subspecies of Bt, including *israelensis*, *morrisoni*, *entomocidus*, *kenyae* and *thompsoni*.

To summarize the current status of Bt systematics, most molecular evidence is in agreement with more classical biochemical and physiological studies which indicate that *B. thuringiensis* and *B. cereus* are the same species, and that the latter becomes the former when it acquires one or more plasmids that express genes for insecticidal proteins. Nevertheless, maintaining Bt as a separate species has practical value because of its insecticidal properties, as does dividing Bt isolates into subspecies based on flagellar antigens. The latter is useful in cataloguing the more than 20 000 isolates of Bt collected to date. However, to understand the insecticidal properties of an isolate, regardless of its subspecies/serotype or other designation, knowledge of the insecticidal protein genes encoded and expressed is required.

Insecticidal Proteins of B. Thuringiensis

Because the â-exotoxin is not permitted in commercial bacterial insecticides in the United States, the two principal active components in commercial Bt preparations are the spore and parasporal body. The endotoxins in the parasporal body account for most of the formulation's insecticidal activity, including initial paralysis followed by death.

The lethal effects of these proteins have been known since their discovery in the 1950s. In the early 1980s, shortly after the development of recombinant DNA techniques, it was discovered that Bt -endotoxins were encoded by genes carried on plasmids. This discovery led to a major research effort in many labouratories around the world aimed at understanding the genetic and molecular biology of these toxins. This effort resulted in the cloning and sequencing of numerous Bt genes, over 100 to date, and characterization of the toxicity of the proteins each encodes.

At the time Hofte and Whiteley wrote their review, a wide variety of confusing names and acronyms were being used to refer to Bt's endotoxin genes and proteins. Computer analyses showed that the nucleotide sequences of most of these genes were quite similar. To standardize the terminology, Hofte and

Whiteley proposed a simplified nomenclature for naming all insecticidal Bt genes and proteins. In this nomenclature, the proteins are referred to as Cry (for crystal) and Cyt (for cytolytic) proteins. Though modified recently, this system is still in use today, and is described briefly here along with its modifications and additions.

The Cry and Cyt nomenclature developed by Hofte and Whiteley was originally based on the spectrum of activity of the proteins as well as their size and apparent relatedness as deduced from nucleotide and amino acid sequence data, and protein gel analyses. At that time, with the exception of a 27-kDa cytolytic protein from *B. thuringiensis* subsp. *israelensis*, all proteins appeared to be related, and probably derived from the same ancestral protein. Therefore Hofte and Whiteley termed the encoding genes '*cry*' genes and the proteins they encode 'Cry' proteins. This designation was followed by a Roman numeral which indicates pathotype (I and II for toxicity to lepidopterans, III for coleopterans and IV for dipterans), followed by an upper-case letter indicating the chronological order in which genes with significant differences in nucleotide sequences were described.

The I and II for lepidopterantoxic proteins also indicate size differences, with the I referring to proteins with a mass of about 130 kDa and the II designating those of 65-70 kDa. Some epithets also include a lower-case letter in parentheses which indicates minor differences in the nucleotide sequence within a gene/protein type. Thus, CryIA referred to a 130 kDa protein toxic to lepidopterous insects for which the first gene (*cryIA*) was sequenced, whereas CryIVD referred to a 72-kDa protein with mosquitocidal activity for which the encoding gene was the fourth from this pathotype sequenced.

The 27-kDa Cyt1A protein first isolated from *B. thuringiensis* subsp. *israelensis* differs from other Bt proteins not only in its smaller size, but also in that it is highly cytolytic to a wide range of cell types *in vitro*, including those of vertebrates. In addition, it shares no apparent relatedness to Cry proteins. Owing to these differences and its broad cytolytic activity, Hofte and Whiteley referred to this as the CytA protein encoded by the *cytA* gene.

In their revision of Bt gene nomenclature, Hofte and Whiteley listed 38 published gene sequences that encoded 13 different Cry proteins and the single CytA protein. Since their publication, the number of Bt Cry endotoxins has more than tripled, and several new *cyt* genes have also been described. As more and more *cry* genes were sequenced and analysed, it was decided to name genes based on their relatedness as determined primarily from the degree of their deduced amino acid identity. As a result, the nomenclature developed by Hofte and Whiteley was modified in the following manner.

The *cry* and *cyt* descriptors have been maintained, but the Roman numerals have been replaced with Arabic numbers to indicate major relationships (90 per cent identity), with higher degrees of identity being indicated by following upper-case letters (95 per cent identity), and minor variations of these alleles being designated by lower-case letters, with the parentheses around the latter eliminated. For example, what was CryIA(c) is now Cry1Ac (and the corresponding gene, *cry1Ac*), a relatively minor change.

However, for some genes/proteins the changes are greater. For example, CryIVD does not cluster with the earlier CryIV (now Cry4) proteins, and is now the taxon Cry11Aa. Although the new designations carry no specific information regarding insecticidal spectrum, primary activity is still inferred because the numbers have been maintained for many of the genes, and there remains a high degree of correlation between relatedness and the spectrum of activity. For example, Cry1 still refers largely to toxicity to lepidopteran insects; Cry2 to lepidopteran toxicity and in some, dipteran activity; Cry3 to coleopteran toxicity, and Cry4 to dipteran toxicity.

Mode of Action of Cry Proteins

Aside from allergenicity, knowledge of the mode of action and structure of Bt's insecticidal proteins are the two interrelated topics that have most bearing on understanding the safety and risks of insecticides or transgenic crops based on these proteins. Although the spore can play an important role in the pathogenicity of *B. thuringiensis* to certain insect species, the Cry and Cyt proteins are responsible for the paralysis and death of most target insects. The three-dimentional structure of three of these is known from X-ray crystallographic studies. Although the mode of action remains to be resolved at a detailed molecular level, these structures along with studies on the effects of the toxins *in vivo* and *in vitro* show how Bt proteins kill insects while being safe for most non-target organisms, especially vertebrates.

In the typical Bt, *B. thuringiensis* subsp. *kurstaki*, for example, the endotoxin crystals dissolve after ingestion upon encountering the alkaline (pH 8-10) juices of the midgut. Dissolution requires the reduction of disulfide bridges that stabilize the Cry molecules in the parasporal crystal. Most Cry toxins are actually protoxins of about 130-140 kDa (*e.g.*, Cry1 and Cry4 proteins) from which an active toxin 'core' in the range of 60-70 kDa is released into the midgut by proteolytic cleavage. Proper activation results in cleavage of 26-29 amino acids from the N-terminus and about 600 amino acids from the C-terminus. Activated toxin molecules pass through the peritrophic membrane and bind to specific receptors on the apical microvillar brush border membrane of midgut epithelial cells, which lies just beneath the peritrophic membrane.

Binding is an essential step in the intoxication process, and in susceptible insects the toxicity of a particular Bt protein is correlated with the number of specific binding sites (*i.e.*, receptors) on microvilli as well as the affinity of the Bt molecules for these sites. However, binding by itself, even high-affinity binding, does not always lead to toxicity, indicating that insertion and probably post-insertional processing in the midgut membrane is required to obtain toxicity. For example, in three different studies it has been shown that the Cry1Ac toxin can bind with high affinity to microvillar membrane vesicles from *Lymantria dispar*, *Spodoptera frugiperda* and *Heliothis virescens*, but with little or no subsequent toxicity. This indicates that insertion and likely post-insertional processing of Cry proteins is essential to intoxication.

In highly sensitive insect species, the microvilli lose their characteristic structure within minutes of toxin insertion, and the cells become vacuolated and begin to swell. This swelling continues until the cells lyse and slough from the midgut basement membrane. As more and more cells slough, the alkaline gut juices leak into the hemocoel where, as a result, the hemolymph pH rises by a half unit or more. This causes the paralysis and eventual death of the insect.

Though this general picture of the mode of action has been known for some time, the details of the insecticidal process at the molecular level remain unresolved, especially the series of events that occur after the toxin binds to a receptor. There is good evidence for an immediate influx of potassium and calcium, in response to which the cell takes in water in an effort to balance these cations. To explain this cationic influx, it has been proposed that Bt molecules insert into the microvillar membrane forming transmembrane cation pores. For pores to form, it is postulated that during membrane insertion, the molecules undergo conformational changes that permit insertion.

After insertion the toxin molecules associate to form a pore composed of six toxin molecules. As more and more Cry molecules enter the membrane additional pores form, leading to the influx of cations and water followed by cell hypertrophy and lysis. For a more in depth discussion of the studies dealing with the mode of action of Bt, the interested reader is referred to the reviews by Gill *et al.*, Knowles and Dow and Schnepf *et al.*

Structure of Cry Proteins

Analysis of *cry* gene sequences showed that the active portion of the Cry toxin molecule (*i.e.*, essentially amino acids 30-630) contains five blocks of conserved amino acids distributed along the molecule, and a highly variable region within the C-terminal half. The variable region was thought to be responsible for the insect spectrum of activity and experimental evidence from recombinant DNA studies was obtained in support of this hypothesis. For example, by swapping the highly variable regions of Cry1Aa and Cry1Ac, the insect spectrum of these molecules could be reversed.

In addition, binding studies showed that in most cases the degree of sensitivity of an insect to a particular Cry molecule was directly correlated with the number of high-affinity binding sites on the midgut microvillar membrane. Studies over the past decade have improved our knowledge of Cry protein molecular biology through determination of the structure of Cry3A, Cry1Aa and Cry2A, identification of regions on the Cry molecule involved in midgut binding and specificity, and identification of several glycoprotein receptors on the insect microvillar membrane. The structure of the Cry3A molecule was the first to be solved, and similarities in conserved amino acid blocks among Cry toxins and conservation in hydrophobicity indicate that this structure is a good general model for Cry toxins.

The solution of the crystal structure of the Cry3A molecule showed that this protein is basically wedge-shaped and consists of three domains. Domain I is composed of amino acids 1-290 and contains a hydrophobic seven-helix amphipathic bundle, with six helices surrounding a central helix. This domain contains all of the first conserved block and a major portion of the second conserved block of amino acids. Theoretical computer models of the helix bundle of this domain show that after insertion and rearrangement, aggregations of six of these domains could form a pore through the microvillar membrane. Domain II extends through amino acids 291-500 and contains three antiparallel â-sheets around a hydrophobic core.

This domain contains most of the hypervariable region and most of conserved blocks 3 and 4. The crystal structure of the molecule together with recombinant DNA experiments and binding studies indicate that the three extended loop structures in the â-sheets are responsible for initial recognition and binding of the toxin to binding sites on the microvillar membrane. Domain III comprises amino acids 501-644 and consists of two antiparallel â-sheets, within which are found the remainder of conserved block number 3 along with blocks 4 and 5.

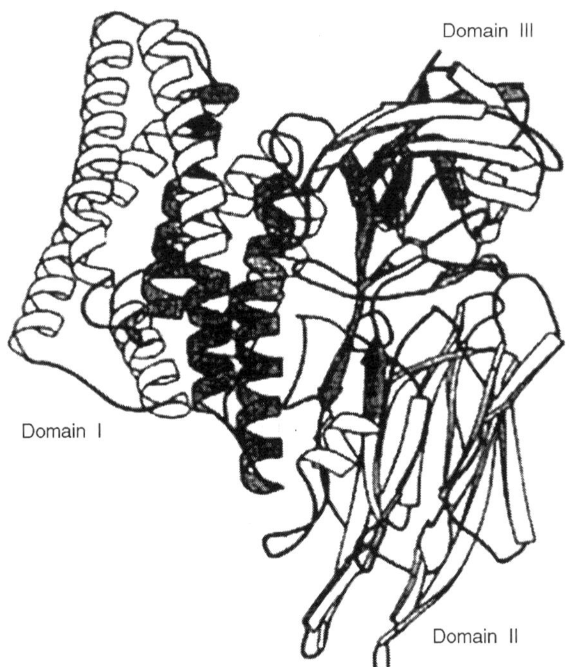

Fig: Three-dimensional Structure of Cry Molecules as Represented by the Structure of Cry3Aa.

The structure resolved by Li *et al.* indicates that this domain provides structural integrity to the molecule. Later site-directed mutagenesis studies of conserved amino acid block 5 in the Cry1Aa molecule suggest that this domain may also play a role in pore formation. Though the structure of the molecule, experiments and computer modeling indicate that binding is attributed primarily to the hypervariable region of this domain, studies by Wu and Aronson also show that positively charged amino acids in domain I are important to binding and toxicity. Moreover, domain III has also been shown to play a role in receptor binding.

Domain I is the pore-forming domain, whereas domain II is responsible for the recognition and binding of the molecule to receptors on the microvilli of the insect midgut epithelium. Domain II therefore accounts for much of the high selectivity and safety of Bt-Cry proteins because mammals and other vertebrates lack receptors for this domain. Domain III assists in stabilizing the structure of the molecule, and also can play a role in binding.

An important aspect of understanding Bt's molecular biology and determining its mode of action at the molecular level is identification of the proteins to which the toxin molecules bind on the microvillar membrane. Sangadala *et al.* contributed to this area with the identification of the first two proteins in the midgut of *Manduca sexta* to which the Cry1Ac toxin binds. They identified an amino peptidase of 120 kDa as a major binding protein.

More recently, cadherin-like proteins and glycoconjugates have also been identified as receptors for Bt endotoxins. These findings indicate that the initial recognition signals on the midgut of a sensitive insect are not ion channels, but one or more proteins or glycoproteins within or which extend from the microvillar membrane. Thus, if these observations hold for other Cry proteins, the midgut receptors are then essentially docking proteins for the toxins, indicating that the toxins do not directly affect ion permeability by binding to ion channels.

These studies of the structure of Bt toxins and their binding properties have identified key regions of the molecule responsible for binding and, putatively, the first binding proteins on the insect midgut epithelium. While further studies are needed to more clearly define the specificity of binding and binding sites, the most enigmatic area of Bt's molecular biology with respect to its mode of action is how the toxin causes toxicity after binding to the membrane. An understanding of this could provide information that might enable the development of insecticidal Cry proteins for use against insects against which we currently have none, such as cockroaches and grasshoppers.

Mode of action and Structure of Cyt Proteins

Several Cyt proteins are known, including Cyt1A, Cyt2A, Cyt1B and Cyt2B, though most of our knowledge of these proteins is based on studies of Cyt1A and Cyt2A, and the crystal structure of the latter toxin. These proteins are highly hydrophobic and all have a mass in the range of 24-28 kDa. They share no significant amino acid sequence to identify with Cry proteins, and are thus unrelated. The first Cyt protein, Cyt1A, was identified as a component of the parasporal body of *B. thuringiensis* subsp. *israelensis*.

Initially, it was thought that Cyt1A played little role in the toxicity of this subspecies to mosquito and blackfly larvae. However, there is now general agreement that it is not only toxic to these and related flies belonging to the dipteran suborder Nematocera, but that it is important as a synergist of Cry4 and Cry11 toxins, and in avoiding the development of insect resistance. While rare in comparison to Cry proteins, since the early work on Cyt1A in *B. thuringiensis* subsp. *israelensis*, Cyt proteins have been reported from mosquitocidal isolates of *B. thuringiensis* subsp. *kyushuensis* (H11a11c), *morrisoni* (H8a8b), *medellin* (H30) and *jegatheson* (H28a28c).

As in the case of Cry proteins, Cyt proteins are synthesized during sporulation as protoxins and assembled into crystalline inclusions that make up a portion of the parasporal body. They are always associated with Cry proteins, but do not cocrystallize with these, instead forming a separate inclusion. The shape of inclusion formed by Cyt proteins varies from hemispherical to angular because they assemble in a spherical parasporal body simultaneously along with inclusions formed by the other Cry proteins with which they occur. However, when produced alone in a recombinant *B. thuringiensis*, the Cyt1A protein forms a bipyramidal crystal with 12 faces.

Based on studies of Cyt1A, after ingestion the 27.3-kDa protoxin molecules dissolve from the crystal under the alkaline conditions present in the midgut of susceptible insects such as mosquito larvae. Proteolytic enzymes then activate the protoxin by cleaving amino acids at both the C- and N-terminus, releasing an active toxin of 24 kDa. This molecule passes through the peritrophic membrane where it then inserts into the microvillar brush border membrane of midgut epithelial cells. Unlike Cry molecules, it appears that Cyt proteins do not require a glycoprotein receptor for binding. Instead, Cyt proteins have a high affinity for the lipid portion of the membrane, specifically for unsaturated phospholipids such as cholesterol, phospholipidcholine and sphigomyelin. After binding to the microvillar membrane, the cells hypertrophy and subsequently lyse.

The specific mode of action of Cyt toxins at the molecular level is not known. The leading hypothesis is that Cyt molecules insert into the microvillar membrane and then assemble into clusters of as many as 18 molecules to form pores that act much like those proposed for Cry toxins. Lysis then results from the influx of cations and water. However, evidence has also been provided that Cyt molecules do not form pores, but rather act like detergents, binding to and perturbing the lipid bilayer and disrupting the structural arrangement of membrane proteins, which would also cause lysis. Unlike Cry proteins, Cyt proteins exhibit cytolytic activity *in vitro* against a wide range of invertebrate and vertebrate cell types, though at higher concentrations. However, toxicity is not observed when ingested *in vivo* by most insects tested, including larvae of lepidopterous species, non-nematoceran flies such as houseflies and fruitflies, and vertebrates such as mice. The mechanism of this *in vivo* specificity is not known, but may involve specific combinations of unsaturated phospholipids. Though Cyt proteins have been thought to be only toxic to nematocerous dipterans, it has recently been shown that at least one beetle species, *Chrysomela scripta*, is sensitive to the Cyt1A protein.

An important property of Cyt proteins, based primarily on studies of Cyt1A, is their ability to synergize the toxicity of the Cry proteins with which they occur. For example, combination of Cyt1A with the Cry4A, Cry4B or Cry11A proteins with which it occurs in the parasporal body of *B. thuringiensis* subsp. *israelensis* results in toxicities three- to fivefold higher than the specific toxicity of Cyt1A or any of the Cry proteins alone. The mechanism of this synergism is not known but probably involves cooperativity of the Cyt and Cry proteins in binding to and/or inserting into the microvillar membrane.

Bacterial Insecticides and Transgenic Crops Based on Bacillus Thuringiensis

More than 30 Bt formulations are on the market in the United States, Europe, Asia and South America. Most of these are complex mixtures of sporulated cells produced by large-scale fermentation and formulation ingredients (stabilizing agents, spreaders, stickers, fillers). The fermentation products contain spores and an array of toxins found in either *B. thuringiensis* subsp. *kurstaki*, which is used to control a wide range of lepidopteran pests, or *B. thuringiensis* subsp. *israelensis*, used to control the larvae of numerous mosquito and blackfly species. Both owe their success to their broad spectrum of activity and the absence of any widespread insect resistance after many years of use. These characteristics are due to the complex composition of the parasporal body, which in each case contains four major proteins, and to the moderate frequency of usage.

With respect to Bt transgenic crops, those on the market are based on full-length or truncated Cry proteins. These crops were constructed by transforming the plants with Bt *cry* genes modified to optimize their expression in plants by changing the codons to reflect plant codon usage. All Bt crops currently on the market produce only a single Cry protein, Cry1Ab, Cry1Ac or Cry3A. However, crops which produce two Cry proteins, such as Bt cotton producing Cry1Ac and Cry2Ab, are being tested in the field. In addition, a variety of insect-resistant crops which produce Bt non-Cry or non-Bt insecticidal proteins are under development, and submissions for regulatory approval have been made in the United States and Australia.

GENETIC ENGINEERING AND GM CROPS

Over the last 30 years, the field of agricultural biotechnology has developed rapidly due to the greater understanding of DNA as the chemical double-helix code from which genes are made. Genetic engineering is one of the modern agricultural biotechnology tools that is based on recombinant DNA technology.

The term genetic engineering, often interchanged with terms such as gene technology, genetic modification, or gene manipulation, is used to describe the process by which the genetic makeup of an organism can be altered using "recombinant DNA technology."

This involves using laboratory tools and specific enzymes to cut out, insert, and alter pieces of DNA that contain one or more genes of interest. The ability to manipulate individual genes and to transfer genes between species that would not readily interbreed is what distinguishes genetic engineering from traditional plant breeding. With conventional plant breeding, there is little or no guarantee of obtaining any particular gene combination from the millions of crosses generated. Undesirable genes can be transferred along with desirable genes or while one desirable gene is gained, another is lost because the genes of both parents are mixed together and re-assorted more or less randomly in the offspring. These problems limit the improvements that plant breeders can achieve, eating time and funds along the way.

In contrast, genetic engineering allows the direct transfer of one or just a few genes, between either closely or distantly related organisms. Not all genetic engineering techniques involve inserting DNA from other organisms. Plants may also be modified by removing or switching off particular genes and genetic controls (promoters).

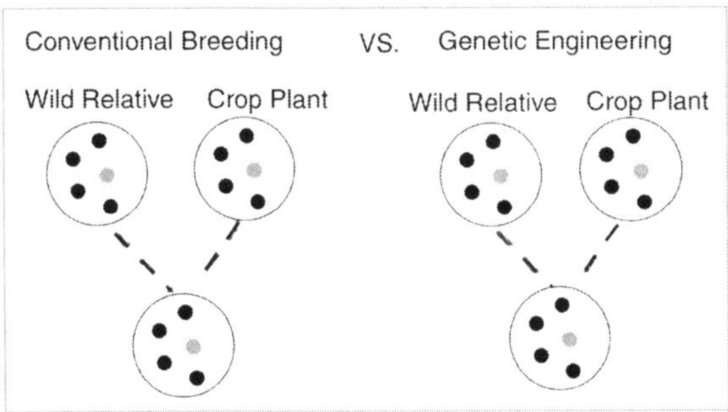

Fig. Conventional vs. Genetic Engineering

Application of Genetic Engineering in Crop Production

Genetic engineering techniques are only used when all other techniques have been exhausted and when:

- The trait to be introduced is not present in the germplasm of the crop;

- The trait is very difficult to improve by conventional breeding methods;
- Tt will take a very long time to introduce and/or improve such trait in the crop by conventional breeding methods.

Modern plant breeding is a multi-disciplinary and coordinated process where a large number of tools and elements of conventional breeding techniques, bioinformatics, biochemistry, molecular genetics, molecular biology and genetic engineering are utilized and integrated.

Development of Transgenic Crops

Although there are many diverse and complex techniques involved in genetic engineering, its basic principles are reasonably simple. It is however, very important to know the biochemical and physiological mechanisms of action, regulation of gene expression and safety of gene and gene product to be utilized. The process of genetic engineering requires the successful completion of a series of six steps.

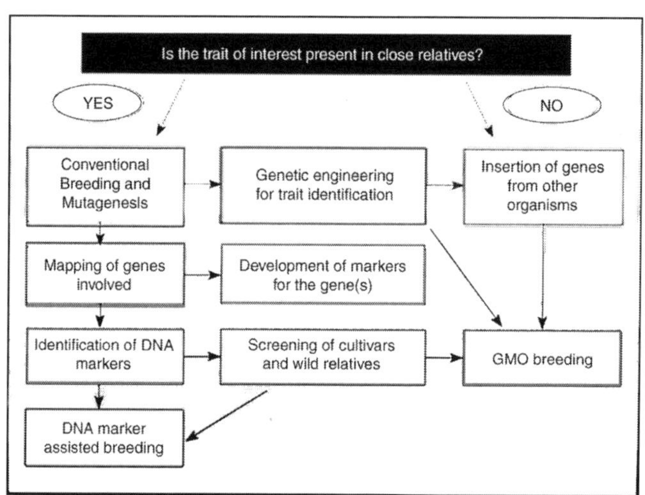

Fig. Integration of Conventional and Modern Biotechnology Methods in Crop Breeding

- *Step 1*: Nucleic acid (DNA/RNA) Extraction Nucleic acid extraction, either DNA or ribonucleic acid (RNA) is the first step in the genetic engineering process. It is therefore important that reliable methods are available for isolating these components from the cell. In any isolation procedure, the initial step is the disruption of the cell of the desired organism, which may be viral, bacterial or plant cells, in order to extract the nucleic acid. After a series of chemical and biochemical steps, the extracted nucleic acid can be precipitated to form thread-like pellets of DNA/RNA.

- *Step 2*: Gene cloning The second step is gene cloning. There are basically four stages in any cloning experiment: generation of DNA fragments, joining to a vector, propagation in a host cell, and selection of the required sequence. In DNA extraction, all DNA from the desired organism is extracted. This genomic DNA is treated with specific enzymes called restriction enzymes cutting it into smaller fragments with defined ends to allow it to be cloned into bacterial vectors. Copies of the vector will then habour many different inserts of the genome. These vectors are transformed into bacterial cells and thousands of copies are produced.

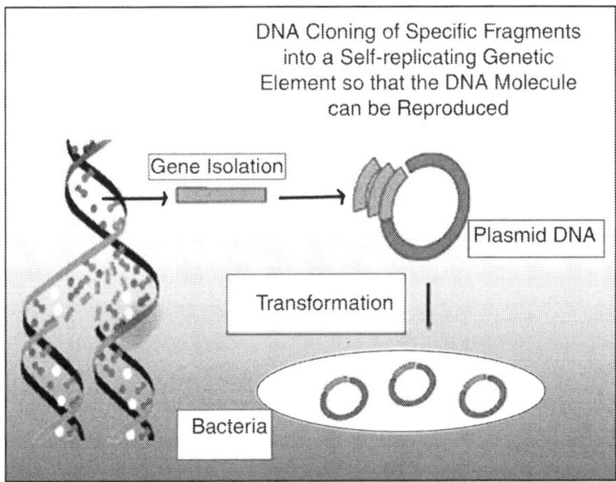

Fig. Gene Cloning

Using information relating to specific molecular marker sequences and the desired phenotype, the vector harboring the desired sequence is detected, selected, isolated and clones are produced. Restriction enzymes are again utilized to determine if the desired gene insert was cloned completely and correctly.

- *Step 3*: Gene Design and Packaging Once the gene of interest has been cloned, it has to be linked to pieces of DNA that will control its expression inside the plant cell. These pieces of DNA will switch on (promoter) and off (terminator) the expression of the gene inserted. Gene designing/packaging can be done by replacing an existing promoter with a new one, incorporating a selectable marker gene and reporter gene, adding gene enhancer fragments, introns, and organelle-localizing sequences, among others. Promoters Promoters allow differential expression of genes. For instance some promoters cause the inserted genes to be expressed all the time, in all parts of the plant (constitutive) whereas others allow expression only at certain stages of plant growth, in certain plant tissues, or in response to external environmental signals. The amount of the gene product to be expressed is also controlled by the promoter. Some promoters are weak, whereas others are strong. Controlling the gene expression is an advantage in developing GM plants.

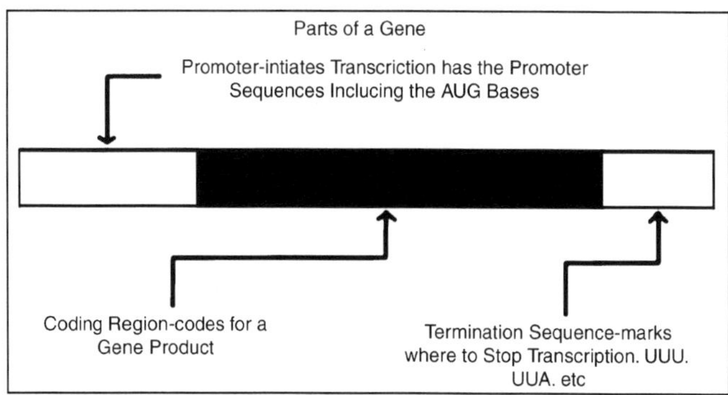

Fig. Parts of a Gene

- *Marker Genes*: Selectable marker genes are usually linked to the gene of interest to facilitate its detection once inside the plant tissues. This enables the selection of cells that have been successfully incorporated with the gene of interest, thus saving considerable expense and effort. Genetic engineers used antibiotic resistance and herbicide resistance marker genes to detect cells that contain the inserted gene. Cells that survive the addition of marker agents to the growth medium indicate the presence of the inserted gene. Although increase in antibiotic resistance in humans and animals is unlikely to occur using antibiotic resistance marker, genes coding for resistance to non-medically important antibiotics are preferred. In addition, alternative types of marker genes have been developed which are related to plant metabolism such as phosphomannose isomerase, xylose isomerase and others.
- *Reporter Genes*: Reporter genes are cloned into the vector in close proximity to the gene of interest, to facilitate the identification of transformed cells as well as to determine the correct expression of the inserted gene. Reporter genes that have been used include: the beta glucuronidase gene (gusA gene) which acts on a particular substrate producing a blue product, hence making the transformed cells blue; the green fluorescent protein (gfp) which allows transformed cells to glow under a green light; and luciferase gene that allows cells to glow in the dark, among others.
- *Enhancers*: Several genetic sequences can also be cloned in front of the promoter sequences (enhancers) or within the genetic sequence itself (introns, or non-coding sequences) to promote

gene expression. An example is the cloning of the cauliflower mosaic virus promoter enhancers in front of the plant promoter.

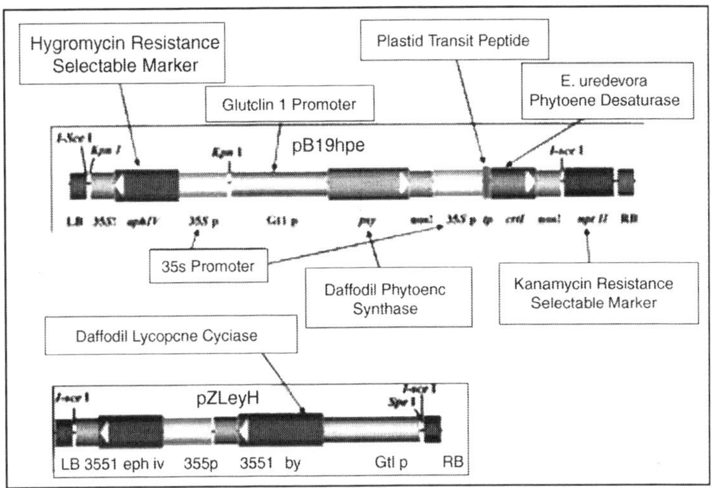

Fig. Components of a Gene Construct Used in Developing Golden Rice

Once the gene of interest is packaged together (with the promoter, reporter and the marker gene), it is then introduced into a bacterium to allow for the creation of many copies of the gene package. The DNA isolated from the bacterial clones can then be used for plant cell transformation using particle bombardment. If however the use of bacteria Agrobacterium tumefaciens is preferred in the plant transformation, the whole gene package should be cloned in between two border sequences (left and right border) of a binary vector.

This will allow processing of the Agrobacterium so that only the transfer DNA (T-DNA) will be incorporated into the plant genome:

- *Step 4*. Transformation The most common methods used to introduce the gene package into the plant cells in a process called transformation or gene insertion, include biolistic transformation using the gene gun and Agrobacterium-mediated transformation.
  - *Particle Bombardment*: Particle bombardment is a mechanical method of introducing the desired gene. The desired genetic sequence is cloned into a plant DNA vector and introduced into the plant using the gene gun or particle gun. As in the common gun, the gene gun uses minute particles of tungsten or gold as the bullet. These particles are coated with the DNA solution and fired to the plant cells through the force of the Helium gas inside

a vacuum-filled chamber. The DNA and the tungsten/ gold particles get inside the cell, and within 12 hours, the inserted DNA gets inside the nucleus and integrated with the plant DNA. The tungsten/ gold particles are sequestered to the vacuole and eliminated later.

Transformed cells are cultured in vitro and induced to form small plants (regeneration) that express the inserted gene.

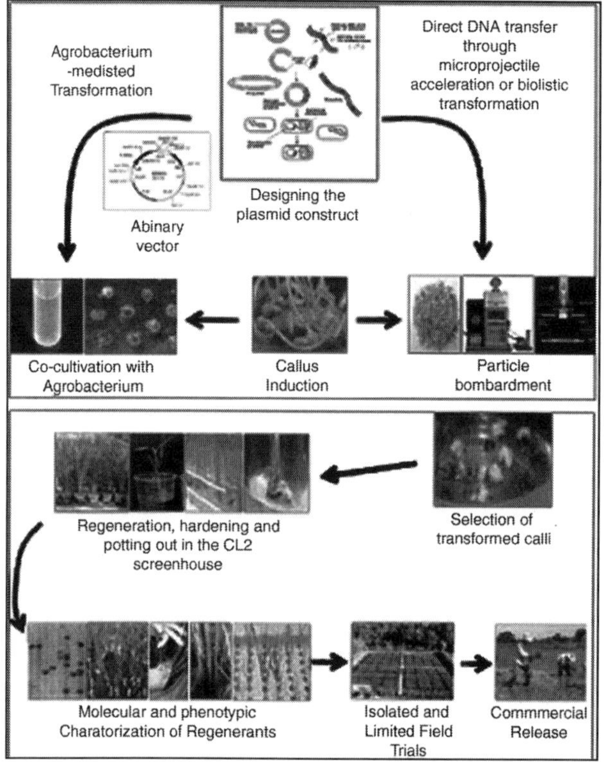

Fig. Genetic Transformation Methods (Biolistics or Gene Gun and Agrobacterium Tumefaciens-mediated Transformation Methods)

Agrobacterium Tumefaciens-meDiated Transformation

The "sharing" of DNA among living forms is well documented as a natural phenomenon. For thousands of years, genes have moved from one organism to another. For example, Agrobacterium tumefaciens, a soil bacterium known as 'nature's own genetic engineer', has the natural ability to genetically engineer plants. It causes crown gall disease in a wide range of broad-leaved plants, such as apple, pear, peach, cherry, almond, raspberry and roses.

The disease gains its name from the large tumor-like swellings (galls) that typically occur at the crown of the plant, just above soil level. Basically, the bacterium transfers part of its DNA to the plant, and this DNA integrates into the plant's genome, causing the production of tumors and associated changes in plant metabolism. Molecular biologists have utilized this biological mechanism to improve crops.

The genes that cause the galls are removed and replaced with genes coding for desirable traits. Plant cells infected with the bacterium will not form galls but produce cells containing the desired gene, which when cultured in a special medium will regenerate into plants and manifest the desired trait. The main goal in any transformation procedure is to introduce the gene of interest into the nucleus of the cell without affecting the cell's ability to survive. If the introduced gene is functional, and the gene product is synthesized, then the plant is said to be transformed.

Once the inserted gene is stable, inherited and expressed in subsequent generations, then the plant is considered a transgenic:

- *Step 5*: Detection of Inserted Genes Molecular detection methods have been developed to determine the integrity of the transgene (introduced gene) into the plant cell. Polymerase chain reaction or PCR is a quick test to determine if the regenerated transgenic cells or plants contain the gene. It uses a set of primers (DNA fragments)—forward and backward primers, whose nucleotide sequences are based on the sequence of the inserted gene. The primers and single nucleotides are incubated with the single stranded genomic DNA and several cycles of DNA amplification is conducted in a PCR machine. Analysis of the PCR products in agarose gel will show if the plants are really transformed when DNA fragments equivalent in size with the inserted gene is present and amplified. Southern blot analysis determines the integrity of the inserted gene: whether the gene is complete and not fragmented, at the correct orientation, and with one copy number. The DNA coding sequence is the probe binding to the single nitrocellulose paper. Autoradiography will reveal the transgenic status of the plant. Northern blot analysis determines whether the transcript or the messenger RNA (mRNA) of the introduced DNA is present and is correctly transcribed in the transgenic plant. The messenger RNA of the transgenic plants are isolated and processed to bind to the nitrocellulose membrane. Labeled DNA is used to bind to the mRNA and can be visualized through autoradiography. Western blot analysis or protein immuno blotting is an analytical technique used to detect whether the transgenic plants produce the specific protein product of the introduced gene. Protein samples are extracted from the transgenic plants, processed into denatured proteins and transferred to a nitrocellulose membrane. The protein is then probed or detected using the antibodies specific to the target protein.

- *Step 6*: Backcross Breeding (if needed) Genetic transformation is usually conducted in elite or commercial varieties which already possess the desired agronomic traits but lacks the important trait of the transgene. Thus, once successfully conducted, the genetically modified plant will be easily recommended for commercialization if it shows stability in several generations and upon successfully passing and fulfilling varietal registration requirements. However, some plant transformations may have been performed in plant varieties which are amenable to genetic transformation but are not important in the target country, or in a variety adapted only in the country where the transformation was conducted. There may also be sterility problems in the transgenic plant. In such cases, conventional plant breeding is performed where the transgenic plant becomes the pollen source in the breeding programme and the elite lines or commercial varieties as the recurrent parent. Backcross breeding enables the combination of the desired traits of the recurrent parent and the transgenic line in the offsprings. The length of time in developing transgenic plant depends upon the gene, crop species, available resources and regulatory approval. It varies from 6 to 15 years before a new transgenic plant or hybrid is ready for commercial release.

COMMERCIALLY AVAILABLE CROPS IMPROVED THROUGH GENETIC ENGINEERING

Transgenic crops which are planted commercially are herbicide tolerant soybean, maize, canola, cotton; insect resistant maize and cotton; and virus resistant squash and papaya. With genetic engineering, more than one trait can be incorporated into a plant and are called stacked traits. These are currently corn and cotton crops with both herbicide and insect tolerance traits. Transgenic crops with combined traits are also available commercially such as the herbicide tolerant and insect resistant maize and cotton.

New and Future Initiatives in Crop Genetic Engineering

To date, commercial GM crops have delivered benefits in crop production, but there are also a number of products in the pipeline which will make more direct contributions to food quality, clean environment, pharmaceutical production, and livestock feeds.

Examples of these products include: rice with higher levels of iron and beta carotene (an important micronutrient which is converted to vitamin A in the body); long life banana that ripens faster on the tree and can therefore be harvested earlier; maize with improved feed value; delayed ripening papaya; papaya ringspot virus resistant papaya; tomatoes with high levels of flavonols, which are powerful antioxidants; drought tolerant maize and wheat; maize with improved phosphorus availability; arsenic-tolerant plants; insect resistant eggplant and rice; edible vaccines from fruit and vegetables; low lignin trees for paper making among others.

FOOD SAFETY

Safety of Genetically–Engineered Food Assessed

Before GE foods and products made from GE crops are approved for use, they have undergone safety testing by the companies or institutions that developed them. Data were reviewed by government regulatory agencies and scientific reviewers based on internationallyaccepted protocols. Frequently, GE foods are also tested by outside groups and the results published in peer-reviewed journals. The process is comparable to safety assessments for pharmaceutical drugs and biomarkers; pharmaceutical companies provide safety data that are subsequently reviewed by the U.S. Food and Drug Administration (FDA) scientists. To date all GE products in the marketplace have undergone full reviews by regulatory agencies regarding safety and content relative to unmodified forms. Submitting the safety data is in the developer's best interests given the legal liabilities incurred should a problem with the food arise following market introductions.

When DNA is Eaten

DNA is chemically identical regardless of its source and is mostly degraded during industrial processing and in the digestive tract. Small fragments can be detected in certain body tissues, such as leukocytes, liver, and spleen. The daily human intake of DNA in food is estimated at 0.1-1g. Estimates of the total daily transgene DNA intake can be calculated assuming 50 per cent of the diet is from GE foods and transgenes represent an estimated 0.0005 per cent of total DNA in food, as 0.5-5ug/day. In July 2007, the European Food Safety Authority released statements on the fate of genes and proteins in food and feed: "After ingestion, a rapid degradation into short DNA or peptide fragments is observed in the gastrointestinal tract of animals and humans" and "To date, a large number of experimental studies with livestock have shown that recombinant DNA fragments or proteins derived from GM plants have not been detected in tissues, fluids or edible products of farm animals"

Changes in the Nutritional Content of Genetically-Engineered Food

GE foods are tested in comparison with conventional counterparts in terms of the nutritional composition: levels of protein, carbohydrate, fat, vitamin, mineral, fibre, moisture, and phytochemicals, and analysed if the composition is substantially equivalent. GE crops and conventional crops should have been grown in comparable conditions to eliminate the effect of the environment in the nutritional composition.

There are also GE crops which are developed to change the nutritional profiles of the foods such as those with increased B-carotene, flavonoids, calcium, folate, and iron availability. According to US-FDA policy, GE foods with altered nutritional traits must be labeled to indicate nutritional differences; one example is VistiveTM, a low-linoleic oil from GE soybeans that can be used instead of trans fat-containing oils. Such crops should be tested for substantial equivalence to compounds unrelated to the introduced trait.

Bt Protein Affect Hsumans

Bt proteins are naturally occurring insecticides produced by the soil bacterium, Bacillus thuringiensis, used to control crop pests such as larvae of butterflies and moths, beetles, and mosquitoes since the 1920s. The crystalline, inactive insecticidal Bt proteins, form bodies inside the bacterium and become active when they are eaten by the target insect larva and cleaved.

The active peptides bind to specialized receptors in the midgut of the insect, creating holes in the gut membrane that cause contents to leak and kill the larvae. The precision of different Bt proteins for their targets resides in the specificity of their tight binding to companion receptors in the insect gut. In recent years, a variety of safety studies were conducted specifically on native Bt proteins to show that they do not have characteristics of food allergens or toxins. Data on CryIAb in maize and cotton and Cry1Ac in tomato, maize and cotton have been carefully reviewed by regulatory agencies in numerous countries, including the U.S., Canada, Japan, UK, EU, Russia, and South Africa. Mycotoxin are toxic and carcinogenic chemicals produced by fungi that gain entry into the holes produced by the larva in corn. The reduction of mycotoxin incidence in Bt corn results in a positive impact in the improvement of corn yield, human and animal health.

Genetically Engineered Foods Cause Food Asllergies

No food is 100 per cent safe, be it conventional, GE, or organic. Allergies are present in the "big eight" which is composed of milk, eggs, fish, shellfish, tree nuts, soybeans, wheat, and peanuts. Since food safety testing conducted on GE foods focuses on the introduced gene and its protein product, it seems unlikely that allergenicity issues related to a commercialized GE food that has undergone strict government health regulatory scrutiny will be greater than that of conventional foods, created by classical breeding and mutation that have not undergone such scrutiny.

Viral Genetic Sequences Inserted in the Genetically Engineered Crops Create a Human Risk

One of the important concerns is the use of the virus-derived promoter which are Introduced sequences in the transgenic plants that regulates how much, where, and when the encoded protein is expressed. This includes the cauliflower mosaic virus 35S which was used in some commercial GE crops, eg. Bt 11, Bt 176, Mon 810 maize, and Roundup Ready soybean. Speculations that the "35S promoter could affect the stomach and colonic lining and cause a growth factor effect with the unproven possibility of hastening cancer formation in those organs" were forwarded earlier without any scientific experimentation.

These speculations have been extensively rebutted by the scientific community because the 35S promoter can be found everywhere in nature. For instance, an estimated 14-25 per cent of oilseed rape in the field is infected with CaMV; similar numbers have been estimated for cauliflower and cabbage. Because of its prevalence in foods, humans have consumed CaMV and its promoters at high levels for decades with no observable effects. The presence of the CaMV promoter in GE plants does not in principle present a different situation. Additionally, DNA in food is rapidly broken down during digestion, giving it little time to interact with the stomach and colonic linings.

Antibiotic Resistance Genes in Genetically Engineered Foods Increase Antibiotic Resistance in Humans and Animal Intestinal Fsslora

To develop antibiotic resistance in microorganisms present in the human and animal digestive tract, there should be a functional transfer of the antibiotic resistance gene, its controlling elements, and its integration in the bacterial chromosome. This is next to impossible, since during chewing, cells in food are broken down. In raw food, as the cells are destroyed, DNA is released and highly active enzymes in the saliva and in the plant start degrading the DNA. This process continues in the digestive tract where other enzymes further breaks down DNA and proteins. In humans, food remains in the stomach for approximately 2 hours, where the remaining DNA is fragmented into small pieces. The antibiotic resistance gene from GE maize was shown not to transfer to gut bacteria in chickens fed with GE maize.

To refrain from using the controversial antibiotic resistance or herbicide tolerance genes as selectable markers, new selection strategies for identifying engineered plants have been developed. These include genes such as phosphomannose and xylose isomerase that facilitate selection by giving transgenic cells a metabolic advantage over non-transgenic cells, as well as other means to excise the marker genes in the commercial product.

Genetic Engineering be Used to Make Pharmaceuticals

Plant-derived pharmaceuticals and vaccines for common diseases such as hepatitis B, pneumonic and bubonic plague, as well as against allergy sufferers, asthma, seasonal allergies and atopic dermatitis have been developed since the early 1990s. Plant vaccines have the advantage of being readily consumed with limited or no processing without the need for cold storage. However, these

GE crops may enter the food supply if not properly handled and monitored. In the USA, where such pharmaceutical crops are cultivated, government regulations are in place. APHIS which regulates the movement and field testing of GE plants requires special steps to prevent plants that produce drugs or industrial enzymes from contaminating food crops:

- Labeling, packaging, and segregating regulated plant materials;
- Reproductive isolation to prevent GE pollen from fertilizing conventional plants;
- Postharvest monitoring to remove volunteer plants;
- Proper disposal of the transgenic materials.

This regulation was further strengthened in 2005 to include the following:

- Exclude field growth without a permit;
- Include crop inspections seven times a year, twice after harvest;
- Increase field isolation distances;
- Use dedicated farm equipment.

The National Corn Growers Association proposed safeguards such as:

- Using plants that are male-sterile or that produce non GE pollen,
- Dedicated production systems that isolate pharma crops,
- Third party verification
- Grower training programmes.

In Sept. 2002, the FDA released a guidance document that recommends multiple strategies to prevent pharma crops from contaminating human or animal feed. This documents suggests that those who are growing drug-producing plants that cross pollinate, such as corn and canola, strengthen containment procedures by growing plants in geographical regions where little or none of those crops are grown for food.

Labeling of Genetically Engineered Foods is not Required by the FDA

Government policy on labeling has been developed differently in many countries. In the USA, the FDA's labeling policy for GE foods is the same as for conventional foods and it assures that consumers are given information about nutritional, health safety or food quality changes in the end product. FDA mandated labels are not used to provide information about the process by which the food is made.

If a GE food is significantly different from its conventional counterpart, the food must be labeled to indicate the difference. Instances where the nutritional profile changes are included, for example if the GE food is created using genetic information from a previously recognized allergenic source, such as peanut, soy, or wheat, or if the new proteins has characteristics of known allergens. For example, oils made from GE soybeans and canola varieties with changes in fatty acid composition must be labeled; foods containing those oils must be labeled and companies producing that oil must use a new name. For example, Monsanto is using the name Vistive TM, to market its low-linoleic acid product from GE soybean oils. If a food contains a new potentially allergy-causing introduced protein, the label must state that the product contains the allergen and name its source.

Organic Foods

Organic farming is a method of agricultural production that does not allow the use of synthetic pesticides, fertilizers or growth enhancers. Food grown under organic certification differ from conventionally produced food by the manner in which they are grown, handled, and processed, but an "organic" label does not guarantee the nature of the product, the food, or ingredient, only its production method.

The important factors for many people who consume organic foods relate to the perceptions that they are healthier, taste better, are better for the environment, have lower pesticide levels and fewer food additives, and are better for animal welfare. However, organic certification does not imply that foods produced using organic methods are more nutritious or safer than those produced without organic methods. Differences reported in nutrient composition between organically and conventionally produced foods are interesting but it is very difficult to control all variables that might affect nutritional quality and ensure that the observed variations are significant and reproducible.

In addition, there are many important nutrients for which no significant differences have been found. Much more research is needed to determine whether food products are reproducible and have a significant impact on human health. Strictly from a nutritional perspective not enough data exist at present to show nutritional benefits from conventionally or organically produced foods that favour consuming either for health benefits. However, if the goal is to promote healthy eating, it is more important for consumers to focus on eating a healthy, balanced diet, rich in fruits and vegetables, than focusing on foods that are produced by particular methods. Convincing epidemiological evidence shows that diets rich in fresh fruits and vegetables, regardless of the methods used to produce them, improve health and are associated with reduced frequency and severity of a number of health conditions.

ENVIRONMENTAL ISSUES RNSECT RESISTANCE TO BT

Resistance of insects against synthetic insecticides and Bt toxins in sprays occur and this will be true for GE crops. To slow this development in GE crops, several strategies have been developed. First generation GE crops produced only one Bt toxin in each plant. Planting refuges of non-Bt crops near Bt crops in the field is the primary strategy of delaying insect resistance. This is based on the idea that insects feeding on plants in the refuge are not selected for resistance. Insect resistance to Bt toxins is recessive. The heterozygous offsprings produced when homozygous resistant insects mate with susceptible insects are killed by the Bt crops.

This high-dose/refuge strategy creates plants that produce Bt toxin concentrations high enough to kill heterozygous insects, making resistance functionally recessive. Insect resistance to Bt toxins can thus be postponed substantially. Another approach is called the pyramid or stacking strategy that combines two or more toxins in a single plant, each with different modes of action. An example is Bollgard II cotton producing Cry1Ac and Cry2b, which targets the same pest in two different ways.

Other approaches to delaying insect development are:

- Mixing seeds of Bt and non-Bt varieties are under small scale experiments
- The use of inducible promoter to drive Bt gene expression only during insect attack.
- Use of modified toxins to kill resistant insects, as exemplified by the use of modified Bt toxin that will not be affected by the mutations in the midgut cadherins.

Cadherins promote toxin oligomerization of Cry1A protein which has alpha helix in the binding site. Modified Cry1A which does not contain the alpha helix are independent of the cadherins and can thus be effective with insects which has developed resistance due to mutated or silenced cadherins To date, the elapsed time before the first cases of field resistance of insects to Bt crops were reported has been longer than what was predicted under worst-case scenarios, suggesting that management strategies may have delayed resistance development. Despite documented cases of resistance, Bt crops remain useful against most target pests in most regions. As insect resistance to Cry toxins currently deployed in Bt crops increases, other strategies to create GE crops resistant to insects are being developed.

Genetically Engineered Crops Cause Adverse Effects on Non-Target Organisms

Effects on GE crops on non-target organisms have been studied with focus on:

- Monarch butterflies and black swallow tails. USA Environmental Protection Agency have concluded based on two studies that Bt corn was not a significant factor in field deaths of monarch larvae, particularly relative to factors such as the widespread use of pesticides and destruction of the butterfly's winter habits.
- Non target soil microorganisms. Studies on four maize varieties with two different Bt proteins (Cry1Ab and Cry3Bb1) versus near isogenic non-Bt varieties reveal that although numbers and types of microbes and enzyme activities differ from season to season among varieties, no statistically significant differences were seen in number of different

microbes, enzyme activities, or pH. Similar results were found comparing Bt and non-Bt cotton, and no Cry2Ab protein was detected in the rhizosphere in the field grown with Bt cotton.

- Non-target arthropods. Studies on foliage-dwelling arthropods on Bt maize expressing Cry3Bb1 compared with those of conventional insecticide treated maize show that there is no adverse impacts on abundance of any non-target arthropods. Insecticide treated arthropods however reduced the number of non-target insects: ladybird beetles, lacewings, and damsel bugs.
- Microbes and non-target water insects. Water sediments and surface water after labeling genomic DNA of GE Bt corn revealed that sediments had more DNA than surface water. In addition, the Cry1Ab protein was not detectable in both samples.

GENETICALLY ENGINEERED CROPS RESULT IN THE POPULATION DECLINE OF OTHER ORGANISMS

Population decline of other organisms has been an ongoing phenomenon since man learned how to domesticate corps. The introduction of modern agricultural technologies including new varieties; competition between local and introduced varieties led to a displacement of local varieties; and displacing local varieties eroded genetic variability of regional crop populations.

Extensive plant breeding in the early 1960s to feed the tremendous increase in the population produced high-yielding varieties of major food crops, resulting in yield increases but also significant displacement of traditional varieties and a concomitant loss in genetic diversity, particularly landraces of cereals and legumes. Recognition of this fact led to establishment of genebanks across the globe with focus on specific crops. One issue on diversity is the gene flow from GE crops to wild and weedy relatives which could render selective advantage of recipients in certain environments. Gene flow can also happen naturally in conventionally bred and commercialized crops.

This is addressed by the adoption of measures needed in cultivating GE crops near centers of origin depending on the nature of the trait and the frequency of its introduction into an ecosystem. Currently, studies on impact assessment of transgenes moving into wild relatives and the potential to change ecosystem dynamics are requested in environmental impact statements before any GE plant is released. It provides insights into the possible outcomes on the environment. Certain impact assessments of some GE crops are also monitored even after deregulation.

Herbicide-tolerant (HT) Crops Lead to Super Weeds

Development of herbicide-tolerant weeds has occurred with both traditionally-bred and GE crops. This phenomenon reduces the effectiveness of certain weed control strategies and decreases weed management options.

Strategies have been developed to minimize the development of herbicide tolerant weeds, such as:

- Use of HT cultivars with resistance genes for herbicides with alternative modes of action that can be used in rotation.
- Use of restriction technologies to prevent gene passage to the next generation through the pollen, *i.e.*, transgenes can be targeted to the cytoplasmic organelles, not in the pollen.
- Rotate the use of HT crops with different modes of action or with non-HT crops.

A few points to consider in using HT crops are: Weeds can also escape herbicide treatment on the basis of application rate, weed age and size, spray volume adjuvants used, water quality and interactions with other herbicides that affect efficacy. Late germination of weeds can also escape herbicide application, thus a second pass of sprays can be done.

Effect of Using GE Crops in Pesticide Usse

Having crops tolerant to herbicides and pest attack increases pest management options and can also reduce the number and strength of pesticide applications. Growth of GE HT crops also allows topical application of herbicide to crops and weeds, which replaces spraying between crop rows and mechanical removal of weeds, both of which can damage crops and result in environmental damage. Reducing mechanical tillage lowers fuel consumption and helps conserve soils prone to erosion and compaction. HT crops can also lead to more flexible herbicide treatment regimes.

The National Center for Food and Agricultural Policy published surveys on U.S. pesticide usage on GE crops. In 2004, HT canola, cotton, maize and soybean as well as Bt cotton and maize showed reductions in herbicide active ingredient (AI) of 25 to 30 per cent.

In a 2006 publication, the USDA National Statistics Service found that from 1996 to 2002, AI use rates for HT cotton and corn, and Bt corn declined as adoption of Bt and HT cotton, corn, and soybeans increased and concurrent shifts occurred towards less environmentally persistent herbicides such as pendimethalin, trifluralin, and metolachlor.

The Environmental Impact Quotient (IEQ) assessment which takes into account the pesticide AI and the environmental impact (EI) of GE crops resulted in significant reductions in the global EI of production agriculture; such that since 1996, the overall EI associated with pesticide use on HT soybean, corn, cotton, canola, and Bt cotton decreased by 15.3 per cent. Cultivation of GE HT crops has also had other positive effects on the environment, *i.e.*, increases in low-or no-till practices and use in combination with integrated pest management schemes, which were made possible because early season pesticide sprays could be eliminated, allowing beneficial insects to establish. Most reports indicate pesticide use and cost decrease following adoption of Bt varieties. In Argentina, numbers of herbicide applications increased with HT soybean but use shifted to more environmentally friendly herbicides. Reduction in pesticide use can also be achieved by using the best methods and tools available, including integrated pest management, biocontrol, organic production methods, and GE organisms to reduce EI while achieving adequate production levels.

Would Bt Crops Need Additional Insecticide Applications

Bt or Cry toxins are toxic to susceptible larvae when cleaved to generate their active form, which then binds to specific receptors in the midgut and creates holes that cause lepidopteran larvae to die.

The first BT GE crops introduced into corn and cotton were targeted to control European corn borer, corn rootworm and cotton armyworm.

Some pests belong to groups insensitive to Bt have to be sprayed to prevent crop damage. With the commercial introduction of corn and cotton varieties with two stacked Bt genes, *i.e.*, Cry1Ac and Cry2Ab in cotton, bollworms and secondary armyworm pests were controlled.

New developments to target different insect pests are: corn with six insect resistant genes against lepidopteran (Cry1F, Cry1A.105, Cry2Ab2) and rootworm (Cry34Ab1 + Cry 35Ab1, modified by Cry3Bb1) pests; the use of a hybrid Cry protein with two binding domains to target lepidopteran and coleopteran pests of potato; use of plant defence proteins such as alpha amylase inhibitors from legumes; use of insecticidal compounds from nematodes, bacterial cholesterol oxidase, avidin, volatile communication compounds, and RNAi approaches targeted to specific insect proteins.

Even with GE approaches, other methods of insect control will be needed, *e.g.*, chemical pesticides, biocontrol, integrated pest management, or organic approaches, because insects are plentiful and ever changing.

Virus–Resistant Genetically Engineered Plants Lead to Novel Viruses

Development of GE crops with resistance to viral diseases has been conducted in squash and papaya using a viral coat protein gene. The USDA APHIS has already deregulated the GE squash allowing commercial production after the virus was shown not to infect wild squash varieties; the resistance gene gave no advantage to wild squash varieties, and the presence of the coat protein gene did not increase viral competitiveness. For GE papaya with the viral coat protein, concerns on viral recombination became a concern since from analyses of viruses, homologous and non-homologous recombination could occur between viruses and between viral genomes and plant genes. Experimental results indicate however that most recombinant viruses are not fully virulent because the new gene combinations are not fully compatible, leaving new hybrids at a competitive disadvantage.

To compete effectively, recombinant viruses must have functional recombinatorial ability, capacity to establish systemic infection, and ability to compete with their progenitors during replication. These requirements place powerful negative selection pressure on newly evolved viruses. Reduced viral replication capacity could also negatively affect recombination frequency in transgenic plants. Large-scale field releases of plants engineered with viral genes are necessary to obtain realistic assessments of the types and recombination frequencies that might occur. Currently, no novel viruses have been reported resulting from GE plants in the field, but likely they would be detected only if their appearance had adverse effects.

At present, the only commercially propagated plants engineered with viral coat protein genes, GE squash, and papaya are grown on small acreages. To minimize the possibility for gene exchange among the viruses, strategies such as RNAi-mediated viral resistance is employed. There is no protein introduced, and the RNAi construct is used to silence a gene from bean golden mosaic virus in Phaseolus vulgaris leading to virusresistant plants.

Genetically Engineered Plants Move to Bacteria in the Field

Horizontal gene transfer is the process of transferring genes among nonsexually related organisms such as from plants to bacteria. Sequence analyses of genes and proteins show that some genes have transferred from plants to bacteria over a very long evolutionary time frame. This transfer can only be simulated in the laboratory using optimized conditions—situations which are difficult to replicate in natural settings. If, however it were to happen in the field, it would be at very low frequencies and the gene would need to provide a selective advantage to survive.

An experiment to determine the persistence of kanamycin resistant bacteria in the soil by increasing the levels of kanamycin in the soil concludes that natural soil conditions rarely would have the selective pressure necessary to keep nptII in the bacterium. Data from this and other studies indicate that homologous recombination and integration of plant genes into competent soil bacteria could occur, but at very low frequencies, and the environmental significance would depend on selective pressure for the trait.

When Pollen Moves From Genetically Engineered Crops to Wild Relatives or Non-genetically Engineered Varieties

Gene flow or the movement of pollen from one plant to another is made possible when the parental plants:

- Flowers at the same time;
- Close enoussgh to allow a vector (insect, wind, or animal) to transfer pollen to receptive females;
- Produce pollen that can result in embryos developing into viable seeds and germinating.

Successful pollination also depends on the longevity of pollen viability, pollen travel distance and the mode of pollination the plant has, whether self or cross-pollinated. Gene flow may present significant economic or environmental risks for either conventionally bred or GE crops on a case-by-case evaluation. Crop-to-wild relative gene flow could result if the plants grow in overlapping regions resulting in new combinations of genes that can improve, harm, or have no effect on the fitness of recipient plants.

Genes can also flow from wild relatives to cultivated crops, introducing new traits into the next generation seeds, but only affect the crop if it is replanted. Planting of GE varieties in areas of genetic diversity of plants needs additional precautions to reduce possible impacts of introgression of GE traits and the potential significant environmental consequences.

To minimize this occurrence, planting of GE crops near wild species should be avoided or other technologies could be used to prevent gene(s) from moving to wild varieties. Gene flow could also occur when compatible plants are present within the vicinity. GE varieties like conventional plants can also persist in the environment. Organic farmers should be aware of these occurrences to be able to adopt the necessary precautions of spatial and temporal isolation.

Organic, Conventional and Genetically Engineered Cropping Systems Coexist

Farmers are used to planting different varieties and planting strategies in order to develop farm products that meet the requirements of the consumers. They are used to planting white and yellow maize, hot and sweet peppers, high and zero erucic acid rapeseed, and still achieve purity standards dictated by certified seed specification. Coexistence strategies must be devised to allow neighbour farmers to farm in an economically viable manner. This can involve alerting each other to their plans and modifying them to accommodate each others' needs. When GE crops are grown next to organic farming operations, certain practices that minimize synthetic pesticide drift can also limit GE gene flow, such as spatial separation of fields, staggered planting dates, and planting varieties with different maturity dates and those that are not sexually compatible.

Other crops-specific methods have been devised to aid coexistence strategies. Gene flow is not only the means for GE to commingle with conventional or organic crops; crops must also be segregated during harvest, shipping and processing. Methods limiting such commingling have in some cases been implemented. With the use of various production methods comes the mixing of permissible inputs and methods, whether with their own farms with products from neighbouring farms, or during harvest and processing. The commingling or adventitious presence (AP) is the unintended occurrence of materials other than specific crops and can include weed seeds, seeds from other crops, dirt, insects, and other foreign materials such as stones or plastics. Different countries have set rules on the degree of AP. In the U.S., for seed crops, rules for AP are specified by the Association of Official Seed Certifying Agencies (AOSCA), where a level of 0.5 per cent seed of other varieties and 2 per cent AP of inert materials is permitted in "pure seed" of hybrid corn.

Venetically Engineered Crops or Organic Farming Lead to More Sustainable Agricultural Production Systems

Sustainable agricultural systems should meet the basic needs of the population while preserving the resources for future generations. The United Nation's Millennium Development Goals to "Ensure environmental sustainability by integrating principles of sustainable development into a country's policies, and programmes to reverse the loss of environmental resources."

This need has been widely accepted and the manner to fulfill this may vary. Conventional farming has led to impressive gains of between 70 and 90 per cent of increases in food production in the last few decades. Unfortunately, these were accompanying environmental impacts as well as sizeable consumption of fossil fuels, unsustainable rates of water use and topsoil loss, and contributions to environmental degradation, air pollution, soil erosion, reduced biodiversity, pest resistance, pollution of lakes and streams, and overuse of surface and ground water. Achieving agricultural sustainability can be addressed through numerous agricultural practices such as: integrated pest management (IPM), biological control, organic methods, and use of GE plants, coupled with selected conventional agricultural methods, can play important roles in future sustainable agricultural practices. Biological control can be a part of an IPM strategy and neither biological control nor IPM specifically excludes the use of GE organisms. Organic production relies on practices, such as cultural and biological pest management, that can include IPM and biological control but excludes the use of synthetic chemicals and GE organisms.

The use of GE organisms can also contribute to sustainable practices by augmenting and replacing certain conventional practices. For example, plants can be created that increase water use, and fertilizer efficiencies, that remediate soil contaminants, increase no-till or low-till practices to help reduce greenhouse gases and produce higher yields without increasing land usage, particularly in developing countries. To achieve true sustainability agriculture must use the best of all practices.

7

Risks Associated with Transgenic Plants and Crops

RISK ISSUES

FOOD SAFETY

To minimize the risk of gene flow from transgenic to adjacent conventional crop fields, federal regulations require buffer strips of conventional varieties around transgenic fields.Food safety issues, except as they impact domestic marketing and exports, are beyond the scope of this publication. Five years ago the major publicized concerns were environmental.

Since then, the environmental community has stalled some transgenic crops. Food safety concerns include:

- Possibility of toxins in food
- Possibility of new pathogens
- Reduced nutritional value
- Introduction of human allergens
- Transfer of antibiotic resistance to humans
- Unexpected immune-system and genetic effects from the introduction of novel compounds

It is in part because of these concerns that domestic consumer demand for organically grown crops continues to increase. There are other marketing problems that reflect religious dietary and general religious (sometimes dismissed as "cultural") sensibilities, as well as ethical/philosophical concerns.

FARM MANAGEMENT ISSUES

In the case of herbicide-tolerant crops, initially farmers hoped to use a single broad-spectrum herbicide for all their crop weeds. It has turned out that they need more than one application in most seasons. By planting insecticidal crops, farmers can eliminate the need to apply pesticides for caterpillar pests like the European corn borer or the cotton bollworm, though they still have to contend with other crop pests.

While these crops offer simplified pest control features, they may complicate other areas of farm management. Farmers who grow both transgenic and conventional varieties of the same crop will need to segregate the two during all production, harvesting, storage, and transportation phases if they sell into differentiated markets or plan to save their own seed from the conventional crops

To minimize the risk of gene flow from transgenic to adjacent conventional crop fields, federal regulations require buffer strips of conventional varieties around transgenic fields. Different transgenic crops require different buffer widths. Because the buffer strips must be managed conventionally, producers have to be willing to maintain two different farming systems on their transgenic fields. Crops harvested from the buffer strips must be handled and marketed as though they are transgenic. Planted refuges-where pest species can live outside fields of insecticidal and herbicide-tolerant transgenic crops-are also required to slow the development of weed and insect pest resistance to Bt and broad-spectrum herbicides. These refuges allow some individuals in the pest population to survive and carry on the traits of pesticide susceptibility. Requirements governing the size of refuges differ according to the type of transgenic crop grown, but a 2006 report in AgBioForum, based on a survey of Indiana farmers, states the requirements are misunderstood by farmers and routinely ignored. For some crops they are unworkable.

Farmers growing herbicide-tolerant crops need to be aware that volunteer crop plants the following year will be herbicide resistant. Such resistance makes no-till or direct-seed systems difficult because volunteers can't be controlled with the same herbicide used on the rest of the crop. In a no-till system that relies on the same broad-spectrum herbicide that the volunteer plants are resistant to, these plants will contaminate the harvest of a following conventional variety of the same crop—a situation farmers tend to avoid for two reasons.

First, the contamination means a following conventional crop will have to be sold on the transgenic market. This leads to the second reason. If farmers grow and market a transgenic crop for which they do not have a technology agreement and did not pay royalty fees, they may face aggressive collection by the company that owns the transgenic variety. Hundreds of U.S. farmers have already been charged with "theft" of a company's patented seed as a result of contamination in the field.

Farmers growing insecticidal crops need to recognize that insect pressure is difficult to predict and may not warrant the planting of an insecticidal variety every year. In a year when pest pressure is low, the transgenic seed becomes expensive insurance against the threat of insect damage. Farmers growing transgenic crops need to communicate with their neighbours to avoid contaminating neighbouring fields and to ensure that buffers are adequate. In Maine, farmers growing transgenic crops are now required by law to be listed with the state agriculture department, to help identify possible sources of cross-contamination when it occurs.

The law also "requires manufacturers or seed dealers of genetically engineered plants, plant parts, or seeds to provide written instructions to all growers on how to plant, grow, and harvest the crops to minimize potential cross-contamination of non-genetically engineered crops or wild plant populations."

Farm management issues common to all transgenic crops include yield, cost, price, profitability, management flexibility, sustainability, market acceptance, and liability. Yield and profitability, as well as market acceptance, are discussed in separate sections below. Liability is discussed under regulation.

CROP YIELD, COSTS, AND PROFITABILITY

Some farmers will get higher yields with a particular transgenic crop variety than with their conventional varieties, and some will get lower yields. Yield variability is related to many factors, including choice of the conventional analog of the transgenic variety, making it very difficult to analyse how any one feature impacts yield. Costs of various inputs are also constantly changing; and the ability of farmers to adjust to changing costs, particularly rapid changes, is limited and affects profitability. However, some yield, cost, and profitability trends do appear to be emerging from the growing body of research data for transgenic crops. As noted in the Wallace Center report, Roundup Ready soybeans were designed simply to resist a particular chemical herbicide, not to increase yields. In contrast, Bt corn and cotton, by resisting insect pests, may result in higher yields from reduced pest pressure.

Yield: Herbicide Tolerant Crops— Soybeans, Cotton, Canola

Herbicide-tolerant soybeans appear to suffer what's referred to as "yield drag." Again, in some areas and on some farms this tendency of Roundup Ready soybean varieties to yield less than their comparable, conventional counterparts varies, but overall, they appear to average yields that are five to ten per cent lower per acre. As described earlier, impaired root development, nodulation, and nitrogen fixation likely account for this yield drag. Drought conditions worsen the effects.

The bacterium that facilitates nodulation and nitrogen fixation in the root zone is apparently sensitive to both Roundup and drought. University of Missouri scientists reported problems with germination of Roundup Ready soybeans in the 2001 crop year. Yields of herbicide-tolerant cotton are reportedly not significantly different from those of conventional cotton. Herbicide-resistant transgenic canola varieties yield less on average than conventional canola varieties. Transgenic canola costs less than conventional canola to produce, but because of its higher yields, conventional canola returns more profit per acre.

Yield: Insecticidal Crops—Corn, Cotton

Insecticidal Bt corn and cotton generally yield higher "in most years for some regions" according to USDA Economic Research Service data from 1996 to 1998. Bt cotton, especially, outpaces yields of conventional cotton by as much as 9 to 26 per cent in some cases, though not at all in others. Yield increases for Bt corn have not been as dramatic. Time will tell whether farmers can expect yield increases or decreases in the long run with these and other transgenic crop varieties.

CHANGES IN CHEMICAL PESTICIDE USE

One of the promises of transgenic technology is that it will reduce pesticide use and thereby provide environmental benefits while reducing farmers' costs. The herbicide-tolerant and insecticidal varieties are designed specifically to meet these goals. Studies estimate a two to three per cent decrease in U.S. pesticide use, but the effects vary widely by crop, region, and year. Increased future pesticide use resulting from the buildup of resistance to heavily used herbicides is a long-term concern acknowledged by the transgenic crop industry. Pesticide use depends on the crop and its specific traits; weather, severity of pest infestations; farm management; geographic location of the farm; and other variables. As a result, conclusions drawn by various studies analyzing pesticide use on transgenic crops remain controversial. According to the Wallace Center report, in a review of the data available up through 2000, crops engineered to contain Bt appear to have decreased the overall use of insecticides slightly, while the use of herbicide-resistant crops has resulted in variable changes in overall herbicide use, with increases in use of some herbicides in some places and decreases in others. The crop for which studies are showing the largest decrease in pesticide use is Bt cotton, with Bt corn resulting in only small changes. Herbicide-tolerant cotton has also resulted in little change in herbicide use. The data for herbicide-tolerant soybeans seems harder to interpret. A recent study of herbicide use data on Roundup Ready soybeans by Charles Benbrook, PhD, former executive director of the National Academy of Sciences Committee on Agriculture and now with the Northwest Science and Environmental Policy Center, concludes that the use of herbicides has actually increased because the weeds have become resistant to Roundup.

While another recent study by scientists in The Netherlands shows a decrease in herbicide use on transgenic soybeans, it is clear that weed resistance to Roundup may lead to increased herbicide use and to the need to shift to more toxic compounds in the future, and this is acknowledged by the industry. American Soybean Association president Tony Anderson agrees that the developing resistance of weeds to herbicides such as Roundup is a problem. The Wallace Center report emphasizes the importance of ongoing monitoring of pesticide use data. If farmers abandon integrated pest management, which utilizes a variety of pesticide and cultural control methods, in favour of the simplified control offered by herbicide-resistant and insecticidal transgenic crops, then early findings of reduced pesticide quantities and toxicity may not hold over the long run. Refer to chapter one of the Wallace Center report for USDA pesticide use data comparisons between transgenic and conventional crops, broken down by crop.

PROFITABILITY

Farmers need to consider all the factors that determine profitability. No single factor can tell the whole story. Transgenic crop seeds tend to be more costly, and farmers have the added expense of a substantial per-acre fee charged by the owners of transgenic varieties. These costs have to be considered along with input cost changes—whether herbicide or insecticide use and costs go down, go up, or stay the same. Market price is another factor. Prices for some transgenic crops in some markets are lower than prices for comparable conventional crops, though rarely they are higher. Farmers need to watch the markets. Some buyers will pay a premium for a non-transgenic product, though as transgenic seeds find their way into conventional transportation, storage, and processing steams, these premiums may disappear along with confidence that "GMO-free" products are in fact truly free of engineered genes. Future availability of conventional seed is another issue. Once farmers try transgenic crops, they have reported becoming locked into the technology, as alternate conventional seed supplies dry up.

Also the potential liability of transgenic plants coming up in a conventional planting the next year is important to farmers. Transgenic seed suppliers aggressively pursue legal cases against any farmer using transgenic seed without having a signed technology agreement.

MARKETING AND TRADE

Brazil, Argentina, and China rank among the top five countries in acreage of transgenic soybeans.Buyer acceptance is a significant marketing issue for farmers raising transgenic crops. Farmers need to know before they plant what their particular markets will or won't accept. Since most grain handlers cannot effectively segregate transgenic from non-transgenic crops in the same facility, many companies are channeling transgenic crops into particular warehouses. Farmers need to know which ones and how far away those are.

Many foreign markets have tended to be more leery of transgenic products than domestic markets, although this may change. World Trade Organization (WTO) directives can force dropping of trade barriers, but consumer acceptance cannot be forced (when choice is possible). Africa is a special case, as authority to accept or reject transgenic products was retained by governments, and several have banned GMOs in any form—even relief grain shipments.

India has now developed its own transgenic industry and is producing transgenic cotton, while actively resisting attempts by others to patent its indigenous crop genetics. Brazil, Argentina, and China rank among the top five countries in acreage of transgenic soybeans, maize, and cotton. Even two European countries—Spain and Romania—are producing transgenic crops for animal feed. Eighty-six countries and the European Union have agreed on implementation steps for the UN's Cartagena Protocol on Biosafety, which came into force in September 2003. A rigourous system for handling, transporting, packing, and identifying transgenic crops was part of the agreement. All bulk shipments of genetically engineered crops intended for food, animal feed, or processing are to be labeled "May Contain LMOs," (Living Modified Organisms) according to the UNEP.

Major producers of transgenic crops, including Canada, Argentina, and the U.S., did not sign the protocol. Trade in transgenic livestock feed is more liberal than trade in transgenic human food. The rapid and widespread dissemination of the Cry9C Bt transgene (StarLinKB], which is not approved for human consumption but was detected in tacos, shows how easily transgenic material can spread from animal feed to human food products. The widespread publicity has resulted in even further resistance on the part of buyers to purchasing transgenic products for human food. According to a report in Britain's *The Guardian*, "No new transgenic crops have been approved by the European Union

(EU) since April 1998, and a defacto moratorium on further approvals has been in place since June 1999."

However, trials of food crops already approved continued, and the European Union officially lifted its moratorium on the introduction of new transgenic crops in 2004, although during the debate over labeling and traceability regulations the moratorium remained in effect. If approved, the new regulations will complicate the export of U.S. farm products to the EU because the U.S. does not require traceability or labeling of transgenic crops. Under the proposed new EU requirements, "all foods and animal feed derived from GMOs have to be labeled and, in the case of processed goods, records have to be kept throughout the production chain allowing the GMO to be traced back to the farm." If approved, the new regulations will complicate the export of U.S. farm products to the EU because the U.S. does not require traceability or labeling of transgenic crops. Spain and Romania rank in the top 14 countries growing biotech crops. Both grow transgenic animal feed crops. Portugal, Germany, France, and the Czech Republic grow small amounts of feed corn (maize)—less than 10,000 hectares (24,700 acres), probably much less.

While the U.S. does not require mandatory labeling of processed food containing transgenic ingredients, the EU, Russia, Japan, South Korea, Taiwan, Australia, New Zealand, and Ecuador do have such requirements, as of 2001. The degree to which the Cartagena protocol will be implemented by other signatories (see above) is unknown. Because many domestic merchandisers of agricultural commodities do not segregate transgenic from conventional crop varieties, it is impossible for them and the farmers that supply them to serve these food markets. Twenty per cent of corn and 35 per cent of soybeans produced in the U.S. are exported, and more than 80 per cent of these crops are used in animal feed. Few, if any, animal feeding trials were carried out before transgenic crops were released. In 2005, grain exports were down 5 per cent overall from the previous year and 26 per cent at Gulf Coast ports (due to the hurricanes). In contrast, in a dramatic increase from 2001, 45 per cent of U.S. wheat is exported. Exports to countries that are resistant to buying transgenic food—particularly Japan and European nations—are dropping, but being supplanted by increased demand from Nigeria and Iraq. Because wheat producers are so dependent on exports, they have vigorously resisted introduction of the first transgenic wheat, originally slated for 2003, now on hold.

The Japanese milling industry has made it clear that it does not want transgenic products. As a result, Monsanto promised not to introduce Roundup Ready wheat until Japan gave its approval. North Dakota and Montana considered legislation that would place a state moratorium on the introduction of transgenic wheat. Recent federal regulation, under the Homeland Security Act, would nullify any such local or state food laws. In addition to national and international policies on the use and importation of transgenic crops, processors and retailers in many countries have set their own corporate policies. Major retail chains in Europe and the U.S. have declared their commitment to avoiding the purchase of transgenic products, both feed and food. But, in the absence of labeling, most have been willing to accept a pervasive presence of transgenic corn, soy, and canola in processed products.

Although European Union rules effectively barring U.S. corn imports have been recently relaxed, since 1997 the European Union ban has cost American farmers access to a $200 million annual market and the U.S. government billions in agricultural price supports.

ORGANIC INDUSTRY

Organic farmers face even bigger marketing and trade risks, since their buyers expect no transgenic contamination. Currently, organic production is process-oriented, not testing oriented—except for exports. The organic industry has a system for segregation, but recent tests for transgenic material in organic products demonstrate that it is not immune to contamination from conventional systems.

New technologies can reliably detect minute amounts of transgenic material. Published reports from Europe and the U.S. confirm a high degree of accuracy for detection methods. European export markets organic farmers might have enjoyed, and those that producers of non-GE conventional crops could have built upon, have proven unstable in the presence of possible transgenic contamination. In 2005 U.S. exports of agricultural products fell below imports, for the first time in 20 years.

SEED TESTING FOR GENETICALLY MODIFIED TRAITS

Selecting the appropriate test for seed will ultimately depend on the end use of the results. Are you looking for the absence or presence of a trait, or do you need quantitative data? The best approach to testing for genetically modified traits is to understand the ultimate use of the tests and then to talk with the laboratory or technologist that will be performing the test.

The technologist will be able to describe the four types of tests commonly used by the seed industry to test for traits:

- Herbicidebioassay
- Immunoassay (ELISA, lateral flow strips)
- Electrophoresis (PAGE, IEF, starch-gel)
- Polymerase chain reaction (PCR)

The technologist can help you select the test that will best fulfill your needs. Five years ago, the Society of Commercial Seed Technologists (SCST) created an accreditation programme for technologists in these four areas. The programme ensures that the technologist is proficient in both the theory and practical application of the genetic purity tests currently utilized by the seed industry.

Using a laboratory with a certified or registered genetic technologist ensures that GM tests are conducted by an experienced person. The SCST Genetic Technology Committee and working groups are extremely active in providing training and education to keep members up-to-date in this rapidly evolving area of seed testing. Hall, Anita, executive director, Society of Commercial Seed Technologists, Inc.

Influence on Public Research

While transgenic crop varieties are generally the property of private corporations, those corporations often contract with public- sector agricultural research institutions for some of their development work. In fact, private investment in agricultural research, including germplasm development, has surpassed public investment in recent years.

With this shift in funding priorities, the following questions become important: Is the private sector unduly influencing the public research agenda? Are corporations directing public research in socially questionable directions while research on, for instance, sustainable agriculture wanes? Are the outcomes of corporate- funded transgenic research and development by our public institutions equitable across the food and agricultural sectors? Is equity even a consideration of our public institutions when they accept this work?

When intellectual property rights (patents) apply to living organisms, making them private property, the free flow of scientific information that has historically characterized public agricultural research is inhibited. What are the implications for the future of agriculture and society of the secrecy that now surrounds so much of what was formerly shared public knowledge? These and other questions need to be addressed by citizens and their public institutions. These issues are of particular concern to farmers and consumers who would benefit from research into alternative technologies that are less costly (in every way), less risky, and more equitable. Equity requires that the economic benefits and risks of technology be fairly distributed among technology providers, farmers, merchants, and consumers.

For long-term sustainability, farmers need research that focuses on farms as systems, with internal elements whose relationships can be adjusted to achieve farm management goals. In contrast, transgenic crop research so far has focused on products that complement toxic chemical approaches to control of individual pest species. These are products that can be commercialized by large agribusiness or agri-chemical interests and that farmers must purchase every year. This research orientation only perpetuates the cost-price squeeze that continues to drive so many out of farming.

Industry Concentration and Farmers' Right to Save Seed

The broadening of intellectual property rights in 1980 to cover living organisms, including genes, has resulted in a flurry of mergers and acquisitions in the seed and biotech industries. According to the Wallace Center report, "Relatively few firms control the vast majority of commercial transgenic crop technologies."

These firms have strategically developed linkages among the biotechnology, seed, and agri-chemical sectors to capture as much market value as possible. However, these tightly controlled linkages of product sectors raise serious issues of market access, product innovation, and the flow of public benefits from transgenic crops.

Unlike Plant Variety Protection—which does not allow for the patenting of individual genes, but only of crop varieties—Intellectual Property Rights prohibit farmers from saving seed and undertaking their own breeding programmes, and prohibit plant breeders from using the material to create new generations of varieties adapted to specific regions or growing conditions. The Intellectual Property Rights have recently been upheld by the U.S. Supreme Court, which relied on a 1795 General Patents statute. By 2000, agri-chemical giants DuPont and Monsanto together owned 73 per cent of the corn seed producers in the U.S. Although some have recently been divested, this kind of corporate control and concentration raises the question of whether there remains enough competition in the seed industry for seed pricing to remain competitive. As additional concentration occurs, how affordable will seed—all seed, not just transgenic seed—be for farmers?

This question takes on added gravity as an increasing number of seed varieties become proprietary and seed production, especially for gardeners, is pushed out of California and Oregon into Nevada and Idaho. Farmers can't save proprietary seed for planting and so must purchase new seed every year. In addition, farmers choosing transgenic varieties must sign a contract with the owner of the variety and pay a substantial per-acre technology fee, or royalty.

The anticipated commercial introduction of transgenic wheat represents a dramatic shift in an industry in which farmers still widely save their own seed. As non-transgenic varieties become contaminated with transgenic ones, even those farmers who choose to stick with conventional varieties will lose the right to replant their own seed. This loss has already occurred in Canada's canola industry, with Monsanto winning its court case against farmer Percy Schmeiser for replanting his own canola variety that had become contaminated with Monsanto's Roundup Ready canola. The adoption of transgenic crop varieties has brought with it an increasing prevalence of contract production. While contract production can lead to increased value and reduced risk for growers, farmers are justified in their concern about their loss of control when they sign a contract with a private company. Issues associated with contract production of transgenic crops must be considered within the broader context of a sustainable agriculture to include ownership, control, and social equity.

Regulation of Transgenic Crops and Apportionment of Liability

Much of the controversy over transgenic crops, both internationally and in the U.S., is in part a result of how the U.S. regulates transgenic crops. The federal government has determined that the commercial products of agricultural biotechnology are "substantially equivalent" to their conventional counterparts and that therefore no new regulatory process or structure is needed for their review and approval. Currently, three federal agencies regulate the release of transgenic food crops in the U.S.: the U.S. Department of Agriculture's Animal and Plant Health Inspection Service (USDA- APHIS), the U.S. Environmental Protection Agency (EPA), and the U.S. Food and Drug Administration (FDA). USDA-APHIS: APHIS looks at how a transgenic plant behaves in comparison with its unmodified counterpart. Is it as safe to grow? The data it uses are supplied largely by the companies seeking a permit for release of the new crop.

Under "fast-track" approval, a process in place since 1997, companies introducing a crop similar to a previously approved version need give only 30 days' advance notice prior to releasing it on the market. According to the Wallace Center report, APHIS staff estimate that by 2000, 95 to 98 per cent of field tests were taking place under simple notification rules rather than through permitting. The Office of Inspector General (OIG) report has called for much stricter tracking-in light of the industry shift to industrial and pharmacrops.

EPA: The EPA regulates the pesticides produced by transgenic crops, such as the Bt in Bt corn and cotton. It does not regulate the transgenic crops themselves. In contrast with its regulation of conventional pesticides, the EPA has set no tolerance limits for the amount of Bt that transgenic corn, cotton, and potatoes may contain. FDA: The FDA focuses on the human health risks of transgenic crops. However, its rules do not require mandatory pre-market safety testing or mandatory labeling of transgenic foods. Initially, the U.S. regulatory process for transgenic food crops required product-by-product reviews. Now, however, to simplify and speed up the process, new products can be approved based on the experience gained in reviewing earlier products. According to the Wallace Center report, the implication is that "some crops might be approved, or disapproved, without actual field testing." The regulatory process, in fact, may not answer most questions about the environmental and human health risks of commercial production of these crops.

Central to the policy of substantial equivalence is the assumption that only the end product of transgenic technology is of concern-not the process of genetic modification. Canada has adopted a similar approach. Europe and other U.S. trading partners, however, have taken a more conservative approach. They focus on the process of genetic modification-the source of many of the environmental and human health risks of greatest concern. How these different approaches play out in reality can be summed up simply. The U.S. and Canada assume a product is safe until it is proven to carry significant risk; the European Union, which follows the "precautionary principle," assumes the same product may carry significant risk until it can be proven safe. The science used by the two approaches is not fundamentally different. The difference is in the level of risk the different societies and political systems are willing to accept.

LIABILITY

Farmers who choose not to grow transgenic varieties risk finding transgenic plants in their fields anyway, as a result of cross-pollination via wind, insects, and birds bringing in pollen from transgenic crops planted miles away. Besides pollen, sources of contamination include contaminated seed and seed brought in by passing trucks or wildlife.

Those farmers whose conventional or organic crops are contaminated, regardless of the route, risk lawsuits filed against them by the companies that own the proprietary rights to seed the farmer didn't buy. Likewise, farmers who grow transgenic crops risk being sued by neighbours and buyers whose non-transgenic crops become contaminated. Because contamination by transgenic material has become so prevalent in such a short time, all farmers in areas of transgenic crop production are at risk. Insurance, the most common recourse for minimizing potential losses because of liability, is not available to the nation's farmers for this risk because insurance companies do not have enough experience for gauging potential losses.

Most of the farmers who have been accused by transgenic seed companies of illegally growing and harvesting their proprietary transgenic varieties have paid fines to the companies rather than go to court to defend themselves. One Canadian grower of nontransgenic canola, Percy Schmeiser, did go to court against Monsanto, and lost. The initial ruling required him to pay Monsanto the approximately U.S. $85,000 value of his crop and $13,000 in punitive damages. The amount was eventually reduced, upon appeal. This farmer's case has implications for other farmers, especially those who traditionally save their own seed for planting the next year's crop.

In another example, a Texas organic farm's corn, assumed to be GE-free, was purchased by a processor who made it into organic tortilla chips. Only after the product had been sold and shipped to European retailers was it discovered to be contaminated with transgenic corn. The processor had to recall its product at a cost of over $150,000. The processor chose not to sue the organic farmer, but could have. The retailer, in turn, apparently did not sue the processor. Had it done so, the liability for the retailer's loss could have fallen on both the processor and the farmer. Until laws or legal precedent clarify the extent of farmer liability, farmers would do well to avoid making assumptions or claims about the purity of their non-transgenic products. (Some export crops, as well as Identity Preserved crops, are now being tested before shipping.) Furthermore, producers of transgenic crops need to take all possible precautions against spreading pollen and seed to their own and others' nontransgenic fields and markets.

A full risk assessment and legal clarification of the distribution of liability among farmers, seed companies, grain handlers, processors, and retailers is needed before farmers can rest assured that transgenic crops won't result in lawsuits against them. For farmers who have adopted transgenic technology—and for those who have not—in the words of Brian Leahy, executive director of California Certified Organic Farmers, "This technology does not respect property rights." In essence, farmers who grow transgenic crops on some of their fields, and farmers who grow none, risk bearing tremendous liability. This situation won't change until farmers either gain legal protection or stop growing transgenic crops entirely.

PRECAUTIONARY PRINCIPLE

The principle of precaution was developed specifically for issues involving ethics and risk. It is described by Katherine Barrett, project director with the Science and Environmental Health Network, as: "... a process for decision-making under conditions of uncertainty. The principle states that when there is reason to believe that our actions will result in significant harm, we should take active measures to prevent such harm, even if cause-and-effect relationships have not been proven conclusively.... It has been invoked in many international laws, treaties, and declarations on a range of environmental issues including climate change, marine dumping of pollutants, and general efforts towards sustainability— [including the 2000 international Cartagena Biosafety Protocol on the transfer of 'living modified organisms'] ... There is growing consensus that the precautionary principle has reached the status of international customary law."

The primary elements of the precautionary principle are identified by Barrett as:

- Avoid harms to the environment that are "irreversible, persistent, bioaccumulative, or otherwise serious."
- Anticipate and prevent potential harms at the source," rather than relying on reactive measures of mitigation, clean-up, or compensation.
- Recognize the limits of scientific knowledge and don't expect a full and conclusive understanding of potential consequences before taking precautionary action, especially when the potential consequences are long-term, unconfined, and broad-scale.
- Shift the burden of proof to the developers of potentially hazardous technologies.

- Finally, some versions of the precautionary principle include analysis of the costs and benefits of precautionary action, though cost-benefit analysis "is not a sufficiently robust decision-making framework to stand alone."

IMPLICATIONS FOR SUSTAINABLE AGRICULTURE

In contrast to the ecological approach of sustainable agriculture, "the current generation of transgenic crops follows a pest management model like that employed for chemical pesticides— through interventions that are toxic to pests," the Wallace Center report points out. This "single-tool" approach is likely to fail in the long run because pests will successfully develop resistance that allows them to thrive.

Standard plant-breeding methods can potentially solve many of the same problems in agriculture that genetic engineers are working on, though there are areas in which genetic engineering can enhance traditional plant breeding. Armed with the map of an organism's genetic code, scientists can test which genes are in a plant to select more easily which ones to cross-breed. "Before we knew where the genes were, we were still breeding in the dark," according to Steven Briggs, head of genomics for Syngenta, a Swiss biotechnology giant, as quoted in the New York Times.

CROSSING OVER IN PLANTS

Plants and animals have two of each gene, on separate but similar chromosomes, analogous to having two copies of the document. The two copies are usually slightly different. By analogy, if the document were The Lord's Prayer, one copy might read The Lord is my shepherd! I shall not want. The other copy might read God is my guide! He provides for me. In sexual reproduction, the two documents are sliced apart and put back together, so that they might read The Lord is my guide! He provides for me. and God is my shepherd! I shall not want. This process is called crossing over. This happens in the sex cells of both parents and then the offspring inherit one "document" from each parent. Crossing over is a major source of genetic variability. Sometimes, however, it produces the same kind of nonsense as in the example "don't be happy! drink the water!".

Transposons

Genes are arranged on the chromosome in order but many genes make a habit of jumping to another part of the chromosome. They are routinely snipped out of the DNA and put back in a different place. These jumping genes are called transposons. By being moved to a different part of the DNA, the functional expression of a transposon changes. A familiar example of a transposon is a gene that determines the colour of the kernels of Indian corn.

Viruses

Many viruses insert their own DNA into the host's DNA document, where it gets copied and becomes part of the genome. More than half of the DNA in the human genome originated as invading viruses. These viruses can carry useful genes from other species into the infected animal or plant. Many human genes originated in other species by this mechanism.

Mutations

With billions of letters in a DNA document, there are always a few copying errors, called mutations. A mutated gene has a changed meaning. It may specify a slightly different protein, or it may switch another gene on or off at a slightly different time.

These changes are usually harmful, but tolerable, especially when we have two copies of a gene and one is unchanged. Occasionally a mutation is either deadly or beneficial. According to the theory of evolution, all of our genes originated, in the distant past, as mutations of other genes, a completely random process.

The critics of genetic engineering mean to leave you with the impression that nature has evolved a finely tuned but fragile system of inheritance but that genetic engineers have no good idea what they are doing. But actually, the genetic engineers make only small well controlled modifications to a genome, whereas nature often makes large and random changes. The likelihood of unexpected effects in transgenic technology is small. Unexpected changes in conventional breeding are virtually certain.

Glyphosate

We earlier mentioned glyphosate, an environmentally benign herbicide. It is biodegradable, lasting approximately two days after use, and its molecules bind to soil so it does not wash into streams or enter groundwater. Even if some madman were to dump it into a waterway, it is 230 times less toxic than the herbicide it replaced. It has been found safe and approved for use in almost every country, including all the countries of the European Union. Yet the anti-transgenic crowd seeks to ban it in Europe using the pretext of uniformizing regulations from one European country to another. What reason is given for banning the safest of all herbicides? Two reasons. It is supposed to be 'implicated' in non-Hodgkins lymphoma, and it is alleged to harm beneficial insects. Examining the evidence for these two claims reveals how some opponents of GMOs will use anything, no matter how shaky, to achieve their purposes. Non-Hodgkins lymphoma patients were asked to recall what pesticides they had been exposed to in the past several years. A statistically insignificant number of them (four) mentioned glyphosate, not surprising since it is so widely used. Even though the investigators considered this association meaningless, it was enough for Greenpeace to demand a ban. The claim about harming insects is even weaker. Several species were exposed to glyphosate and the control group were not exposed. There was similar mortality in both groups.

But there was mortality. This is enough for Greenpeace to claim that glyphosate kills beneficial insects. Greenpeace ignores a follow-up study by the International Organization for Biological Control, which concluded that glyphosate was exceptionally safe. One might have expected the IOBC to be hoping that it would find something wrong with glyphosate, since its charter is to popularize biological controls. But IOBC's scientists are ethical. The real reason for banning glyphosate is that it is made by Monsanto and used as the herbicide for the herbicide-resistant transgenic soybeans.

Labelling Transgenic Food

Once the opponents instigated a doubt about the safety of genetic engineered foods, and remember that there has not, in ten years, been even the slightest evidence of that, the next step was to agitate for labelling. Why, after all, shouldn't consumers have a choice? I agree people who want to avoid transgenic food should be able to make that choice. But labelling can take two forms. One could label food which is not transgenic, or one could label food which is transgenic. I can't imagine anyone objecting to labelling foods which are transgenic-free. But the anti-transgenic demand is adamant for the other choice. We can say that the difference is between those who would label the non-transgenic food with a smiley face and those who would label transgenic food with a skull and crossbones. Labelling sounds so reasonable. How could anyone oppose it if they didn't have something to hide? But labelling has a down side. It costs.

The cost is not in the ink to print the label. It is in keeping the products separate. We are talking now mostly about soybeans and corn. These grains are harvested by the hundreds of tons, shipped in railroad cars, stored in grain silos, sold in futures contracts. Approximately 70 per cent of processed foods in the US have some ingredient derived from biotechnology.

This includes most milk and cheese, sugar, soy products, corn and corn sweeteners, vegetable oils, etc. 63 per cent of US grown soybeans are now transgenic. Neither buyers nor sellers distinguish between transgenic and conventional soybeans. Your tofu is a mixture of both kinds. (In my local supermarket, there is only one brand of tofu available and the manufacturer has recently decided to use only organic soybeans, which are specifically labelled non-GMO. At the same time, the package size changed from sixteen to fifteen ounces and the price was raised by $.25.) To keep the two kinds separate, we would need, at a minimum, to have two separate distribution channels, two storage systems, two futures markets. In my opinion, the pressure for skull and crossbones labelling is really a pressure to increase the cost of the genetic engineered food. So far, the transgenic food has had only producer advantages — it is cheaper to produce. Take away its cost advantage and it is no better than the conventional foods.

Never mind that corn with the Bt trait, one third of the US crop, is grown without pesticides. Never mind that the transgenic soybeans prevent soil loss and global warming. The Economist magazine estimated that segregation of grains would add a 25 per cent premium to the price of some processed foods, like packaged cereals. This is probably an over-estimate, but we should remember that these costs will fall inordinately on people with low income. But if this were the only problem with labelling, the food industry giants would quickly adopt grain segregation and labelling. They are being forced into grain segregation anyway by the labelling regulations of nations that import US grain.

The industry is really concerned that the labels will make it easier for groups like Greenpeace to boycott their products. Stop the scare campaigns and the resistance to labelling will disappear overnight. If you doubt that demands for skull and crossbones labelling can be used to purposely create a disadvantage, imagine what your reaction would be if someone were to propose that products assembled by hispanic workers must be so labelled. The public demand for labelling would be justified by a "right to information", but just below the radar screen you would not be surprised to hear that hispanic workers might be illiterate, or might be illegal immigrants, or might be drug users.

But you would recognize that the demand for labelling was meant to disadvantage hispanics. We have established a precedent in this country about labelling. The government does not mandate labelling without a very good reason. The main exception is when safety (skull and crossbones label) is involved. Once the government decides that, say, a certain chemical pesticide is safe, nobody can require the food grown with that pesticide to be labelled. When a segment of the consuming public wants a label (smiley face) about a trait that it cares about, the market provides such a label and it is reflected in the cost of the product.

For example, there are people who prefer to eat food grown with no chemical pesticides. They buy "organic" food. Everyone understands that the label "organic" means that the food was grown without pesticides. Everyone also knows that they have to pay more for the food labelled organic. Similarly, orthodox Jews have certain religious rules about what they eat. These include a requirement that cows and chickens must be slaughtered in a particular way. This is a requirement for the Kosher label. Jews do not expect all meat to have a label detailing how the animal was slaughtered. Even in Israel, the people who want the label pay for the privilege, and kosher meat is often quite a bit more expensive that ordinary meat.

Environmental Scare Stories

We mentioned earlier a concern that some genetic engineered crops could cross breed with wild relatives. To the pseudo-environmental groups (those who like to call themselves green even though much of what they advocate would harm the environment), this absolutely rules out any permissible use. Their argument goes as follows: "We don't have any idea what plants could cross breed with what other plants, and we don't have any idea what effect the unusual genes would have in the wild plants. Therefore we should take no chances." In fact, hybridization between different species is extremely uncommon and it has never been observed between distantly related species. A tomato won't cross with a potato, even though they are both members of the nightshade family.

Yet the GE opponents would have us believe that just about any plant can hybridize with any other. And they have quotes from PhDs to prove it. Actually there is exactly one controlled experiment showing a hybridization between a genetic engineered crop and a wild species. The crop was a transgenic canola (oilseed), with that same herbicide tolerance gene we have encountered before.

Here is what PhD Jeremy Bartlett, of the John Innes Plant Research Centre of Norfolk, England, wrote to the Manchester Guardian. He said we don't know what plants will hybridize with what other plants. He said that there is a documented example of transgenic canola hybridizing with a wild mustard.

He went on in the same letter to talk about the possibility of gene transfer to soil organisms. Dr. Bartlett is a PhD in biology. He surely knows that there is little basis for speculating that plants will pass genes on to soil organisms. But let's give him the benefit of the doubt where a speculation is concerned. Still he must have known that the documented case he was referring to was observed at a test plot of the John Innes Research Centre, his own institution.

The transgenic canola was planted in the centre of the plot and various other species were planted at various distances to measure the rates of hybridization. There was only one case observed. The canola hybridized with the wild mustard. Jeremy Bartlett means for you to think that if an oilseed can hybridize with a wild mustard, then anything can hybridize with anything. Fortunately, the John Innes Centre puts its research reports on the world wide web, so I read the actual report. Guess what? Canola is a hybrid itself, a cross between Brassica rapa and Brassica napus, two closely related wild plants in the same genus. Brassica rapa is wild mustard! So we don't have evidence for a transgenic crop hybridizing with just about anything else imaginable, beyond the capability of man to anticipate. We have evidence of a plant crossbreeding with its closest relative, placed in the test plot because it was so likely to crossbreed, and it was the only case observed.

When the biotechnology opponents write or talk about this, they always say wild mustard and canola, never Brassica rapa and napus. Dr. Bartlett means to play with our minds, to mislead us. Hybridization with wild plants is a concern, but it is a managable concern.

RESISTANCE TO PLANT VIRUS DISEASES

PROTEIN

One successful application of genetic engineering in crop improvement is in the production of virus-resistant plants. This has been achieved by genetic transformation of plants with novel resistance genes based on nucleotide sequences derived from the viruses themselves or from virus-associated nucleic acid as well as genes from other sources.

The demonstration that the expression of tobacco mosaic virus (TMV) coat protein (CP) in transgenic plants protects the plants against virus infection has led to explorations of using other viral or virus-associated genes in producing genetically engineered resistance against in plant virus. Transgenic plants expressing Sat-RNAs have been obtained in a number of labouratories and have been tested in greenhouses and fields. Recent studies with defective interfering RNA or DNA protection and sense-antisense-ribozyme RNAmediated protection offer much promise for a broad range virus resistance in plants.

Our labouratory has successfully produced satellite RNA-mediated resistance to cucumber mosaic virus (CMV) in tobacco and tomato, CP-mediated resistance to rice stripe virus in rice and fusion protein (CP and nuclease) mediated high resistance to tobacco mosaic virus in tobacco. One way to confer more effective and durable field resistance to virus disease in transgenic crop is to use a combination of multiple resistance genes.

ENGINEERING RESISTANCE TO PLANT-VIRUS DISEASES

Since 1981, Sat-RNAs have been used as a biological control agent (BCA) of diseases caused by CMV on a large scale in China. The results show that the BCA is effective in controlling diseases caused by virulent CMV strains in many crops and, in addition, induces resistance to certain types of fungal diseases. The protective effects of CMV Sat-RNAs have been demonstrated in a number of greenhouse and field tests. These studies have provided the basis for engineering plant resistance to virus infection by introducing the Sat-RNA as a transgene.

In our labouratory, the cDNA of an attenuating Sat-RNA monomer from CMV strain 1 was synthesized and constructed into the plant expression vector pRok2. The Sat-RNA cDNA was introduced into different plant species including tobacco and tomato.7 When transgenic tobacco and tomato plants were tested in the greenhouse for resistance against challenge infection by CMV, they showed a significant decrease in disease index compared to nontransformed plants.

In an experimental scale field test conducted in the spring of 1990, the transgenic tomato plants inoculated artificially with CMV gave about 50 per cent higher fruit yield than the non-transformed control plants and showed a decrease in disease index. However, there was no significant difference in leaf yield between the satellite transgenic and the control tobacco plants, although the disease index of transgenic tobacco was about 10 per cent lower than that of the control.

Due to an insufficient level of Sat-RNA in the plants at an early stage of infection, the transgenic tobacco plants showed only a weak resistance to CMV, which directly affected the leaf yield, since it is the early leaves that are harvested in tobacco production.

Although Sat-RNAs confer some extent of resistance to virus infection, field tests carried out in China from 1990-1992 show that the resistance of the transgenic plants is not strong enough to fully protect tobacco and tomato plants from the damage caused by natural virus infections.

SAT-RNA IN A MULTIPLE GENE STRATEGY OF ENGINEERING RESISTANCE

One way to obtain more effective and durable resistance to virus diseases in transgenic plants may be to use a combination of multiple resistance genes. In this strategy, different genes that produce blockage or interference at different stages of a virus infection are used to obtain engineered resistance.

Transgenic lines of tobacco expressing both CP and Sat-RNA of CMV were obtained through transformation using a chimeric vector containing two expression cassettes of the two genes under the control of the CaMV 35S promoter. When challenged with CMV, the virus concentration of challenging CMV in transformed plants that express only Sat-RNA was about 10-20 per cent of that in non-transformed plants, while in CP + Sat-RNA transformed plants it was only 4-5 per cent. A comparison of disease incidence and disease index of CP + Sat-RNA, CP alone or Sat-RNA alone transformed tobacco plants. The resistance of CP + Sat-RNA transformed plants was about twice as strong as that of Sat-RNA alone or CP alone transformed plants. Since the CP level expressed in CP + Sat-RNA transformed plants was similar to that in transformed plants that contained only the CP gene, the two-fold increase in resistance of CP + Sat-RNA transformed plants was not due to higher levels of CP expression. On the other hand, the level of Sat-RNA accumulation in the CP + Sat-RNA transformed plants was lower than that in the Sat-RNA alone transformed plants.

This may be due to the effect of the CP gene conferring resistance at an early stage in CMV infection, reducing the replication of CMV as well as Sat-

RNA. Preliminary data from a field test of 1992 confirmed the enhanced resistance of the CP + Sat-RNA transformed plants to CMV. The CP + Sat-RNA transformed plants show stronger resistance to natural CMV infections than either the CP alone or Sat-RNA alone transformed plants.

TRANSGENIC TOBACCO PLANTS RESISTANT TO CUCUMBER MOSAIC VIRUS

Since *Nicotiana tabacuum* is an amphidipoid species, the tetraplold transgenic plants will segregate extensively in their progenies. It needs to be propagated several generations before obtaining a homozygous line. In order to speed up this process, a procedure for a fast production of homozygosity of transgenic plants was devised in our work.

Leaf discs of diploid tobacco plant derived from anther cultures of a commercial cultivar NC-89 were transformed with a chimeric vector of CP and Sat-RNA genes of CMV.

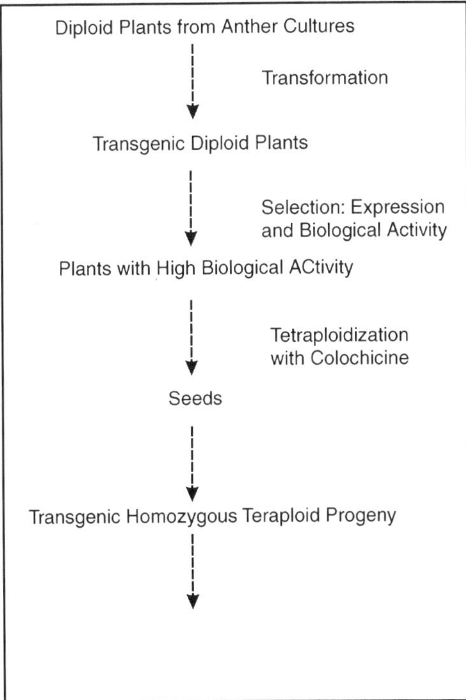

Fig. A Procedure for fast Homozygosity Generation in Transgenic Plants.

The kanamycin resistant diploid plants were selected by their expression of CP and Sat-RNA. Then the transgenic plants were tested for their resistance to CMV after inoculation with high concentration of CMV (50-100 ìg/mL). Only the plants with high resistance to CMV were tetraploidizated with colchicine.

The transgenic homozygous tetraploid progeny can therefore be obtained after only one generation.

A certain amount of transgenic seeds will be produced within one growing season. After propagation of the transgenic seeds once in Hainan Province a large scale of field test can be achieved in the main tobacco producing area in the second year.

RESISTANCE TO RICE STRIPE VIRUS IN TRANSGENIC INDICA RICE PLANTS

Rice stripe disease occurs in rice growing areas of China, Japan, Korea and the Commonwealth of Independent State, and causes significant reduction in rice yield. The coat protein (CP) gene of a Chinese isolate of rice stripe virus (RSV-C) was synthesized, cloned and sequenced, which is very similar to the Japanese strain. Cell suspension cultures were initiated from embrogenic calli of rice Annong S-1 (indica rice) by inoculating yellowish, compact and embryogenic calli derived from seeds into a suspension culture medium containing proline and maltose.

After being cultured at 26°C in the dark for about half a year, finely-dispersed and embryogenic suspension culture was established. Before bombardment, the suspension culture was evenly applied onto three-layerfilter-paper discs in a petri dish. Cells were bombarded with 1 ìm diameter tungsten particles coated with DNA of the expression vector pROK2 containing the CP gene under control of the 35S promoter. The plasmid DNA was absorbed to tungsten particles using a calcium-spermidine precipitation procedure. 2.5 ìL of the coated particles was loaded onto macroprojectile and each dish with the suspension culture was bombarded three times under a partial vacuum.

Following bombardment, the suspensions were cultured in a modified N6 medium. Two days later the suspensions were transferred to the same medium containing G418 (40 ìg/mL), which were subcultured every week. Being subject to G418 selection for two months, white and fast-growing colonies were emerged from the browny cultures. Green plants could be regenerated when the resistant calli were transferred to a differentiation medium. In one experiment ten plantlets regenerated from G418 resistant calli were tested for their transgenic nature by Southern blot analysis using a32P-dCTP-labelled CP gene as a probe. The genomic DNA extracted from the selected as well as the control plants were digested with EcoR1 and BamH1 and then hybridized with the probe. Two plants showed two hybridization bands of 0.6 kbp and 0.7 kbp corresponding to the CP gene. Western blot and ELISA further analyses demonstrated that CP (32 kDa) was expressed in the transgenic rice plants. Sixteen transgenic thus obtained and 100 non-transformed rice plants were inoculated with RSV-carrying leaf hoppers, 24 days post inoculation feeding, most of the untransformed plants developed chlorotic stripes and chlorosis in the young leaves, the disease index (DI) being 53 per cent. After 40 days post inoculation feeding, most of the control plants showed severe symptoms, being stunted and the leaves developing chlorotic and brown necrotic streaks, with the DI being 88 per cent. In contrast, the transgenic rice plants subject to the inoculation feeding developed no or mild symptoms. After 24 and 40 days post inoculation feeding, the disease index of the transgenic plants was 15 per cent and 25 per cent respectively.

INCREASED RESISTANCE TO TOBACCO MOSAIC VIRUS

Coat protein-mediated resistance (CPMR) has been proved to be successful against several families of plant viruses. Recently CPMR was obtained by expressing a truncated CP of tobacco virus. Our work has indicated that expression of a fusion protein of TMV-CP and Staphylococcal nuclease (SN) in transgenic tobacco plants significantly enhances the CPMR to TMV. TMV-CP or SN gene alone and CP'!SN or SN'!CP fusion protein genes were obtained by PCR, inserted downstream of 35S promoter of a binary vector pROK2 and transformed into tobacco leaf discs.

No SN gene alone transformed tissues could be regenerated, but the tissues containing CP'!SN or SN'!CP fusion genes grew normally and could be regenerated into plants. The transformed nature of the regenerated tobacco plants were confirmed by Southern blot hybridization and PCR analysis. It was demonstrated that the fusion proteins of CP'!SN and SN'!CP were expressed in the transgenic plants by Western blot analysis. CP'!SN- or SN'!CP-transformed plants exhibited higher resistance to TMV than CP alone transformed ones. 50 per cent of CP'!SN- and SN'!CP-transformed tobacco plants were symptomless for three months after inoculation with 1 ìg/mL TMV.

Nuclease activity test of the fusion protein expressed in transgenic plants was carried out *in vitro*. Fusion proteins of SN and CP were extracted from the transgenic plants and were obtained by immunoprecipitation with antiserum of TMV. The precipitated proteins were infiltrated with 2 per cent SDS and then with water containing excess amount of TMV-CP, and reacted at 30°C with the incubation mixture of 25 mM Tris-HCl (pH 8.8), 10 mM $CaCl_2$ and 40 ng 32P-5'-end-labelled oligdeoxynucleotide. The result showed that no nuclease activity was observed. No obvious difference in structure and infectivity of virion isolated from SN-CP transgenic and non-transgenic plants has been found.

The mechanism of resistance to virus by expression of a fusion CP was unknown. We speculate that the fusion CP is somehow dysfunctional and more effective to disrupt the normal virus-host relationship than CP does alone. The disruption may be at the level of virus movement throughout the plant or the fusion CP is encapsidated into virion to generate defective virus particle.

RISKS ASSESSMENT OF RELEASING GENETICALLY ENGINEERED PLANTS

Some transgenic organisms, such as micro-organisms and animals will be grown under controlled conditions and only released to the environment accidentally. However transgenic plants will be grown on a large scale in the open environment. There is concern that transgenic plants may produce risks.

Spread of Transgene by Pollen

Because genetic engineering will often result in the introduction of genes not previously present in a species, there is concern that such genes may spread to other plants and to weeds by crosspollination. It is important to have information on pollen transfer distances and the range of species that can be pollinated. To obviate this problem trial release of transgenic plants have often required the removal of flowers to prevent pollen spread as for the transgenic tobacco production. However, if such plants are to be used for the seeds or fruits, unlimited spread of pollen will occur.

Releasing of Antibiotic Resistance Genes

Antibiotic resistance genes are often used as a selective marker in transformation procedure. In the case of antibiotic resistance, it will be a problem only if it is inherited by pathogenic micro-organisms which are being controlled by the antibiotic in question. The risk of a pathogen inheriting and being able to express an antibiotic resistance gene derived from a plant is at least a million-fold lower than such resistance being obtained from other micro-organisms.

Pollution of Honey

A further problem that might be encountered is the presence of cloned gene products in pollen. This could harm the bee larvae which are fed pollen, pollutes honey with minute amounts of the gene products, and makes the pollen more allergenic. Most gene products will not be found in or on pollen in other than minute amounts and thus contamination should not be a problem.

Transgenic Plants Becoming Weedy

Potentially the greatest risk is that certain types of transgenic plants might become weedy, for example, the acquisition of herbicide resistance genes which will make them less easy to control in subsequent generations. On the one hand herbicide resistance genes are used as selective markers in many labouratories; on the other hand large companies are attempting to encourage growers to use more herbicides, and are using genetic engineering for purely commercial reasons.

Safety of Application Sat-RNA for Virus Diseases Control

Safety concerns about the use of Sat-RNAs remain. Even though virus replication is inhibited in transgenic plant, a small amount of Sat-RNA can still be encapsidated into helper virus particles and transmitted to other plant species, which it may cause more severe diseases. Moreover, because the exacerbating and attenuating Sat-RNAs can differ in only a few nucleotides, there is always a possibility that a symptom attenuating Sat-RNA can mutate into an exacerbating one.

During a decade of using Sat-RNAs as biological control agents in China, a number of facts concerning the safety of using Sat-RNA in agriculture have been realised. In spite of the large scale application of the BCAs on different crops, no necrogenic or other harmful Sat-RNA variant was found in a systematic field survey in which random plant tissue samples were analysed by temperature-gradient gel electrophoresis (TGGE) to detect Sat-RNA sequence variants.

Secondly, the efficiency of aphid transmission of CMV in the presence of the Sat-RNA is reduced to about 10 per cent of that in the absence of the Sat-RNA.

The reduced transmission efficiency, due to the low concentration of CMV when the Sat-RNA is present, lowers the chance that the Sat-RNA will be released in the field. Finally, it is possible to maintain the benign property of the Sat-RNA by periodic and detailed quality control of the BCA stocks.

It is our hope that the use of Sat-RNA as a viral control strategy will become increasingly safe as our knowledge of Sat-RNA structurefunction relationship increases. Indeed, it has recently been found that the Sat-RNA of CMV can lose its ability of being encapsidated into CMV CP by changing a few nucleotides; this in turn excludes the possibility of Sat-RNA transmission (D. Baulcombe, personal communication).

Safety of viral coat Protein-mediated Resistance

Transencapsidation of viral nucleic acid by transgenically expressed CP may induce heteroencapsidated virion, which might be transmitted by new vectors and to new host plants. The CP of plum pox potyvirus has been shown to confer aphid-transmissibility on a non-transmissible isolate of zucchini yellow mosaic virus.

However, recent studies showed that intentionally truncated, antisense or nonexpression (-AUG) CP genes could provide measurable protection or even complete immunity against the appropriate parent virus; this in turn excludes the possibility of heteroencapsidation of viral nucleic acid by transgenically expressed CP. In conclusion, safety, from a public view, was ensured by biologically containing the transgenic plants so that they would not induce risk.

The emphasis on biological containment has meant that for release to be permissible it is necessary to demonstrate beyond reasonable doubt that the transgenic plants are safe. But nothing is completely safe, and safety *per se* is not always desired; for example, plant species without any toxic metabolites would be exquisitely sensitive to pests.

PRODUCTION OF TRANSGENIC PLANTS

To produce transgenic plants, the crucial steps are the introduction of the cloned genetic material into the plant cell nuclear and the facilitated integration of the cloned gene into the plant chromosome. It is interesting that the best method for doing so uses a system that already exists in nature, the system by which the plant pathogenic bacterium Agro-bacterium tumefaciens injects a portion of its plasmid DNA into plants and inserts it into the plant become. The Agro-bacterium system is an excellent system for introducing foreign genes into the chromosomes of plants. The transfer into intact plant cells occur at high frequency, the T-DNA is usually integrated into the plant chromosomes at high frequencies without undergoing structural alterations, and the cells that have received the T-DNA can be selected easily by using antibiotic resistance such as the neomycin resistance maricer. Finally the transgenic plants produced in this manner are quiet stable for at least several generations. Consequently almost all existing transgenic plants with potentially desirable traits have been obtained by this method.

HERBICIDE RESISTANCE PLANTS

The advantages of making crop plants resistant to herbicides are obvious. Although there is a fear that such resistant may eventually increase the use of herbiude chemicals, there are also reasons to except that these transgenic plants will promote the use of sager, more biodegradable herbicides, perhaps in smaller amounts one example involves the herbicide glyphosate which inhibits 5-enol-pyruvylshikimate 3-phophate synthase-an enzyme involved in the biosynthesis of aromatic amino acids.

This enzyme has been purified from crop plants and sequenced and DNA Probes corresponding to its amino acid sequence have been synthesized. These probes were used to isolate CDNA for the enzyme from the CDNA library of a plant cell line known to overproduce 5-enol-pyruvylshikimate 3-phophate synthase. The CDNA was then cloned behind the strong CaMv promoter, and the promoter gene complex was introduced into plant cells via a disarmed Ti plasmid vector. The transgenic plants produce a much higher level of the target enzyme and thus are significantly more resistant to glyphosate. These results are encouraging because glyphosate has very low toxicity to animals is rapidly degraded in soil.

INSECT RESISTANT PLANTS

Bacitlus thuringiensis is used in the biological control of caterpillars because its sporulating cells contain toxic proteins. The gene for one toxic protein was cloned behind promoters that are effectively expressed in plants and was introduced into plant cells via a Ti plasmid vector, thus producing plants toxic to caterpillar. The major problem with this approach has been the low level of expression of the toxin protein in plants, which is presumably related to the fact that the gene come from a bacterium,. Nevertheless, the method has produced tomato and tobacco plants that proved quiet resistant to caterpillars in field tests. Cotton however, is often attacked by insects that are more resistant to the B. thuringiensis toxin and the low level of expression of this toxin in transgenic cotton plants provided the plants with little if any protection. Recently, the coding sequence of the toxin was altered extensively to replace codons that are rarely used in plants as well as to preclude the formation of a strong secondary structure in MRNA. When cotton plants were provided with this modified gene, their production of the bacterial toxin increased 100-fold and they showed impressive resistant to common bepidopteran insects than damage unmodified plants.

Some seeds contain high concentrations of protease inhibitors, which are thought to interfere with the digestive process in insects. The cloning of cowpea trypsin inhibitor genes, for example and their transfer to tobacco plants have resulted in good resistance to a wide variety of leaf-eating insects.

VIRAL RESISTANT PLANTS

The method most frequency used for producing virus-resistant plants sprang from the observation that, of ten times, plants infected by nearly avirulent virus are thereafter resistant to super infection by a related, highly virulent one. Thus, deliberate infection with avirulent virus strain has been used to produce protection in crop plants. The method is not totally safe, however, because the avirulent strains many mutate to produce strains that are significantly pathogenic. Most plant viruses are positive strand RNA viruses covered by coat protein sub units. When the virus enters an injured plant cell the replication process begins, starting with the progressive uncoating of the virus from the 5-end of the RNA. Tranagenic plants, whose genomes contain introduced tobacco mosalc virus (TMV) coat protein genes and which continuosly synthesize the coat proteins, show resistance to virus infection. The phenotype of these plants is consistent with the idea that resistance is a result of interference with the uncoating of the virus particle. Although the plant cells are resistant to infect TMV particles, they remain sensitive to the TMV RNA or to partially uncoated TMV particles.

The coat protein gene is usually put behind a strong promoter such as the 35S promoter of CaMV and is introduced into a plant's genome via the Agro-bacterium Ti system. Transgenic plants showing significant resistance to TMV alfalfa mosaic virus, cucumber mosaic virus, tobacco streak virus and tobacco rattle virus have already been produced in this manner. In a recent experiment two virus and potato virus Y, were introduced simultaneously into a commercially important potato cultivar, and one of the resulting trangenic plants provide quiet resistant to both viruses under field test conditions.

TRANSGENIC CROPS

The application of recombinant DNA technology has revolutionized plant breeding in recent years. Genetic transformation in plant breeding greatly increases the gene pool and broadens the scope of genetic changes that modern breeders may draw upon. The first generation of genetically modified (GM) crop plants hold great promise in carrying new or improved agronomical traits that require less intensive farming methods or which have improved quality traits. These GM crop plants have been produced to improve farming production systems, aiming, for instance, to reduce negative impacts on the environment, such as fertilizers, herbicides and pesticides.

In general they are of little direct interest to the consumer. However, the genetic manipulation of the primary and secondary metabolism of plants offers distinct possibilities in the development of nutritionally improved products, raw materials with added pharmaceutical value, and crops with real health benefits for the consumer. These are second-generation GM plants. Recently, several prototypes in molecular plant breeding with novel traits for (non-) food and industrial applications have been introduced into the market, and many more GM crops have the prospect of commercial exploitation in the near future.

The safety and wholesomeness of these GM crop plants raises questions beyond those posed by conventional foods. In this context, traditional plant breeding of most food crops has conformed to the standard 'generally recognized as safe' (GRAS). In some instances safety is assumed to be assured during performance and quality testing by breeders (*e.g.*, glucosinolates and erucic acid in rape seed). However, the process is largely focused on the comparison of agronomic and phenotypic characteristics, such as colour, maturity, yield or disease resistance.

Products with an unusual taste or that have harmful effects following ingestion or skin contact have normally been rejected from the breeding programme. However, the safety of conventional plant varieties has usually been presumed from the assumption that prudent consumers will avoid those species once it is known that they cannot consume them without adverse health effects. When the commercial production of GM plants started, it sparked a debate centred on issues associated with concerns on environmental, human and animal safety matters. Many people questioned whether the 'simple' GRAS approach, based on historical and empirical experiences, is sufficient to guarantee food safety of the GM crop plants.

It is also asked whether the risks of chronic exposure of humans and animals to GM crops are adequately covered, and whether large-scale breeding of GM plants will have detrimental effects on ecosystems.

FOOD SAFETY ISSUES

Although the variety registration of existing crop plants has not resulted in adverse effects in humans, the conventional assessments carried out by plant breeders was not considered to be sufficient to ensure food safety of GM crop plants. As a result, Europe issued a regulation on novel foods and novel food ingredients that sets the legal framework for the market introduction of genetically modified organisms (GMOs). This came into force on 16 May 1997.

The Novel Foods Regulation establishes a system of mandatory pre-market approvals of novel foodstuffs, including GM plants. The accompanying EU guideline 97/618/EC gave a clear indication of the types of data that are needed to form the basis of an assessment. In particular, four central safety issues have been set forth related to food and feed safety:

- The nutritional and toxicological consequences of inserted gene products;
- The potential of pleiotropic (unintended) effects in the host organism due to the insertion event;
- The allergenicity of expressed proteins and novel foodstuffs;
- The potential of gene transfer to human and animal gut flora.

This means that in transgenic insect-resistant Bt tomatoes encoding the C-terminal truncated Bt2 gene derived from a *Bacillus thuringiensis* (Bt) strain, *IAb5*, the novel protein has to be evaluated on its own merit, while the safety and wholesomeness of the 'remaining' transgenic tomato is assessed separately. Whereas, for example, in antisense RNA exogalactanase tomatoes, which have improved rheologic characteristics due to downregulation of the endogenous exogalactanase activity, the safety assessment is primarily focused on the 'remaining' novel fruit. In general, there is basic agreement on the safety issues to be addressed. One particular area of concern is the possibility of unexpected or unintended metabolic perturbations due to genetic modification that may alter, for instance, levels of nutrients and health-influencing components. The use of rDNA techniques does not necessarily result in fundamental changes in crop plants compared to the food produced by conventional breeding methods. However, it should be emphasized that a uniform international agreement for the evaluation of GMOs is urgently needed. For example, an agreement on how to establish in practical terms the similarities or differences between the 'remaining' GM crop plant and the traditionally bred plant seems to be much more difficult to achieve. This chapter is devoted to the description of advanced strategies for the evaluation of potential unintended effects in GM crop plants. Pleiotropic (unintended) alterations in agronomic traits or composition may arise from insertion mutagenesis or as a result of metabolic effects of the novel gene product(s). The challenge is to gain the ability to discriminate between metabolic alterations due to somaclonal variations, unintended effects or natural diversity. Thereto, emphasis is given to a platform of technologies designed to identify changes in the molecular machinery of crop plants. As is the case for detection methods for GM crops and derived products, there are needs and criteria for sampling strategies, statistical models and manufacturing standard reference matehers recently.

UNINTENDED EFFECTS

Strategies for assessing the food safety and wholesomeness of novel foodstuffs are currently in the exploration phase. Above all there is a need for uniform and harmonized quantitative methods for the testing of potential unintended effects in whole foods, such as GM crop plants are. The safety of, for example, synthetic chemicals and food additives is established by assessing separate elements of the compound in question.

Although very successful in case of single chemicals, the evaluation of a whole food product may not be a simple addition of individual assessments of the constituents. Traditional testing protocols involving labouratory animal 90-day feeding trials with (whole) food products are far from ideal. This type of animal feeding study with complex food matrices is complicated by the likelihood of nutritional imbalances leading to dietary problems, confounding factors, an insensitivity for specific endpoints, and the impossibility of using large safety margins, if any.

Therefore, in a global context several regulatory, research and industry-based bodies have proposed alternative concepts for the safety evaluation of genetically engineered foods and food ingredients. In particular the Organisation for Economic Co-operation and Development (OECD), the Food and Agriculture Organization (FAO) and the World Health Organization (WHO) have developed concepts and principles for the safety evaluation of GMOs and derived products. Their reports concluded that risk assessments should be directed at demonstrating that a GM crop plant or derived food product is as safe as a traditional or previously authorized product. Accordingly the EU Competent Authority incorporated these general principles into their Novel Foods Regulation and accompanying guidelines. In this context, novel foods and food ingredients that are considered to be 'no longer equivalent' to traditionally bred plants are subject to labelling as defined in the Council Regulation 1139/98.

Substantial Equivalence

The term 'substantial equivalence' appears in the EU Novel Foods Regulation as a major principle in the risk assessment. It emphasizes that, within the context of the Regulation the evaluation of the safety and nutritional value of a GM plant should be focused on a comparative analysis with conventionally bred products.

OECD's Group of National Experts on Safety in Biotechnology enunciated this comparative approach towards the safety assessment of transgenic food plants as the concept of substantial equivalence. In their report, it is assumed that existing food organisms possess a long history of safe use (*i.e.*, GRAS). Consequently, the most practical approach to the determination of safety is to consider whether the GM plants are substantially equivalent to analogous food product(s), if such exist. Once substantial equivalence of a novel foodstuff has been established, it provides assurance of safety that is equal to or better than that of its comparators.

Further clarification of the concept of substantial equivalence originates from a WHO sponsored workshop in 1994. It has been concluded that the establishment of substantial equivalence is not a safety assessment *per se*, but a dynamic, analytical exercise. Thus, it may mean that analysis of, for instance, gene expression patterns, global changes in protein expression and/or differences in metabolic capabilities (the metabolome) of the novel product in comparison with conventionally bred products should be performed in order to examine equivalence.

This strategy has worked satisfactorily for the assessment of the first-generation crop plants for which sufficient background knowledge is available. However, relatively less experience has been gained with the safety and wholesomeness evaluation of novel food plants for which no substantial equivalence can be established. Even in the case of single gene modification, potential alterations in metabolic pathways are difficult to identify and, consequently, the implications of the genetic modification process on the metabolism of plants are poorly understood. Moreover, data on the mechanisms by which plants regulate, for instance, their response to environmental stress factors and other conditions are remarkably scanty.

Critical Nutrients and key Toxicants

Application of the concept of substantial equivalence appears in practice to lead to various interpretations. Different requirements for risk assessment with regards to the potential of unintended effects exist as a result of genetic modification. Actual approaches entail a consideration of the characteristics of host and donor organism, phenotypic properties and toxicological evaluation and include comparative determinations of any changes in critical (anti-)nutrients and key toxicants for the food source in question. In general, the analysis of an expanded spectrum of components is unnecessary, but should be considered if there is an indication from other traits that there may be an unintended secondary effect of genetic modification.

However, compositional analyses based on single parameters using a list of crop-specific critical (anti-)nutrients and key toxicants as a minimum requirement of screening for possible variations in the content has its limitations. Limited data are typically available for the less important food components, as in less well-known crops most critical (anti-)nutrients and natural toxicants will be unknown.

Moreover, the toxicants to be assessed may be partly influenced by the cellular function of the product encoded by the inserted gene. Furthermore, the natural variation of critical comparators may mask the evidence of secondary effects due to genetic modification. Thus, there is no criterion that determines whether a difference between a GM plant and its natural counterpart is outside the literature values for the ranges of parameters that express natural diversity.

Relevant information on indicators of unexpected effects may not have been observed during the development of an edible transgenic crop species. In the future other compounds could become of importance, such as anti-oxidative, oestrogenic and anti-carcinogenic compounds. In determining critical nutrients and/or key toxicants, differences in consumption patterns and practices of processing and consumption in various geographical settings and cultures must also be recognized. It is our opinion that conclusions about relative safety and wholesomeness based on a list of selected single components may not be equally valid in a global context and in all times. Therefore, there is an urgent need to set up generic approaches using platform technologies to evaluate undesirable metabolic perturbations in GM food crops.

A Post-genome Challenge

In our view the identification of unintended effects encompasses the ability to interpret and use innovations in molecular genetics research such as genomics, proteomics and metabolomics on a crop-by-crop basis. This will establish a whole data package directed towards a holistic view on possible unintended side effects due to genetic modification. In essence, the focus is shifting towards molecular characterization to understand functional activity. Vital to this approach will be informative profiles of molecules at different integration levels, *e.g.*, mRNAs, proteins including post-translational modifications and molecules at the level of primary and secondary plant products and metabolites that might be useful benchmarks for the detection of unintended effects.

The relative importance of these various 'profiling' techniques in establishing data sets for assessing unintended effects is not indicated by their order and will vary from species to species. Alterations in, for example, expression levels, post-transitional modification, interactions or chemical composition do not *per se* imply that the product is less safe or even unsafe. Such a platform of technologies may determine whether unintended effects exist between the GM plant and its control(s) or whether there is substantial equivalence apart from certain well-defined expected differences.

Measures of the relative activities of various molecular constituents associated with metabolic capabilities in non- and modified plant cells under different conditions will provide new insights into how metabolism and genetic modification are orchestrated. But it should be emphasized that the fields of plant genomics and proteomics are still in their infancy.

CDNA MICROARRAYS

Over the past few years, the DNA microarray technology has emerged as a powerful high-throughput method for the analysis of gene expression. At present, microarrays for gene expression studies generally consist of cDNAs, which are physically deposited onto small glass surfaces. The major advantage of DNA microarray technology over conventional gene-profiling techniques is that it allows small-scale analysis of expression of a large number of genes in a sensitive and quantitative manner. Furthermore, it allows comparison of gene expression profiles under a large number of different conditions.

The cDNA microarray technology has already been successfully introduced in disciplines ranging from cell and developmental biology to drug development and pharmacogenomics. In the following a number of technical aspects of the technology as well as its potential value for the safety assessment of genetic modification of food plants will be discussed. The cDNA microarrays are produced by deposition onto a solid surface of purified PCR products corresponding to specific genes.

Typically a few nanolitres of DNA solution (100-500 μg/ml) are spotted using a micro-dispensing robot. Essential information on robot construction and protocols can be found on the Internet. In general, with respect to the spotting devices, a higher density of spots is accomplished when using a passive dispenser (> 2500 DNA spots/cm^2). Thereto, glass slides covered with a positively charged layer (*e.g.*, poly-l-lysine) or carrying reactive groups are most commonly used.

To analyse the expression of genes in a plant cell or tissue sample, polyA+ RNA (typically 0.5-2.5 μg is required for each reaction) is purified and labelled by reverse transcription using an oligo-dT primer incorporating a fluorescently labelled nucleotide. The labelled cDNA pool is then hybridized to the microarray. The required hybridization conditions (*i.e.*, sample concentration, stringency of hybridization) and parameters, such as detection level and the correlation between transcript concentration and hybridization have been determined for each experimental setting.

The intensity of a hybridization signal, which is a measure of the relative expression level of each gene, can be read by charge-coupled device (CCD) cameras or confocal/non-confocal laser scanners. RNAs extracted from two different samples can also be simultaneously analysed with a microarray by labelling each with a different fluorescent label using either Cy-3 or Cy-5. At RIKILT-DLO we are currently testing whether the analysis of differential gene expression using DNA microarrays is an informative strategy for the safety evaluation of GM plants.

To address this question, cDNA libraries are constructed and enriched by a subtractive cloning approach for cDNAs preferentially expressed in either green or red tomato fruit. These subtracted cDNAs, as well as control cDNAs representing known tomato genes (*i.e.*, sequence information derived from public sequence databases), have been spotted on arrays and are used for comparison of gene expression profiling of control and genetically modified tomatoes. In the near future we will construct more defined arrays containing genes of toxicologically relevant pathways, such as those involved in the synthesis of natural plant toxins (*e.g.*, the glycoalkaloids á-tomatine and á-solanine).

MRNA FINGERPRINTING

A very promising strategy to evaluate the equivalence of genetically modified crops is gene expression profiling. Conventional methods for the analysis of differential gene expression include Northern blotting, S1 nuclease protection, comparative expressed sequence tag (EST) sequencing, differential display and serial analysis of gene expression. More recently, Kok *et al.* have designed a method of detecting altered gene expression by means of mRNA fingerprinting or reverse transcription polymerase chain reaction (RT-PCR). Based on the original concept of Liang and Pardee specific subsets of the mRNA population of a tomato plant were amplified. This enables the expression of different genes in the GM crop and genes in the parental line to be compared and any empirical information about significant alterations to be established.

The method is still under development and requires further validation, but has been shown to be reproducible and able to detect important differences in gene expression. Some of the most significant remaining problems are that the method largely depends on individual skills based on experience and is time-consuming, even in its most simplified form.

PROTEOME PROFILING

The introduction of Bt genes of bacterial origin may theoretically influence the nature of post-translation modification (*e.g.*, glycosylation) in the transgenic tomato fruit. This is because as well as the impact of the protein moiety, for example, the type of glycosylation pattern may alter due to inactivation of glycosidases and glycosyltransferases in the host cell. In addition, glycosylation of the inserted Bt gene product in the plant might result in altered protein activity and stability.

The importance of structurally elucidating these oligosaccharides, especially the *N*-glycans, can be attributed to their widespread functions in the cell. These functions include correct folding, biological activity and stability of proteins as well as involvement in plant development. Moreover, *N*-glycans represent antigenic epitopes by themselves. Immunological as well as biochemical studies revealed the allergenic potency for the á1,3-fucosylation of the proximal *N*-acetylglucosamine residue. Additionally, also the β1,2-

xylosylation of the β core mannose is suspected to be an allergenic epitope. Recently, there is increasing evidence that these *N*-linked oligosaccharides may have a substantial impact on the immunogenicity of glycoproteins, as *N*-glycans are suspected to play a certain role in the adverse food reactions of hypersensitized patients. At the Institute for Agrobiotechnology, Austria, this issue was tackled by developing a generic technology for isolating, purifying and characterizing *N*-linked oligosaccharides in plant tissues.

Our low-temperature acetone powder method appeared to be most suitable for the isolation of (glyco-)proteins from plant tissue (*e.g.*, tomato fruit). The resulting (glyco-)protein fraction was proteolytically digested and corresponding glycopeptides purified by cation-exchange chromatography. The oligosaccharides (*N*-glycans including á1,3-fucosylated oligosaccharides) were released from the peptide backbones by *N*-glycosidase A (PNGase A) and fluorescently labelled with 2-aminopyridine.

Subsequently, pyridylaminated glycans were separated and analysed structurally by repeated two-dimensional high-performance liquid chromatography (2D HPLC) using a size-fractionation and a reverse-phase column in combination with an exoglycosidase treatment. MALDI-TOF (matrix-assisted laser desorption ionization time-of-light) mass spectrometry was applied for further confirmation of the observed structures. Sixteen different *N*-glycosidic structures could be detected in tomato fruit. The two most abundant glycans in all the tomato samples showed identical properties to those of the major *N*-linked oligosaccharides of horseradish peroxidase (MMXF: Man(3)Xyl-FucGlcNAc(2)) and pineapple stem bromelain (MOXF: Man(2)Xyl-FucGlcNAc(2)), respectively, and accounted for about 65-78 per cent of the total glycan content. Oligomannosidic glycans occurred in only small quantities (3-9 per cent). The majority of the *N*-glycans found was β 1,2-xylosylated and carried an á1,3-fucose residue linked to terminal N-acetylglucosamine.

A structural element was identified and shown to be an IgE-reactive determinant, which contributes to cross-reactivities among non-related glycoproteins, and is present in a wide variety of plants extracts. It is proposed that this structural element might also be of importance in some cases of food and pollen allergy.

The carbohydrate profiling of, for example, populations of green and red-ripe fruit of non-modified and modified Bt tomato plants (expressing the *cryIA5* gene) revealed no differences in the nature of *N*-glycans, nor were there significant differences in the relative amounts of glycan. Interestingly, the relative amounts found were quite similar across seasons and comparable to the commercial variety Notoro used as an extended control. As far as the *N*-linked oligosaccharides were involved it was not possible to detect significant alterations in the á1,3-fucose content of Bt tomatoes due to genetic modification.

PROFILING OF THE METABOLOME

Among other factors, identity of a complex GM crop plant with its traditionally bred parent entails metabolism and the level of (un-)desirable substances contained therein. Therefore, we would like to build up an overall picture of components associated with different metabolic capabilities in the (non-)modified plant tissue. A multi-compositional analysis of the so-called metabolome may overcome the test limitations of, for example, single critical nutrient or key toxicant analyses and problems related to their limits of natural diversity.

An attractive alternative is the utility of high-resolution proton (^{1}H) nuclear magnetic resonance (NMR) spectroscopy (^{1}H-NMR) in combination with separation techniques (liquid chromatography (LC)). Recently, we have shown that the approach of chemical fingerprinting using off-line LC-NMR can obtain information on possible changes in complex plant matrices due to environmental effects as long as GLP-like conditions for sample handling, data acquisition and automation of data handling are maintained. In principle, it will be possible to screen all the different low molecular weight components (MW $<$10 kDa) present in the metabolome of the plant tissue.

TOXICOLOGICAL PROFILING

The significance of detected compositional alterations may need to be further explored by *in vitro* toxicity assays in order to screen for potential adverse effects and for mechanistic studies. It may even be that the latter approach leads to a redesign, refinement or even replacement of (sub-)chronic rodent feeding trials. *In vitro* systems derived from organs and tissues from animals and humans, and various types of cultured recombinant cell lines, are successfully used for screening for the toxic potential of single compounds. However, their ability to screen for the potential toxicity of whole foods or extracts thereof has been insufficiently explored up till now, and has been examined by us within the framework of a tiered safety evaluation of GM plants.

The validity of using *in vitro* test assays for comparative testing of whole food products, food ingredients and extracts thereof depends on the sensitivity and selectivity of the toxic response induced upon exposure. However, since these models are not yet validated, they should be considered as early warning or alert-systems to identify potential changes in toxicity of the engineered product only.

GENOTOXICITY ENDPOINTS

The mucosa of the gastrointestinal tract is clearly a primary, potential site of action for the effects of novel foods. Therefore effort has been concentrated on the development of a battery of intestinal cell systems (*e.g.*, IEC-6, IEC-18, Caco-2 and INT407 cell lines) with different endpoints for initial cyto- and genotoxicity testing. The choice of a suitable method of sample preparation was considered of utmost importance in assaying vegetables.

Freeze-dried tomato fruit was extracted (10 per cent suspensions, w/v) in water, 0.9 per cent saline or chloroform:methanol (2:1, v/v) immediately before incorporation into the culture medium. It was encouraging that aqueous and chloroform/methanol extracts of, for example, red-ripe tomatoes up to a maximum of 10 per cent suspensions (w/v), had little or no general toxicity once the pH (> 6.0) and osmotic pressure had been taken into account. No toxicity to the intestinal cells Caco-2 and IEC-18 under short-term conditions was detected in red-ripe, non- and antisense RNA exogalactanase tomato using alpha-glutathione S-transferase (á-GST) lactate dehydroyrenase (LDH) leakage, reduction of 3-(4,5-dimethylthiazol-2-yl)-2,5 diphenyltetrazo-lium bromide (MTT conversion), neutral red (NR) uptake and total cellular protein as *in vitro* endpoints.

On the other hand, green tomato fruit displayed a pronounced toxic and cell-damaging effect. No statistical significant differences were found between transformants and isogenic control lines in terms of their activity in the assays. Treatment of tomato tissue with pepsin (ratio pepsin to tomato protein 1:250, w/w) did not result in an increase of or loss of inherent cytotoxicity. Supplementary investigations with naturally occurring plant components, for example, quercetin, β -carotene, á-lycopene, genistein showed that only steroid glycoalkaloids, such as á-tomatine, exhibited a rapid toxic effect, whereas á-tomatidine was ineffective. It was noted that the *in vitro* models did show general matrix-related toxic effects.

For example, á-tomatine diluted in a matrix of extracted red-ripe tomato tissue (final concentration of 5-10 per cent w/v) showed an increase of dose-response-dependent cytotoxicity in human Caco-2 cells (*e.g.*, IC50 7.9 ± 2.1 mg/L versus IC50 33.9 ± 5.3 mg/L of á-tomatine only). Using the COMET assay, suspensions of freeze-dried tomato fruit up to a maximum of 10 per cent (w/v) did not exhibit DNA-damaging effects in a range of gastrointestinal cells derived from rats (IEC-6 and IEC-18) and humans (Caco-2, INT407). Genetic modification did not result in increased genotoxicity, but fruit extracts were able to suppress the DNA-damaging effects of the known genotoxins H_2 0_2 and N-methyl-N2 -nitro-N-nitrosoguanidine (MNNG).

Although these endpoints are important toxicological measures they do not indicate the molecular mechanisms of toxic damage. So far, pronounced adverse effects could only be observed with glycoalkaloids. The screening for subtle differences in toxic responses of other plant constituents appeared to be limited. However, therefore, the use of critical endpoints for such studies other than cytotoxicity should be clearly defined.

ASSESSING TRANSCRIPTION FACTORS

There is a demand for *in vitro* models that not only determine the (geno-

)toxic potency of plant components or complex mixtures, but also provide information on the mechanism by which that compound exerts its toxic effect. As an *in vitro* profiling system the eukaryotic stress gene assay (*i.e.*, CAT-Tox(L)) was assessed.

This utilizes human HepG2 cells stably transfected with chloramphenicol acetyltransferase (CAT) reporter constructs. The aim of the CAT-Tox(L) assay is to collect and quantify molecular responses to cellular stress and toxicity that have a transcriptional component in their regulation. The assay employs an ELISA method to measure the transcriptional activity of the promoter or response element constructs.

The advantage is that many of the stress responses occur before any measurable cytotoxicity, thus allowing monitoring of stress pathways at sublethal levels. Since 14 unique transcription responses can be measured simultaneously, the CAT-Tox(L) provides comprehensive cellular stress profiles. The assay generates specific stress gene induction profiles for a variety of stressful and toxic plant components such as genistein, diadzein and sinigrin. Aqueous extracts of red-ripe, non-and antisense RNA exogalactanase tomato fruit dosed up to 0.05 g per assay did not cause any cytotoxicity to the HepG2 transfectants nor were there molecular responses to cellular stress or toxicity.

Extracts of non-modified and modified green tomato fruit showed cytotoxicity and induced the construct containing the xenobiotic response element (XRE). This gene-induction profile was not comparable to that of, for example, the reference molecules dioxin, 3-methylcholanthrene, benzo[a]pyrene or β -napthoflavone. These polycyclic aromatic hydrocarbons were characterized with a co-induction of cytochrome P450 1A1 and glutathione-*S*-transferase *Ya* subunit- CAT fusion constructs.

There was substantial evidence that the systemic fungicides fenarimol and bupirimate did not induce any of the genes in this assay. There was no promotor in the assay that could be linked to universal cellular stress. A-tomatine caused severe cytotoxicity at high concentrations and yet did not induce any of the genes (*i.e.*, causing membrane damage). However, it was observed that á-tomatine diluted in a matrix of red-ripe tomato tissue (final extract concentration 5 per cent w/v) elicited an increased cytotoxicity concomitant with a dose-dependent response profile involving XHF (collagenase involved in inflammatory reactions sensitive to mitogens and cytokine IL-1). So far, the approach of the CAT-Tox(L) model shows potential but there is a need for validation based on molecular mechanisms of plant constituents. Although the induction of these stress genes represents cellular perturbations, the mechanistic validations need to be extended before any conclusions can be drawn on the value of these findings. Certainly, the development of a database to explore relationships between a wide variety of compounds stressful to plants and meaningful stress gene profiles may serve as molecular fingerprints unique for a particular component.

CHEMICAL FINGERPRINTING

The principles for establishing a chemical fingerprint were directed towards the detection of alterations in compositions in ^{1}H-NMR spectra arriving from two populations, for example, non-modified counterpart(s) and an engineered one. In this way the unprocessed parts of the plant used for human consumption were fractionated and analysed, *i.e.*, in tomato comparison on the fruit level. Genetically modified tomato varieties, such as the antisense RNA exogalactanase fruit, have been studied using this innovative technology.

The spread of measurements of the individual profiles was comparable for the GMO and its control(s). By subtraction of the ^{1}H-NMR spectra the differences of the means were calculated at 99 per cent confidence intervals between the GM crop average (n = 8/batch per line) and its isogenic control average (n = 8/batch per line). The difference of the means was defined here as all differences of the means of the amplitudes of individual NMR signals obtained from spectra of various populations.

The differences of the means were used to quantify how the transgenic crop plant differed from its parental line. The normalized 99 per cent confidence intervals of the normalized means of fractions A-E after eightfold independent replicates demonstrated that differences in amplitudes exceeding at least 20 per cent could normally be detected. This is an adequate sensitivity to allow statistical evaluation, and is in line with recommendations of the Nordic Council. They proposed that if the average value of a parameter differs by more than 20 per cent an explanation should be sought. With the purpose of establishing those differences that may arise from metabolic effects linked to genetic modification the default statistical analysis used a mixed effects randomized block model. Random block effects such as variations in the stage of cultivation, location, logistics or climate had a significant impact on the overall chemical fingerprints of cultivars.

For example, chemical fingerprints varied much more between field sites and seasons than between replicates in one plot. A tomato background mean (*i.e.*, crop mean value) was established by constructing an extended range of non-modified tomato references for comparison with natural variability, and to assess the impact of external factors (*i.e.*, variations in season, climate, etc.). The crop mean value was used to quantify how the differences of the means between the transgenic crop plant and its control differed from the natural diversity of commercial counterparts.

Representative examples of typical baseline correction and peak selection results obtained from the ^{1}H-NMR spectra of antisense-exogalactanase RNA tomato fruit using software as described in Lommen *et al.*. A global compositional description per fraction is as follows: fraction A, polar compounds such as fruit acids, glucose and other monomeric and oligomeric sugars, amino acids and TCA-cycle substrates; fraction B, large variety of aromatic, aliphatic and sugar-like compounds; fraction C, primarily non-tomatine glycoalkaloids and aromatic compounds; fraction D, tomatines and aromatic/indole-like compounds and fraction E, primarily lipids and carotenoids. Typically the detection limits in the various fractions are estimated in the range of mg/kg wet weight tomato tissue.

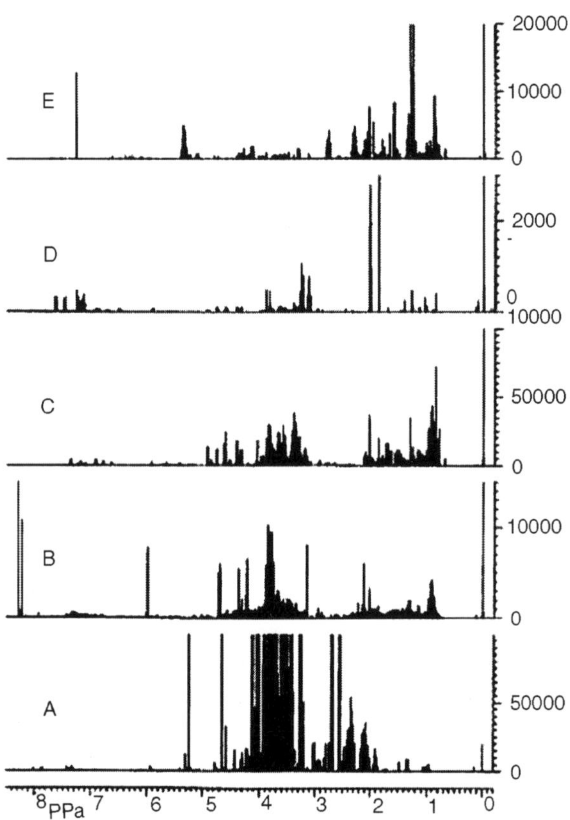

Fig. Chemical Fingerprinting Analysis using off-line LC-NMR (vertical scale in arbitrary units).

To differentiate between compositional changes either due to genetic modification, genetic variability or environmental variations the chemical fingerprints were studied in a hierarchical approach. Hence, the fingerprints of the GMO are compared to those of:

- The isogenic control line grown side-by-side under identical conditions in one plot;
- The isogenic control line grown side-by-side under identical conditions at multiple sites;

- An extended range of commercial varieties of that crop;
- The influence of downstream processing.

It was recognized that 45 per cent of all compounds of the control line harvested in different years varied considerably in concentration. Most differences were found in fractions A, C and D. A similar degree of impact on composition was observed in the case of processing temperatures (*e.g.*, 100°C versus 70°C) on tomato juice. These findings were compatible with the protease activity surviving the 70°C treatment.

But in fruit harvested from various individual plants, divided into two populations, approximately 95 per cent of all constituents were practically identical in concentration. Also the determination of the differences of the means between transformant tEG to its isogenic control bred side-by-side in the same plot showed that 249 out of 3000 amplitudes (8 per cent) varied in concentration. Assignments by NMR indicated that, for example, citric acid was present in fraction A of the control at 1.4 times the levels found in the transformant. However, in fraction A of the transformant glutamic acid and/or glutamine showed an increase of 1.3 times the levels found in the isogenic control. On the other hand, one single aromatic compound, á-lycopene, appeared to be present in fraction E of transformant tEG at 2.5 times the concentrations found in the parental line.

No alterations were observed in ß-carotene levels. These effects may be the result of the antisense downregulation of the *tEG1A* gene in tomato. In our view the natural diversity must be taken into consideration, however, when interpreting the biological relevance of any significant difference found in the differences of the means between the GMO and its parental line. The parental line was extended by entering non-isogenic commercial references (*i.e.*, crop mean value).

Although limited in the range of lines, the comparison of the crop mean value to the differences of the means showed that the contribution and magnitude of differences (*e.g.*, citric acid, glutamic acid/glutamine, á-lycopene) diluted out. It was recognized that in the agricultural practice the conditions of culturing, processing and storage of appropriate references will be mostly unknown and they may have a non-isogenic genotype.

Notwithstanding these complicating random block effects we recommend that, for example, the chemical fingerprint should not be limited to the transgene and its isogenic parent bred under identical circumstances only, but should also include a sampling through multiple sites and a comparison with an extended range of controls of that crop. Otherwise it would be a too limited approach towards the establishment of the claim of substantial equivalence and it would neglect the ranges of natural variations in that crop. In future, if many more lines have to be compared with each other this approach may benefit from multivariate statistical methods.

GENE EXPRESSION PROFILING

A second potential application of DNA microarray technology in safety evaluations concerns screening for biological effects of (fractions of) plant compounds that have been found by other profiling techniques to be present or modified in the genetically modified crop plant when compared to the parental line and other references. At RIKILT-DLO we are testing the feasibility of this strategy by analysing defined food compounds for possible effects on human intestinal gene expression.

To this end, human intestinal cell lines are exposed for different time periods to various amounts of crop plant components and the mRNA isolated. The labelled mRNA is hybridized to DNA microarrays containing unknown intestine-specific genes, as well as already identified control genes known to be intestine specific or to be functional indicators (for toxicological processes, apoptosis, etc.). As a source for unknown, intestine-specific genes we use cDNA libraries from human intestine biopsies and human intestinal cell lines which have been enriched by a subtractive cloning approach.

Having established the usefulness of this approach for the identification of biological functions of defined plant compounds, these cDNA microarrays will be exploited to screen, for example, GM crop plants for possible effects on human health. The concept of substantial equivalence is broadly accepted as a basis for the risk assessment of GM crops and derived novel foods. Data requirements for establishing substantial equivalence should be based on identifying unintended metabolic perturbations in the engineered crops that have been caused by genetic modification. A generic approach cannot be envisaged, however. In all cases the collation of data should be directed by the type of genetic modification and related consequences. It is recommended that unintended effects should be identified on a case-by-case basis, since such effects are dependent on the genotypic and phenotypic parameters of the new gene products involved. The application of innovations in molecular genetics research will help to define the conditions under which the new food products can be marketed. The present approach has generated a sound scientific basis for the evaluation of the potential of unintended effects in GM crops and derived food products. The results may assist regulators and legislators to test and, if necessary, to improve the applicable regulations. Moreover, the technologies described and the results obtained may serve as a general framework for the risk assessment of GM crop plants, and may contribute to a better understanding by the public of recombinant DNA techniques in plant breeding and their implications.

It is concluded that a parallel approach focusing on a comparative analysis using informative profiles of molecules at various integration levels (*e.g.*, mRNA, protein, metabolites) including a toxicological profiling of relevant plant matrices, offers good prospects for the identification of hazards related to unintended effects. The choice of comparators and of external parameters to support a claim that a GM crop plant presents no unintended effects compared with its traditional counterpart should be based on sound scientific judgement. It is recommended that data are generated on the unprocessed part of the plant, such as the tomato fruit or potato tuber, used for human and/or animal consumption. Furthermore, data banks should be set up containing information on natural variations of essential plant or other product constituents (*i.e.*, crop mean values).

The chemical fingerprinting technique, using a combination of off-line liquid chromatography and proton-NMR imaging, appears to be a powerful screening method for the detection of secondary effects in the metabolome that may be applied on a routine basis. The carbohydrate profiling technique is useful for the detection of unexpected changes at *N*-glycan levels; in particular it is applicable to the detection of post-translational modifications in newly expressed proteins as well as in the whole GM crop plant. DNA microarray technology is rapidly evolving and much effort is being put into improving spot density, reduced production time, and increased reproducibility and sensitivity.

Improving the latter is critical because it makes it possible to use smaller amounts of starting material. It will depend on technical aspects such as the quality of the scanners and spotting machines, the development of fluorescent dyes with improved characteristics (*e.g.*, narrow excitation and emission peaks; high level of photon emission; resistance to photo-bleaching), supports with reduced background and more target sequence binding capacity. DNA microarrays generate a huge amount of complex hybridization data and major challenges are to develop more advanced computer software to help to find statistically significant correlations within, and between different experiments, and to link the data to sequence information and submitted expression profiles available in public databases. Application of these technologies in combination with other measurements of, for instance, the performance and quality may limit or even replace animal feeding studies aimed at the detection of unintended effects.

Together, these techniques, known as functional genomics, will enable us to answer questions about what happens to expression and protein composition when a plant cell changes metabolic state, for example, due to genetic modification or environmental factors. It is foreseen that this holistic view on the cellular machinery will be achieved soon, as the coincidence of genome sequencing with improvements in the analysis of expressed proteins is of growing importance and continues to be successful. Eventually, microarrays for genomic studies will probably also be used to evaluate the numerous constituents of GM crop plants. Some constituents might not cause obvious changes in cellular behaviour or morphologic appearance but could cause subtle metabolic alterations that would show up when the mRNA content was interrogated by an array.

It is likely that microarrays might be used to evaluate the life cycle of a plant much more precisely and to understand the complex metabolic control systems of plants. The ultimate test for GM crop plants will be acceptance by the public in large. It is therefore extremely important that industry, the scientific community and the regulatory authorities are able to respond to queries from the public by providing transparent information on the criteria for safety assessment. Improved safety assessment at the molecular level should refine and complement strategies by anticipating the potential risks and sources of unintended effects in GM plants.

Bibliography

Alaka Pande and V.G. Rao: *A Compendium of Fungi on Legumes from India*, Scientific Publication, Delhi, 2001.

Anil Prakash: *Fungi in Biotechnology*, CBS Publication, Delhi, 2001.

Awani Kr. Singh: *Encyclopaedia of Infection Fungies*, Anmoln Publication, Delhi, 2010.

Chandra Deo Narain Singh: *Advanced General Pathology of Animals*, IBDC Publication, Delhi, 2010.

D. Gopalakrishna Rao: *A Text Book on Systemic Pathology of Domestic Animals*, IBDC Publishers, Delhi, 2010.

Deepa H. Dwivedi, S.K. Dwivedi and S. Gupta: *Agro-Animal Resources of Higher Himalayas*, Satish Serial Publication, Delhi, 2010.

Elle Matthews: *Animal Adventures in Africa : Serengeti Safari*, Unicorn Books, Delhi, 2010.

G K Ghosh.: *Biopesticide and Integrated Pest Management*, APH Publication, Delhi, 2009.

J A Bierens De Haan: *Animal Behaviour*, Reprint Publication, Delhi, 2004.

K R Nair.: *Integrated Production and Pest Management*, Gene Tech Books, Delhi, 2007.

Lata Bhattacharya: *Animal Biochemistry*, Discovery Publication, Delhi, 2010.

M.M. Ranga: *Animal Behaviour*, Agrobios Publication, Delhi, 2002.

Mahendra Kumar: *Animal Byproducts Utilization Through Semi-Moist Rendering*, Daya Publication, Delhi, 2007.

N. Venugopal Rao, T. Uma Maheswari, P. Rajendra Prasad, V. Govardhan Naidu and P. Savithri.: *Integrated Insect Pest Management*, Agrobios Publication, Delhi, 2003.

Opender Koul, G.S. Dhaliwal, S.S. Marwaha and Jatinder K. Arora.: *Biopesticides and Pest Management (2 Vols-Set)*, Campus Books International, Delhi, 2003.

P Ramadass: *Animal Biotechnology : Recent Concepts and Developments*, MJP Publication, Delhi, 2008.

Parmeshwar Singh.: *Encyclopaedia of Agricultural Pests and Their Management*, Anmol Publication, Delhi, 2010.

Prasum Tyagi: *A Textbook of Animal Physiology*, Dominant Publication, Delhi, 2010.

R.K.S. Kushwaha: *Fungi in Human and Animal Health*, Scientific Publication, Delhi, 2004.

R.P. Srivastava.: *Mango Insect Pest Management*, International Publication, Delhi, 1997.

S. Ignacimuthu, s.j. and S. Jayaraj.: *Recent Trends in Insect Pest Management*, Elite Publication, Delhi, 2008.

S.C. Dwivedi and Nalini Dwivedi.: *Integrated Pest Management and Biocontrol*, Pointer Publication, Delhi, 2006.

Sanjay Kumar Singh: *Fungi*, Campus Books Publication, Delhi, 2008.

Udit Narain and Mukesh Srivastava: *Fundamentals of Fungi*, Advance Publication, Delhi, 2006.

V K Rastog: *Encyclopaedia of Fung*i, Anmol Publication, Delhi, 2007.

Vaishali J. Patil and T.V. Sathe.: *Insect Predators and Pest Management*, Daya Publishing House, Delhi, 2003.

William J.A. Payne and R. Trevor Wilson: *An Introduction to Animal Husbandry in the Tropics*, Wiley, India, 2013.

Index

J

K

L

M

N

O

P

Q

R

S

T

V

Y

Z